# Food Waste Management and Sustainability

# Food Waste Management and Sustainability

Guest Editor

**Tom Dugmore**

MDPI

Basel • Beijing • Wuhan • Barcelona • Belgrade • Novi Sad • Cluj • Manchester

*Guest Editor*
Tom Dugmore
Department of Chemistry
University of York
York
UK

*Editorial Office*
MDPI AG
Grosspeteranlage 5
4052 Basel, Switzerland

This is a reprint of the Special Issue, published open access by the journal *Sustainability* (ISSN 2071-1050), freely accessible at: https://www.mdpi.com/journal/sustainability/special_issues/314HCSLDD2.

For citation purposes, cite each article independently as indicated on the article page online and as indicated below:

| Lastname, A.A.; Lastname, B.B. Article Title. *Journal Name* **Year**, *Volume Number*, Page Range. |
|---|

**ISBN 978-3-7258-7789-8 (Hbk)**
**ISBN 978-3-7258-7790-4 (PDF)**
**https://doi.org/10.3390/books978-3-7258-7790-4**

# Contents

# About the Editor

**Tom Dugmore**

Tom Dugmore—Department of Chemistry, University of York is a Lecturer in the Department of Chemistry specialising in teaching Green Chemistry, particularly at postgraduate level on the MSc in Green Chemistry and Sustainable Industrial Technology. He is a graduate of the University of York at both MChem and PhD level, specialising in Green Chemistry throughout before spending time in industry in the Waste Management sector, returning to the University in 2013 and taking up a lectureship in 2019.

A Fellow of the Higher Education Academy, he works both within the Department and across the wider university in implementing greener principles into the curricula through development of interdisciplinary undergraduate option modules and promotion of the MSc principles locally and across the Higher Education sector. He is also involved in schools and outreach programmes, promoting the work of the Green Chemistry group in the wider community and helping inspire the next generation of chemists.

His own work focuses on the valorisation of food waste into useful chemical products and materials in line with the United Nations' Sustainable Development Goals, particularly Target 12.3—to halve per capita global food waste at the retail and consumer levels and reduce food losses along production and supply chains, including post-harvest losses, by 2030.

*sustainability*

MDPI

*Review*

# Food Loss and Waste: A Review of Voluntary National and Voluntary Local Reviews of the United Nations

**Emi Kameyama [1,*], Sabine O'Hara [1] and Marian Stuiver [2]**

[1] College of Agriculture, Urban Sustainability and Environmental Sciences (CAUSES), University of the District of Columbia (UDC), 4200 Connecticut Ave NW, Washington, DC 20008, USA; sabine.ohara@udc.edu
[2] Data and Knowledge Hub for Healthy Urban Living, Utrecht University, Yalelaan 2, 3584 CM Utrecht, The Netherlands
* Correspondence: emi.kameyama@udc.edu

**Abstract**

Food loss and waste (FLW) presents a major global challenge and is explicitly referenced in the several of the United Nations Sustainable Development Goals (SDGs), most notably in Target 12.3. The growing literature on FLW addresses a number of strategies to mitigate the problem; however, much of this work prioritizes technical solutions while studies focused on institutional solutions are less prevalent. Our study offers a systematic review of the FLW solutions discussed in the Voluntary National Reviews (VNRs) and Voluntary Local Reviews (VLRs) of the United Nations. We apply the Institutional Analysis and Development (IAD) framework pioneered by Nobel Laureate, Elinor Ostrom to our analysis of FLW solutions in the VNRs and VLRs to identify prevalent patterns as well as gaps evident in the reviews. We base our systematic review of FLW solutions on the VNRs and VLRs rather than on the peer-reviewed literature since government-reported FLW initiatives play a key role in guiding individual and collective actions toward achieving SDG 12.3. Using a systematic review approach, we identified 165 publicly available VNRs and 116 VLRs that report FLW activities and were published in English between 2016 and 2024. We then applied the IAD framework categories to the 281 identified Reports. Our findings suggest that nearly 60% of the Reports focused on technical solutions while 40% reported socially oriented solutions. The socially oriented solutions were primarily directed toward consumers and less frequently toward businesses and policymakers. These findings suggest that FLW solutions can be meaningfully advanced by placing more emphasis on the role of informal institutions in in advancing awareness and social engagement to address FLW. A stronger emphasis on social rather than technical solutions may also facilitate the development of local and regional partnerships that advance FLW objectives across all levels of governance, from local to global.

**Keywords:** food loss and waste; UN sustainable development goals; institutional analysis and development

## 1. Introduction

Approximately 40% of global food production is lost or wasted along the food supply chain. Food loss and waste (FLW) therefore represents a critical issue associated with economic costs exceeding $1 trillion annually [1]. FLW entails not only the loss and waste of food products but also of critical resources that went into producing the food products. Among these are losses in freshwater, nutrients, fossil fuel, and labor. These direct and indirect losses create negative consequences for the environment in general and the climate in particular. In 2022, FLW generated an estimated 10% of global greenhouse gas (GHG) emissions [1] and caused the eutrophication of freshwater bodies and coastal waters and their ecosystems [2]. At the same time, nearly 30% of the world's population experienced moderate to severe food insecurity [3].

Not surprisingly, reducing FLW is articulated in the 17 United Nations Sustainable Development Goals (SDGs). Goal 12, for example, calls for responsible consumption and production, and SDG 12.3 aims to "halve per capita global food waste at the retail and consumer levels and reduce food losses along production and supply chains, including post-harvest loss" [4]. The FLW reduction target of SDG 12.3 foresees the use of two sub-indicators to measure progress: (1) a Food Loss Index (12.3.1a) and (2) a Food Waste Index (12.3.1b). These indexes are managed by the Food and Agriculture Organization of the United Nations (FAO) and the UN Environment Programme (UNEP). According to the FAO, "food loss refers to the decrease in edible food mass at the production, post-harvest and processing stages of the food chain, mostly in developing countries. Food waste refers to the discard of edible foods at the retail and consumer levels, mostly in developed countries" [5]. Food loss is generally attributed to limitations in harvesting techniques, insufficient storage and cooling facilities, a lack of infrastructure, and socioeconomic factors [6–8]. Food waste is typically the result of consumer related matters like a lack of behavioral considerations, insufficient food preparation coordination, and inadequate planning [9]. In 2021, more than 13% of the food produced globally was lost, and 19% of all food was wasted at the retail and consumption stage [10].

Projections suggest that SDG 12.3 will not be met. The list of factors that have contributed to the lack of progress in meeting FLW reduction goals include the COVID-19 pandemic, worsening climate conditions, and various geopolitical conflicts. FAO Director-General Qu Dongyu has stressed the need for collective action to scale up efforts to reduce FLW while also reducing GHG emissions [11].

The growing academic literature on FLW explores a range of opportunities to improve FLW reduction efforts. Some of these solutions target environmental objectives [12], while others focus on the food supply chain [13], specific groups of foods [14], definitional elements of FLW [15], and circular economy approaches that provide an alternative to the standard take-make-dispose approach that has shaped much of today's food system [16]. Scholars generally agree that FLW transformations are complex and will require multifaceted solutions, such as the "co-evolution of both physical infrastructure systems that produce, deliver, and manage food and waste, and the human infrastructure, from front-line workers to governance and institutions" [17] (p. 1).

Much of the growing literature on FLW is focused on technical solutions [18]. However, an emerging body of work notes that solutions limited to technical considerations will be insufficient as FLW also reflects social, cultural and governance contexts [18–20]. This points to the importance of institutions that shape FLW. According to Durkheim, an institution can be defined as the beliefs and modes of behavior codified by a collectivity such as society at large or a sub-group [21]. North defines institutions as the human designed rules, procedures and organizations that structure political, economic, and social interactions, and consist of both informal constraints and formal rules [22]. Formal institutions include laws, policies, and regulations, while informal institutions include norms, mindsets, and accepted practices [23,24]. Giddens suggests that institutions are "standardized modes of behavior which play a basic part in the time-space constitution of social systems" [25] (p. 96). They are not static, but are constantly reformed through people's actions. These actions can shape informal institutional mindsets and formal institutions as norms (informal) shape policies (formal) and vice versa [24,26]. Governments are "the institution at

https://doi.org/10.3390/su18020833

the forefront of balancing the interests of society, its sub-groups, and individuals" [24]. An analysis of government actions and institutional characteristics is therefore critical to understanding the complexities of FLW. Yet while the importance of formal and informal institutions is well-established [27], an institutional approach to FLW has only recently been explored in the academic literature [28].

The objective of this study therefore is to analyze FLW solutions through an institutional lens using the Institutional Analysis and Development (IAD) framework pioneered by the Nobel Laureate, Elinor Ostrom. The IAD framework elucidates formal and informal institutional factors impacting collective initiatives [29]. In applying the IAD framework, our objective is to identify prevalent patterns as well as research gaps in reported solutions to FLW (Figure 1). We base our analysis on a systematic review of FLW activities reported in the Voluntary National Reviews (VNRs) and Voluntary Local Reviews (VLRs) submitted to the United Nations. These Reviews have proven to be key resources for understanding strategies to address SDGs and assess whether and how SDG targets are met at the national and local levels since governments are key actors in shaping FLW-related solutions [30,31]. With almost global participation, VNRs are increasingly seen as a vital resource that can aid our understanding of how countries are approaching the SDGs and SDG targets [31,32]. We therefore conduct a systematic review of VNRs and VLRs and apply Ostrom's IAD framework to examine how national and local/regional governments have addressed FLW between 2016 and 2024. Our analysis focuses chiefly on initiatives that address FLW. The results of our review indicates that most of the proposed strategies to address FLW are technical in nature while institutional and social-oriented solutions are underrepresented. This confirms similar findings in the academic literature and suggests that FLW solutions can be advanced by increasing research on the role of institutions, and especially on informal institutions that support FLW conscious mindsets.

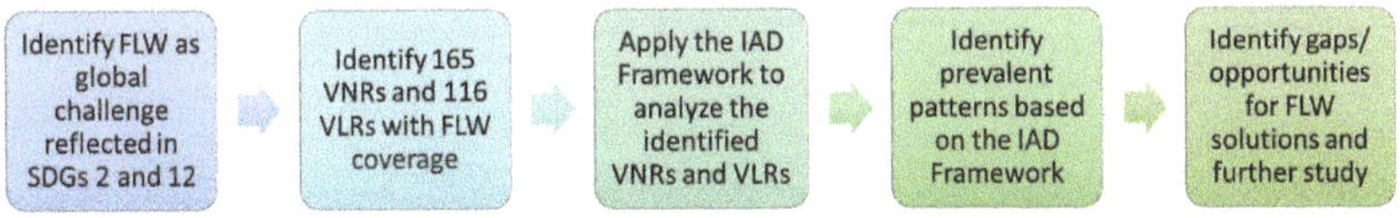


**Figure 1.** An overview of the research process.

## 2. Materials and Methods: Conducting a Systematic Review of FLW Solutions Through the Lens of the Institutional Analysis and Development (IAD) Framework

Our methodology consists of two components. We conduct a systematic review of the FLW activities reported in publicly available VNRs and VLRs. To structure the activities reported in the VNRs and VLRs, we apply the IAD framework to assess the types and prevalence of the reported FLW activities. This approach allows us to distinguish between technical vs. institutional activities.

### *2.1. Analytical Approach: The IAD Framework*

The IAD framework, developed by the 2009 Nobel laureate Elinor Ostrom provides a tool for analyzing complex problems by breaking them into smaller, practical elements, allowing outcomes to be evaluated without losing the complexity of real-life decision-making scenarios [23,28]. The framework has been widely used to analyze collective action cases to sustain common pool resources such as forests, pastures, scarce freshwater sources, and fisheries [33,34]. It "can be used to investigate any broad subject where humans repeatedly interact within rules and norms that guide their choice of strategies and

https://doi.org/10.3390/su18020833

behaviors... and can lead one out of the path dependency of existing patterns of practice when their accompanying ways of thinking have not yielded solutions" [29] (p. 41).

As shown in Figure 2, the IAD framework involves several elements. Its core element is "the identification of an action situation and the resulting patterns of interactions and outcomes, and evaluating these outcomes" [35]. In action situations, participants "observe information, select actions, engage in patterns of interaction, and realize outcomes from their interaction" [36]. Action situations are typically influenced by three sets of exogenous variables or contextual factors: (1) biophysical conditions, (2) community attributes, and (3) rules-in-use [35,36]. Biophysical conditions include technological and infrastructure attributes, including physical and material resources. Community attributes include values, beliefs, and preferences, including cultural norms and common understandings accepted by the community. Rules-in-use refers to both formal rules such as laws, policies, and procedures, and informal ones such as norms, standard operating practices, and habits [23]. Patterns of interactions are shaped by the choices actors have available to them and by the three types of exogenous variables that form the context of all actions (see Figure 2). The IAD framework proposes six evaluative criteria to assess the outcomes of actor-context dynamics: (a) economic efficiency, (b) fiscal equivalence, (c) redistribution equity, (d) accountability, (e) conformance to values of local actors, and (f) sustainability [35].

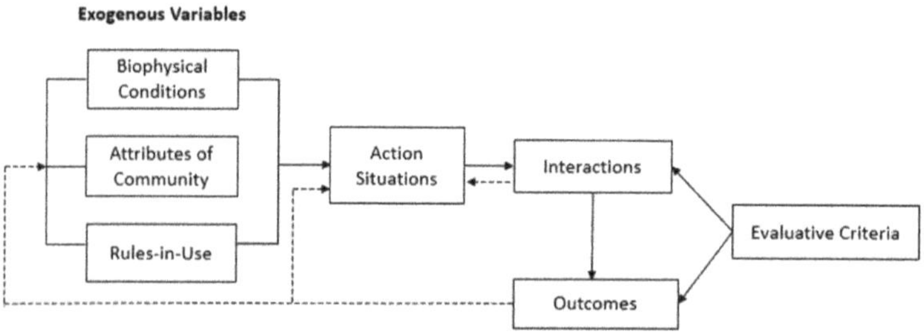


**Figure 2.** Institutional Analysis and Development Framework. (Source: Adapted from Ostrom [35]).

Other conceptual frameworks relating to the role of institutions include social-ecological systems that analyze the interactions among four core subsystems of governance, users, resource system, and resources units [37,38]; the sociological institutionalism framework which emphasizes causal structures rooted in culture and interpretation [39]; the social-technical transitions framework which proposes four main pathways of (1) transformation, (2) reconfiguration, (3) technological substitution, and (4) de-alignment and re-alignment [40]; and the multilevel governance framework which focuses on the growing complexity of authorities across local, regional, national, and supranational levels of engagement [41] and the involvement of both public and private sector actors [42]. In this study, we apply the IAD framework as it offers a widely recognized concrete analytical tool to structure our analysis of the complexities of FLW.

Several articles have used the IAD framework to analyze FLW solutions. For example, a recent study has applied the IAD framework to understand critical insights on household food waste recycling in Hong Kong [28] and another assessed community-based composting in Viet Nam [43]. The IAD framework has also been used to analyze waste management issues, such as campus sustainability initiatives in the United States [44]; recycling and recovery efforts in Brazil, Indonesia, and Nigeria [45]; households waste disposal in China [46]; and larger food systems related topics like agri-food cooperative models [45], agri-environmental climate schemes [47], and urban agriculture [48]. A summary of these recent contributions can be found in Table 1.

https://doi.org/10.3390/su18020833

**Table 1.** Recent Uses of the IAD Framework.

| Topics | Authors | Application Cases and Countries |
|---|---|---|
| Food Waste Recycling | Chua and Yau [28]<br>Nguyen and Watanabe [43] | Household food waste recycling in Hong Kong<br>Community-based composting in Viet Nam |
| Solid Waste Management | Chikowore and Kerr [44]<br>Oh and Hettiarachchi [45]<br>Zhang and Zhao [46] | Waste systems on campus in the United States<br>Waste recycling and recovery in Brazil, Indonesia, Nigeria<br>Households waste disposal in China |
| Food Systems | De Herde [47]<br>Sattler et al. [48]<br>Vidal Merino et al. [49] | Agri-food cooperative models in Belgium<br>Agri-environmental schemes in six European countries<br>Urban agriculture in Ghana, Uganda, and South Africa |

A number of articles have also applied the IAD framework as a structural element to analyze and synthesize information in complex decision scenarios [33,50,51]. Yet to date the IAD framework has not been applied to systematically examine reported FLW activities at the national and regional/local levels. Our study seeks to address this research gap.

### 2.2. *A Systematic Review of FLW Solutions Reported in VNRs and VLRs*

We apply the IAD framework to systematically analyze FLW activities reported in the VNRs and VLRs submitted to the United Nations (UN). VNRs are the primary government reports presented at the UN High-Level Political Forum on Sustainable Development (HLPF) which was established at the 2012 UN Conference on Sustainable Development [52]. The 2030 Agenda for Sustainable Development encourages UN member states to "conduct regular and inclusive reviews of progress at the national and sub-national levels, which are country-led and country-driven" [53] to share knowledge and experience. VLRs are designed to complement VNRs by reporting on the progress cities and regions have made toward meeting the SDGs [54]. The VLRs thus acknowledge the fact that local communities have a significant role to play in advancing the SDGs overall [30].

The first HLPF after the adoption of the 2030 Agenda for Sustainable Development was held in 2016. Subsequent HLPFs have been held annually in July, and each forum includes an in-depth review of several SDGs. SDG 12 on responsible consumption and production, which includes FLW objectives, was reviewed in 2018 and 2021 and is scheduled to be reviewed again in 2027 [55]. All UN member countries with the exception of Myanmar, Haiti, and the United States have submitted VNRs [56], and many countries and cities have submitted their VNRs and VLRs multiple times since 2016.

A growing body of work examines different themes across VNRs and VLRs. For example, linkages between SDGs 2 (zero hunger) and 6 (water) were examined based on 195 VNRs reported by 159 countries [57]; coverage of SDG 4 (education) was examined based on 33 VNRs from the 2019 HLPF [58]; and an analysis of city-level SDG reviews was based on 151 VLRs published between 2018 and 2022 [59]. The coverage of FLW in VNRs and VLRs however, has not yet been addressed.

#### 2.2.1. The Systematic Review Process

Our systematic review of VNRs and VLRs used a two-step approach. In the first step, we identified publicly available VNRs and VLRs that cover FLW activities [53,54]. Our analysis included all full VNRs and VLRs published in English between 2016 (the first year VNRs were released) and 2024. Reports published in other official UN languages such as Arabic, Chinese, French, Russian, and Spanish were not included (see Table 2). We then identified the VNRs and VLRs included in our analysis by conducting a full-text screening process using a string of keywords, including "anaerobic digestion," "compost," "food," "food loss," "food waste," "organic waste," "postharvest," "urban," "waste," and "SDG

https://doi.org/10.3390/su18020833

12.3." VNRs and VLRs that discussed specific waste management issues such as plastics, e-waste, medical waste, hazardous waste, etc., without any specific reference to food waste were excluded. The VNRs and VLRs in our sample used diverse reporting structures and varied in depth and length [60]. Many reports presented chapters according to UN Goals and discussed their FLW activities in considerable detail. Goals 2 and 12 were particularly frequently mentioned. Other reports presented information according to their own national or local priorities. We identified the VNRs and VLRs in our sample based on our keyword search regardless of the level of detail reported or the structure used.

**Table 2.** Inclusion and Exclusion Criteria.

| Inclusion Criteria | Exclusion Criteria |
|---|---|
| Full VNR and VLR reports | Executive summaries and duplications |
| Published between 2016 and 2024 in English | Non-English written reviews and published in 2025 |
| Discussions related to FLW | Solid waste management without references for food waste |

In establishing our sample, we applied the PRISMA protocol (Preferred Reporting Items for Systematic Reviews and Meta-Analyses) to our review of VNRs and VLRs [61,62]. Figure 3 illustrates our review process. At the national level, a total of 410 VNRs published as of 21 July 2025 were identified [53]. We eliminated reports that referenced expressions of interest for the 2025–2029 reporting cycle ($n$ = 44), VNRs published in languages other than English ($n$ = 104), and a partial report ($n$ = 1). Our sample thus consisted of 261 VNRs written in English and published between 2016 and 2024. Our full-text analysis through keyword search revealed that 63% of the VNRs (165 of 261) included FLW coverage. We extracted data from these 165 VNRs and quantitatively and qualitatively synthesized the reported FLW activities.

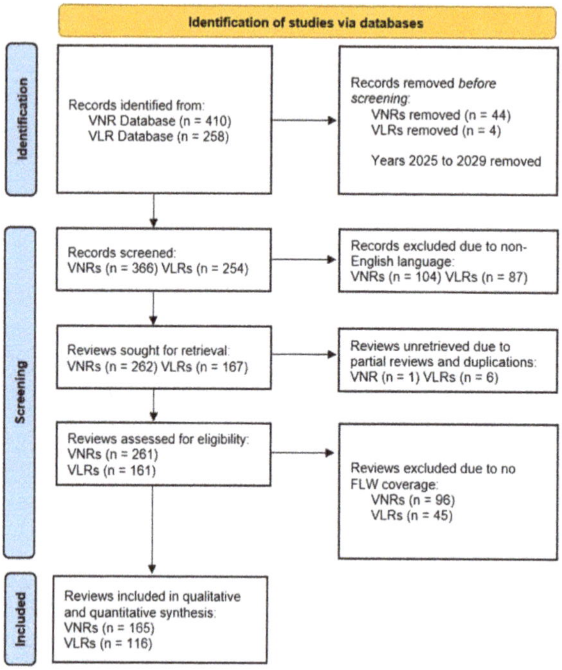


**Figure 3.** PRISMA 2020 flow diagram for the inclusion of VNRs and VLRs with FLW coverage.

https://doi.org/10.3390/su18020833

At the regional and local levels, a total of 258 VLRs were publicly available as of 21 July 2025 [54]. Of these, 116 VLRs met our inclusion criteria of full VLRs with FLW coverage, published in English between 2016 and 2024. The full-text screening revealed that 72% of the VLRs (116 out of 161) reported FLW activities. Data were extracted from these 116 VLRs to synthesize results across all of the VLRs in our sample (see Supplementary Materials for a list of all VNRs and VLRs studied).

We sorted our systematic review results into eight world regions derived from the reported regional groupings of the UN SDG indicators [63] whereby we separated Northern America from Europe to yield the following regions: (1) Sub-Saharan Africa, (2) Northern Africa and Western Asia, (3) Central and Southern Asia, (4) Eastern and South-Eastern Asia, (5) Latin America and the Caribbean, (6) Oceana, (7) Europe, and (8) Northern America. Our approach made it possible to analyze the FLW coverage in VLRs submitted from five U.S. cities (Los Angeles, New York, Orlando, Pittsburgh, Sacramento) and the State of Hawai'i, even though the U.S. has not submitted VNRs during the timeframe of our systematic review.

2.2.2. Applying the IAD Framework

In the second step, we organized the coverage of FLW in the VNRs and VLRs in our sample based on Ostrom's IAD framework categories (see Figure 2). The categories included in our application of the IAD framework are:

① Action situations
② Biophysical conditions (exogenous variable)
Rules-in-use (exogenous variable)
Attributes of community (exogenous variable)
③ Patterns of interactions
④ Outcomes based on technical evaluative criteria
Outcomes based on social evaluative criteria

We mapped the FLW activities reported in our sample into relevant IAD categories as shown in Figure 4. The numbers in Figure 4 indicate the sequencing used to identify prevalent IAD patterns in the coverage of FLW activities reported in the VNRs and VLRs. These patterns also correspond to activities found in the published literature that synthesized the IAD framework, such as work by Chua and Yau [28] and Fouladvand et al. [33]. For the purpose of our analysis, 'action situations' refers to national and local/regional governments' reports of their respective FLW activities including key actors that are discussed in the VNRs and VLRs in our sample. This includes public sector actors, private sector actors like consumers, farmers and fisheries, nongovernmental organizations (NGOs and INGOs), and educational institutions. Our analysis indicates that the most frequently discussed FLW solutions reported are based on the well-established food recovery hierarchy of waste reduction before recovery and recycling. This hierarchy frequently serves as a framework to guide FLW strategies [12,64]. This implies that FLW reduction strategies (i.e., minimizing food surplus) are considered preferable to recovery strategies (i.e., through food donation, redistributing, and upcycling), and the lowest preference is attributed to recycling strategies (i.e., using food waste as animal feed, in anaerobic digestion, and composting). Disposal in landfills or incineration of food is considered a strategy of last resort [65].

In the biophysical condition category, we identified VNRs and VLRs that discuss technological and infrastructure-related solutions to address FLW. Key FLW reduction strategies include solutions like postharvest interventions through improved storage, better transportation, and updated infrastructure [7]. Also mentioned are improved packaging and digital solutions such as meal planning apps [64]. Urban solutions include strategies that reduce the distance between food production and consumers and

https://doi.org/10.3390/su18020833

thus the need for a robust food storage and distribution infrastructure [14,66,67]. Also included are more technical food production methods in urban environments like hydroponics and aquaponics production and solutions that advance a more localized food systems [68–70]. Technological and infrastructure-related solutions also include food waste recovery, including the establishment and operation of food banks and surplus food redistribution efforts such as the distribution of so-called ugly vegetables that do not meet standard consumer expectations, and upcycling technologies for "remaking wasted food into new food products for human consumption" [71] (p. 16). Biophysical solutions mentioned also include repurposing food for animal feed, converting food waste into energy through anaerobic digestion, and composting organic waste with various technologies and facilities.

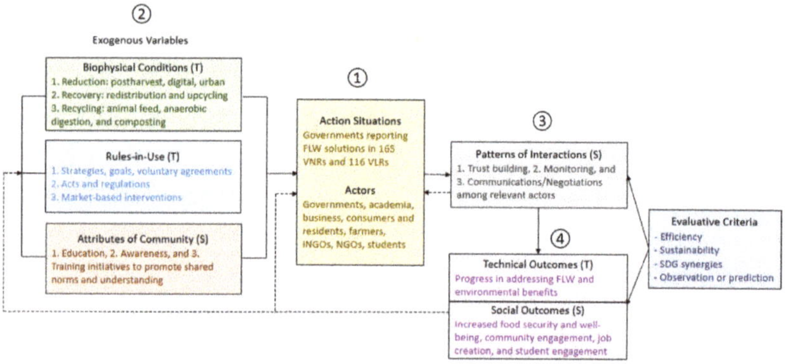


**Figure 4.** Application of the IAD Framework to analyze FLW activities reported in VNRs and VLRs. Note: (T) indicates technical strategy and (S) denotes socially oriented strategy. The numbers indicate the sequencing used to identify prevalent IAD patterns in FLW coverage in the VNRs and VLRs.

In the rules-in-use category of the IAD framework, we identified formal institutional solutions like regulations, legislation, guidelines, market-based interventions, and similar formal institutional strategies that seek to reduce FLW. Some of these rule-in-use strategies also reflect technically oriented solutions like strategies, for example, that reflect the European Union's FUSIONS (Food Use for Social Innovation by Optimizing waste prevention Strategies) initiative. Yet these FLW strategies typically rely heavily on policy approaches, regulations and market-based interventions [72]. Frequently, the formal institutional measures can be divided into carrot and stick approaches that rely on encouragement and punishment [73]. This includes strategies like banning food waste from landfills and establishing mandatory targets and rewards for waste reduction, composting and landfill bans on food waste, but also market-based instruments such as subsidies, grants, and tax incentives. However, few studies have proven the effectiveness of taxes and fees in reducing food waste [28,74].

In the 'attributes of communities' category of the IAD framework, we identified various FLW education, awareness, and training initiatives targeting various actors Community attributes such as norms, values, and culture are shaped by the actions and motivations of individuals within the community, as well as the interactions and behaviors directed toward the community [33,75]. Given the general lack of awareness of the significance of FLW, a growing number of studies examine the potential of educational interventions in schools to reduce household food waste [76], efforts to reduce consumer food waste at household level [77,78], and the need for improved training to address uncertainties regarding when to discard food at the retail level [17,79]. Education and awareness programs can also foster shared norms and values around FLW, and training activities can influence

https://doi.org/10.3390/su18020833

community attributes by fostering knowledge and skills that lead to positive behavioral changes both at the individual and community (social) level.

According to the IAD framework, 'patterns of interactions' can lead to effective improvements [23]. We therefore included the category in our analysis of IAD variables identified in the VNRs and VLRs in our sample. Several reports mention interactions among relevant actors. Reducing FLW requires many collaborative efforts involving a range of actors. According to Ostrom, patterns of interaction facilitate cooperation as opposed to conflict and may include (1) reciprocity, trust, and reputation building, (2) monitoring and sanctioning behavior, (3) frequency and type of negotiation, communication, or contestation, and (4) whether people tend to free-ride or contribute to collective efforts [80]. We found patterns of trust building (mutual exchange and strategic partnerships), monitoring activities (data reporting and measurement), and communication and negotiation strategies (education, public awareness and stakeholder engagement in policy development) among a range of relevant actors at the local, regional, national, and global levels. Engaging local communities in collaborative efforts with individuals who share common informal institutional mindsets can also unlock viable pathways for more effective formal institutional frameworks like local and regional policies and can create positive spillover effects [24].

The IAD framework category we termed 'outcomes' captures reported activities and strategies that apply identified criteria and report technical and socially oriented outcomes as compared to activities. Outcomes are categorized according to the following criteria: efficiency, sustainability, and SDG synergies. While outcomes can be either observed (for empirical analysis) or predicted (for theoretical analysis) [43], we analyzed the VNRs and VLRs based on outcomes that were observed and anticipated. Among the more technical outcomes were those with clearly stated quantitative objectives such as the amount of postharvest loss prevented and the amount of food waste prevented, recovered, and recycled. We also found sustainability objectives discussed in the reviewed VNRs and VLRs, such as linkages between FLW and environmental benefits, for example. These connections are also increasingly examined in the academic literature particularly with respect to the impact of FLW strategies on GHG emissions [64]. In addition, we identified VNRs and VLRs that discuss interdependencies across the SDGs that can help achieve synergies between Target 12.3 and other SDGs. We also identified VNRs and VLRs that discuss socially oriented outcomes. According to Ngan et al. [81], common descriptions of social sustainability include characteristics like a strong civil society, human well-being, and social engagement. We found both VNRs and VLRs that discuss the social implications of reducing FLW including improved food security and well-being, community engagement, job creation, and the engagement of students [82,83].

Our systematic data analysis and sorting of FLW coverage in the Reviews based on the IAD framework indicates that multiple IAD categories are addressed in the majority of VNRs and VLRs in our sample. This aligns with qualitative content analysis approaches that emphasize the importance of deriving categories from the data [84]. To validate our findings, we triangulated our data with other sources including FLW solutions described in the peer-reviewed literature, gray literature (government and NGO reports), and project websites. This strategy is widely recommended to enhance the credibility, trustworthiness, and rigor of qualitative research (see for example, Morgan [85]). Finally, we sought to minimize bias by giving equal weight to VNRs and VLRs and by not favoring specific IAD categories referenced in the Reviews. This approach addresses potential concerns about the single-source dominance or preferential treatment of certain categories, which can distort qualitative findings. The combination of systematic coding, inductive anal-

https://doi.org/10.3390/su18020833

ysis, triangulation, and explicit bias-mitigation seeks to ensure rigor in our qualitative content analyses.

## 3. Systematic Review Results

Our analysis of VNRs and VLRs according to the identified IAD framework categories revealed that nearly 60% of the reviewed reports focused on technical solutions and outcomes, and approximately 40% covered socially oriented solutions (see Table 3). Figure 5 presents a summary of FLW coverage in the reviewed VNRs and VLRs based on the IAD framework for the eight UN world regions. The rules-in-use category dominated in the VNRs, while VLRs most frequently reported solutions relating to the biophysical conditions category. An exception are reports from Europe where rules-in-use focused re-ports dominate both at the national (VNRs) and local level (VLRs).

**Table 3.** FLW coverage in the reviewed VNRs and VLRs based on the IAD Framework.

| | Technical Oriented Solutions | | | Socially Oriented Solutions | | |
|---|---|---|---|---|---|---|
| | Biophysical N | Rules N | Total N (%) | Community N | Interactions N | Total N (%) |
| VNRs | 87 | 120 | 207 (57%) | 67 | 87 | 154 (43%) |
| VLRs | 90 | 78 | 168 (58%) | 61 | 61 | 122 (42%) |
| | Technical Oriented Outcomes N (%) | | | Socially Oriented Outcomes N (%) | | |
| VNRs | 56 (56%) | | | 44 (44%) | | |
| VLRs | 55 (59%) | | | 38 (41%) | | |

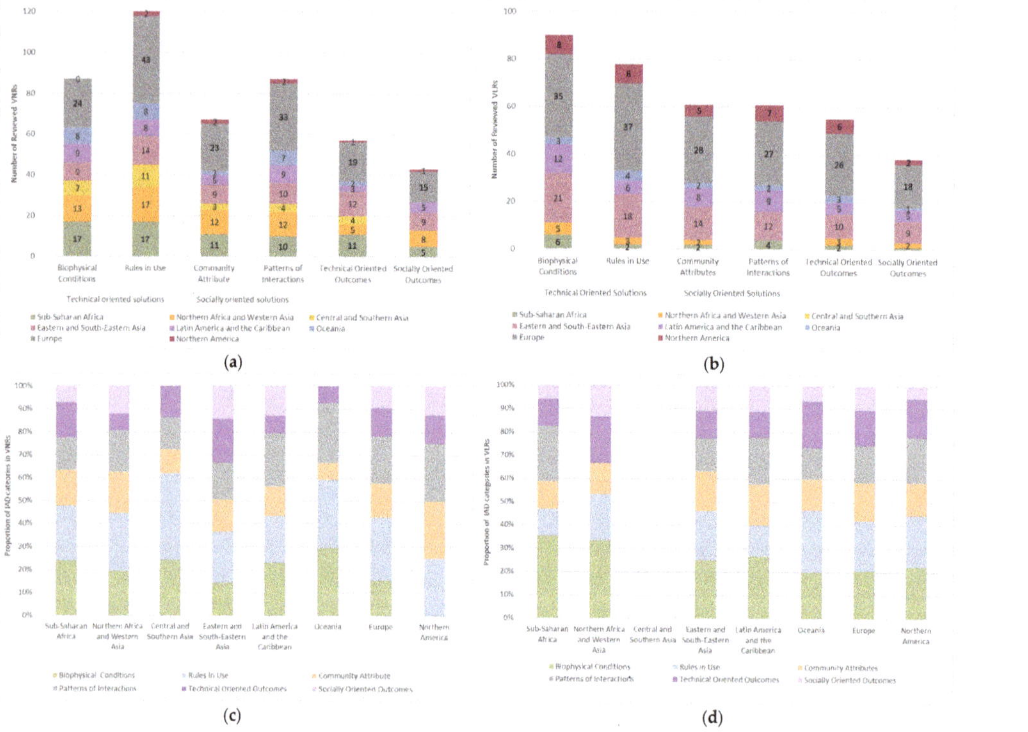


**Figure 5.** FLW coverage in the reviewed (**a**) VNRs and (**b**) VLRs by World Region based on the IAD framework, and the proportion of the IAD categories in the reviewed (**c**) VNRs and (**d**) VLRs within each world region.

https://doi.org/10.3390/su18020833

In terms of the regional distribution of the reports, roughly one third of VNRs came from Europe (34%), followed by Sub-Saharan Africa (21%), Northern Africa and Western Asia (12%), and Eastern and South-Eastern Asia (12%) as shown in Figure 6a. Reports from Europe (44%) also constituted the largest number of VLRs followed by reports from Eastern and South-Eastern Asia (23%), Latin America and the Caribbean (11%), and Northern America (8%). These findings are consistent with a review of the academic literature reported by Oliveira et al. [16], who indicate significant interest in addressing FLW among European scholars. As Figure 6b shows, the number of local reports addressing FLW issues increased particularly dramatically between 2017 and 2021 (an increase of 1200%) and local reports now exceed the number of national reports with FLW coverage.

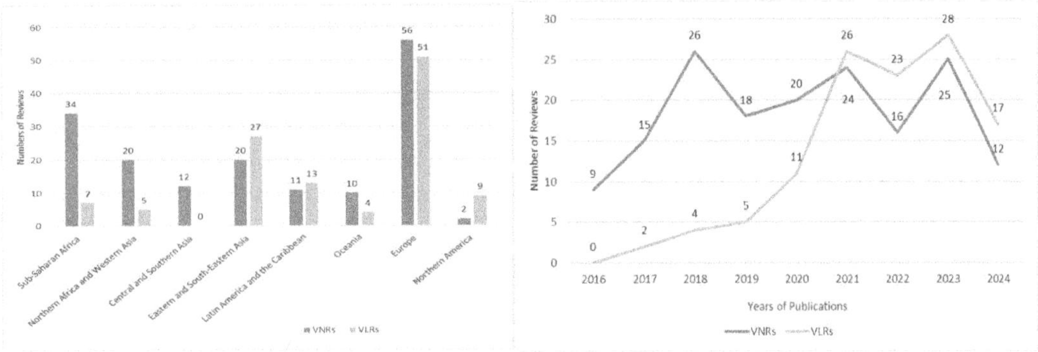


(a) (b)

**Figure 6.** FLW Publication Trend (**a**) by world region and (**b**) by year.

Some regional differences in both VNRs and VLRs stand out and are discussed in the following section which provides an analysis of FLW coverage based on the identified IAD categories.

### 3.1. *Reports Focused on Biophysical Conditions*

87 of the 165 reviewed VNRs and 90 of 116 VLRs (53% and 78%, respectively) discussed technological- and infrastructure-related FLW solutions. A significant proportion of the reports came from the local level across all world regions (see Figure 5).

#### 3.1.1. VNRs and Biophysical Conditions

Of the 87 VNRs in our sample, 52% discussed solutions related to FLW reduction (45 VNRs), 45% focused on food waste recycling (39 VNRs), and 32% on food waste recovery (28 VNRs) (Table 4). Because some of the reports cover more than one type of solution (reduction, recovery, and recycling), the percentages add up to more than 100%. The majority of the reported FLW reduction strategies (76% or 34 of 45 VNRs) discussed postharvest interventions, with a focus on modernizing storage facilities for various types of commodities such as seeds, grains, and fruits and vegetables. These reports came mainly from Sub-Saharan Africa and from Small Island Developing States (SIDS), with the Island States discussing primarily reducing losses of fishery products. The second most frequently mentioned improvements were FLW reductions through improved mechanization and transportation infrastructure. 13% of the reports discussed food waste reduction technologies such as smart packaging and the use of coatings to extend the shelf life of fruits and vegetables, and another 13%, like the VNRs from Egypt, Guyana, Mauritius, and Seychelle, for example, discussed efficient production practices such as bio-farming, hydroponics, and aquaponics.

https://doi.org/10.3390/su18020833

**Table 4.** Biophysical conditions discussed in the reviewed VNRs and VLRs.

| | VNRs N | VLRs N |
|---|---|---|
| FLW Reduction | 45 | 7 |
| Food Waste Recovery | 28 | 33 |
| Food Waste Recycling | 39 | 70 |

Over 30% of the reports focused on food waste recycling technologies came from Europe. Almost two thirds of these VNRs discussed composting in different contexts such as at schools, residential facilities, etc., while 36% focused on anaerobic digestion and 10% on utilizing food waste as animal feed thus reducing reliance on animal feed imports (see for example a report from Japan). Reports focused on food waste recovery primarily reported establishing or operating food banks and other food donation efforts (21 out of 28 VNRs or 75%). More than half of these recovery focused reports (57%) came from European countries. A far smaller percentage of the VNRs (14% or 4 VNRs) focused on upcycling technologies. Interestingly, the upcycling reports came from different world regions and were quite region specific, including repurposing shrimp shells into organic spices and fertilizers in Egypt and producing biopolymers from banana and cocoa crops in Samoa.

#### 3.1.2. VLRs and Biophysical Conditions

Among the 90 VLRs, food waste recycling received the largest coverage with 78% or 70 of the local reports. This was followed by food waste recovery (37% or 33 VLRs) and FLW reduction (8%). Composting played a major role among these reports (77% or 54 of 70 reports). Reported solutions include home, community, and commercial composting systems across all seven world regions, followed by anaerobic digestion (21% mainly from Europe) and waste-to-energy projects (including from Cape Town). Two VLRs (eThekwini and Helsingborg) discussed the production of biochar from food waste. Food waste recovery and redistribution solutions were addressed in 52% of the reports (17 out of 33) and came mainly from Europe, including from Finland (Stadin safka), the Netherlands (Too Good To Go), and Portugal (Fruta Feia). 42% of the reports (14 of 33 reports) discussed food banks while some distinctly regional reports focused on upcycling shrimp shell waste (Indonesia) and repurposing used cooking oil (Malaysia). Only 8% discussed FLW reduction through community-based urban agriculture solutions that address both food access and food waste.

### 3.2. *Rules-in-Use*

The vast majority of both VNRs (73%) and VLRs (67%) discussed solutions based on rules-in-use strategies. Approximately 30% of the VNRs and more than 22% of these VLRs come from Europe (Figure 5).

#### 3.2.1. Rules-in-Use in VNRs

Among the 120 VNRs that discussed rules-in-use solutions, the vast majority reported on various FLW reduction plans, followed by a limited number of market-based interventions (17%) and regulations (16%) (see Table 5). The rules-in-use based FLW reduction strategies include developing a dedicated national food waste action plan or national roadmap to reduce FLW (29%). The rules-in-use focused reports come predominantly from Europe. However, the relative importance of reported rules-in-use strategies as a percentage of the overall strategies reported from a world region, indicates the strongest commitment to rules-in-use solutions in Central and Southern Asia (38%) and Oceana (30%). Postharvest management strategies come (19%) mainly from Sub-Saharan Africa,

https://doi.org/10.3390/su18020833

and 14% of the reports discussed a goal of halving food waste by 2030 which aligns with SDG 12.3. Several reports (11%) also addressed FLW reduction strategies as part of broader food security and food systems strategies, including the EU Farm to Fork Strategy. Five VNRs from Europe (Denmark, France, Greece, Norway, and the UK) discussed voluntary agreements, such as the Courtauld Commitment launched by the Waste and Resources Action Programme (WRAP) and a No Food Waste Certification program originating from Greece. 31% (32 out of 104 VNRs) of the rules-in-use focused national reports discussed food waste recycling strategies as part of a national waste management strategy (59%), a circular economy strategy (15%), and a zero-waste masterplan (13%).

**Table 5.** Rules-in-use strategies reported in VNRs and VLRs.

| | VNRs N | | | VLRs N | | |
|---|---|---|---|---|---|---|
| | **Strategies** | **Regulations** | **Market-Based** | **Strategies** | **Regulations** | **Market-Based** |
| Reduction | 73 | 6 | 8 | 40 | 2 | 1 |
| Recovery | 7 | 6 | 6 | 0 | 1 | 0 |
| Recycling | 32 | 5 | 6 | 28 | 5 | 3 |
| Disposal | 0 | 4 | 0 | 1 | 2 | 0 |

Key market-based interventions discussed in the national reports include free food waste audits for small businesses in the United Kingdom, subsidies for retailers to promote imperfect produce in Western Asia, and a Value Added Tax exemption to encourage food donations mainly in Europe. Regulatory measures include food waste reduction legislation in Japan and Romania, postharvest regulations such in Ghana and Jamaica, food donation regulations, and an outright ban on food waste in landfills in Kazakhstan.

#### 3.2.2. Rules-in-Use in VLRs

Of the 78 VLRs that reported rules-in-use strategies, 88% (69 VNRs) focused on planning documents, followed by regulations (13%) and market-based interventions (5%). The large majority of the local reports (40 out of 69 VLRs or 58%) discussing FLW reduction plans came from Europe. Reported strategies include local food reduction plans, green city strategies, and local action plans to implement Agenda 2030. 41% (28 out of 69 VLRs) of the local reports discussed food waste recycling strategies. These reports too came predominantly from Europe, followed by plans from Eastern and South-Eastern Asia and North America, including the Aloha+ Challenge from Hawai'i, L.A.'s Green New Deal, and zero waste strategies from Orlando, Florida, Pittsburgh, Pennsylvania, and New York City, New York. Noteworthy are also local reports emphasizing circular economy principles (22%) and zero waste strategies (14%). Other rules-in-use solutions that address regional solutions within national boundaries include the California Senate Bill 1383 that introduces organic waste recycling from the state of California, the Catalan Law 3/2020 which focuses on preventing FLW in the Catalan region, and the use of sensors to address food waste disposal violations from the city of Suwon in South Korea.

### 3.3. *Attributes of Communities*

A total of 67 out of 165 VNRs (41%) and 61 out of 116 VLRs (53%) discussed the community at-tributes category of our IAD analysis, which was underexplored across the eight world regions compared to the biophysical and rules-in-use categories.

#### 3.3.1. Community Attributes in VNRs

Of the 67 VNRs in this IAD category, 60% (40 VNRs) discussed raising awareness, while 34% (23 VNRs) focused on education, and 31% (21 VNRs) on training activities

(Table 6). Raising public awareness among consumers played a significant role (83% or 33 out of 40) among reports from Europe, Western Asia, and Eastern and South-Eastern Asia. These countries reported on national campaigns such as the WRAP 'Love Food Hate Waste' campaign in Australia and the UK. 10% discussed raising awareness among students, including a holistic approach in Europe that promotes healthy eating, locally sourced food, and food waste reduction. In terms of education, 52% (12 of 23 VNRs) of the reports focused on consumers and 30% on students, mainly in Europe through programs such as the EU Platform on Food Losses and Food Waste and Boroume at School in Greece. Regarding training, 62% (13 out of 21) of the VNRs in this category discussed farmer training emphasizing improved postharvest practices across Sub-Saharan Africa and Western and South-Eastern Asia. Only 7% (5 out of 67 VNRs) of the reports discussed awareness, education, and training activities for businesses, such as educating restaurants on ways to reduce stocking expiration sensitive products in Saudi Arabia, and only 6% (4 out of 67) discussed education and training for policymakers such as training programs on food storage and fumigation in Jamaica.

**Table 6.** Promoting community attributes discussed in the reviewed VNRs and VLRs.

| | VNRs N | | | | | | VLRs N | | | | | |
|---|---|---|---|---|---|---|---|---|---|---|---|---|
| | **Business** | **Consumers** | **Farmers** | **Policy Makers** | **Students** | **All Actors** | **Business** | **Residents** | **Farmers** | **Policy Makers** | **Students** | **Teachers** |
| Awareness | 1 | 33 | 2 | 0 | 4 | 1 | 3 | 29 | 0 | 2 | 1 | 0 |
| Education | 1 | 12 | 2 | 1 | 7 | 2 | 3 | 16 | 0 | 1 | 12 | 1 |
| Training | 3 | 0 | 13 | 3 | 2 | 2 | 3 | 3 | 0 | 0 | 0 | 1 |

3.3.2. Community Attributes in VLRs

Among the 61 VLRs in this category, 54% discussed education, 52% covered raising awareness, and 16% discussed training activities. Prominent themes in the education focused reports include initiatives for local residents that focus on food waste recycling in cities across Malaysia and the United States for example. 36% of the VLRs discussed education initiatives for students such as European initiatives that promote sustainable school meals with seasonal local produce and vegetarian options that reduce the carbon footprint of food and food waste. Awareness raising initiatives focused primarily on residents (90%), and 82% of the VLRs (9 out of 11) from Europe focused on FLW reduction while 70% of VLRs (7 out of 10) from Eastern and South-Eastern Asia focused on raising awareness for food waste recycling. A number of VLRs from Europe also discussed public awareness campaigns, such as reducing food waste by eating more plant-based foods, consuming less meat and processed foods, and opting for shorter food chains along with organic, local, and sustainable food choices that promote more sustainable food systems. Training initiatives at the local level focused on composting training, for example in Turkey and South Africa. While 14% (9 of 61) of the VLRs discussed education, awareness, and training initiatives for businesses, only 5% (3 VLRs) discussed education and awareness for policymakers.

*3.4. Patterns of Interactions*

87 of the 165 VNRs (53%) in this category and 61 of the 116 VLRs (53%) discussed a diverse range of interactions among relevant actors at the global, national, and local levels that were distributed across all world regions.

3.4.1. Patterns of Interaction in VNRs

Among the 87 VNRs in this category, 28 different patterns of interactions were reported including among major stakeholders across government, INGOs, businesses, and NGOs as key actors at the national level (Figure 7a). Most of the reports in this category (14%)

discussed interactions between governments, while 13% discussed government-INGOs interactions, 8% government-business interaction, and 8% government-business-NGOs interactions. 50% of the government-government interactions focused on international initiatives, for example, between Fiji and Japan and between Slovenia and Kosovo and among regional players in five Indian Ocean islands. INGOs also played a key role in collaborations with national governments, including UN agencies such as the FAO, UNEP, WFP and regional entities like the EU and the Pacific Regional Environment Programme (SPREP) to support SIDS such as Micronesia, Samoa, and Solomon Islands. Of the 87 VNRs in this category, 53% discussed interactions related to trust building such as through technical partnerships and national networks, while 36% focused on communications and negotiations such as convening FLW campaigns and developing strategies, and 26% on monitoring activities such as developing measuring and reporting tools to address FLW.

(**a**) (**b**)

**Figure 7.** Patterns of interactions discussed in (**a**) VNRs and (**b**) VLRs. Categories are not mutually exclusive.

### 3.4.2. Patterns of Interaction in VLRs

The 61 VLRs that mentioned 'patterns of interaction' identified 25 different patterns among relevant stakeholders. NGOs played a stronger role at the local level compared to public sector actors whereby 18% of the reports discussed interactions between governments and NGOs at the subnational level, followed by government-business interactions (13%) and government-government interactions (11%) (see Figure 7b). A number of VLRs discussed city-to-city partnerships at the international level, such as collaborations between the city of Buenos Aires and the Metropolitan Region of Santiago, between the cities of Jakarta and Osaki, Japan, and between the cities of Hamburg and Dar es Salaam. Several VLRs also discussed local and regional participation in the C40 Good Food City Declaration and the Milan Urban Food Policy Pact which address sustainable food systems objectives. Some reports also emphasize the coalitions that formed within some states and cities, such as Zero Waste Hawai'i and various Circular Economy Hubs such as the Basque Country

https://doi.org/10.3390/su18020833

in Spain and Hunter region in Australia. Of the 61 VLRs in this category, 64% discussed interactions relating to trust building such as partnerships around FLW networks and technical cooperation, 41% described communications such as providing information to residents, and only 8% interactions relating to monitoring activities at the local level.

### 3.5. Technically Oriented Outcomes

56 of the 165 VNRs (34%) and 55 of the 116 VLRs (47%) in our sample discussed technical outcomes, which represents a modest percentage compared to the technical solutions described in Sections 3.1 and 3.2 above. VLRs from Northern Africa and Western Asia show the highest proportion of reports focused on technically focused outcomes (20%). VNRs from Eastern and South-Eastern Asia also report comparatively high levels of activity in this category (see Figure 5).

#### 3.5.1. Technically Oriented Outcomes in VNRs

Of the 56 VNRs, 38% (21 VNRs) discussed the environmental benefits of reducing FLW, 32% (18 VNRs) covered progress in food waste recycling, 32% (18 VNRs) discussed FLW reductions, and 9% (5 VNRs) improving food waste recovery. Two-thirds (14 out of 21) of the reports focused on environmental benefits discussed predicted benefits, such as the potential for reducing GHG emissions in Iceland and the UAE, for example, and promoted the water-energy-waste nexus through initiative like the co-digestion of food waste and sewage sludge in Singapore. The remaining one-third (7 out of 21) discussed observed benefits, such as objectives to enrich soils and reduce water pollution through composting in Lesotho, North Macedonia, and Slovenia.

As for food waste recycling, 89% (16 out of 18) of the VNRs discussed outcomes like food waste recycling goals in Singapore from 12% in 2012 to 16% in 2017, generating 7.3 million kWh of electricity annually from potato waste in Turkey, and converting the equivalent of 33,000 tons of commercial food waste into energy in Australia. Biomethane production from food waste contributes to SDG 12.3 along with SDGs 7 (clean energy) and 17 (partnerships), while 83% (15 out of 18) of the reports discussed reducing food loss, such as postharvest losses of food crops primarily in Sub-Saharan Africa (i.e., millet and sorghum losses in Namibia) and South-Eastern Asia (i.e., rice losses in Viet Nam), reduced cold chain related losses in Thailand, and positive trends in reducing food losses measured through the Food Loss Index in Kenya. Only 11% discussed progress in reducing food waste, such as a reduction in food waste per inhabitant in Norway.

#### 3.5.2. Technically Oriented Outcomes in VLRs

Among the 55 VLRs that mentioned outcomes related initiatives, 51% (28 VLRs) discussed progress in food waste recycling, 38% (21 VLRs) discussed environmental bene-fits, 23% (13 VLRs) improvements in food waste recovery, and 22% (12 VLRs) FLW reductions. Regarding food waste recycling, 89% (25 out of 28) discussed observed outcomes and emphasized the amount of organic waste collected or processed for composting. Regarding environmental benefits, 67% (14 out of 21 VLRs) of the reports discussed observed outcomes like $CO_2$ reductions, diverting 3500 tons of $CO_2$ emissions from food waste prevention in Jakarta, reducing around 5500 tons of $CO_2$ emissions from a biogas plant in Bad Köstritz, Germany, and reducing food-related GHG emissions from approximately 9800 tons of $CO_2e$ in 2018 to less than 5400 tons of $CO_2e$ in 2021 in Ghent, Belgium. Other environmental benefits such as reduced water usage at the local level were not discussed in VLRs.

In terms of food waste recovery, all 13 VLRs in this category discussed the amount of surplus food collected, donated, and distributed in various units such as kilograms, tons, pounds, and number of meals. A few VLRs also reported progress in reducing the amount of food waste per capita in the city of Amsterdam and food waste per day per

https://doi.org/10.3390/su18020833

capita reductions in the District of Seodaemun-gu, South Korea. Numerous VLRs also highlighted the benefits of addressing Target 12.3 with respect to other SDGs, such as goals number 2 (zero hunger), 3 (good health), 6 (clean water), 11 (communities), and 14 (life below water).

### 3.6. *Socially Oriented Outcomes*

44 VNRs (27%) and 38 VLRs (33%) discussed socially oriented outcomes, which represented a modest percentage compared to the socially oriented solutions described in Sections 3.3 and 3.4 above. VLRs from Northern Africa and Western Asia and from Eastern and South-Eastern Asia report the highest mentions of activities in this category with 13% of their reports referencing socially oriented outcomes and outcomes measures (Figure 5).

#### 3.6.1. Socially Oriented Outcomes in VNRs

Of the 44 VNRs that discussed social outcomes associated with FLW related initiatives, 61% (27 VNRs) discussed improved food security and well-being, 20% (9 VLRs) discussed greater community engagement, 18% (8 VLRs) covered employment opportunities, and 11% (5 VLRs) addressed increased education and student engagement. These benefits mainly focused on the number of people who benefitted from FLW initiatives. Regarding food security and well-being, 67% (18 out of 27) of the reports highlighted positive impacts of food recovery efforts in supporting vulnerable populations and reducing food waste, such as the Food Bank Viet Nam objective that served more than 3 million people between 2016 and 2021. Other examples included the construction of keyhole gardens, which benefitted over 115,000 people in Lesotho, and upcycled food waste such as the whey-based drink Piena Spēks that offers a range of potential health benefits while addressing food waste in Latvia.

Examples of community engagement efforts discussed in the reports include the number of partners engaged in food banks that foster connections among community members, and employment opportunities like the Chito Shrimp initiative reported from Egypt that benefited over 250,000 Egyptians by repurposing shrimp shells. Other reported social outcomes include the increased level of student engagement. Examples include initiatives from Slovenia where 150,000 children participated in the School Fruit and Vegetable Scheme initiative and from Greece where 24,000 students participated in a Boroume School program, contributing to several SDGs such as 3 (well-being), 4 (education), 12 (consumption), 13 (climate), and 15 (life on land).

#### 3.6.2. Socially Oriented Outcomes in VLRs

Of the 38 VLRs in the social outcomes category, 47% discussed increased food security and well-being, 37% covered community engagement, 18% discussed education and student engagement, and 16% discussed employment opportunities. Prominent (67%) in the food security and well-being group of reports are those focused on food recovery benefits for vulnerable populations and surplus food reductions. Other examples included shared refrigerator projects in the city of Suwon, South Korea that promote food sharing with neighbors, and rooftop allotments in Barcelona, Spain with increased access to fresh produce while reducing food waste. Community engagement played a stronger role in VLRs, including at BuurtBuik in the Netherlands where FLW initiatives are fostering community connections and preventing social isolation. Zero Waste Schools in Hawai'i engaged students in composting programs which received national awards and Kodomo shokudo (children's cafeterias) promoted food education for children and intergenerational partnerships by engaging children and older adults in Kitakyushu, Japan. Food distribution, upcycling, and composting programs also created employment opportunities, and contributed to multiple SDGs such as 1 (no poverty), 2 (zero hunger), 3 (good health),

https://doi.org/10.3390/su18020833

8 (decent work), 9 (industry), 11 (communities), and 17 (partnerships) in addition to Target 12.3.

## 4. Discussion and Future Directions

The IAD framework we applied to the 165 VNRs and 116 VLRs in our sample used three sets of exogenous variables (biophysical conditions, attributes of community, and rules-in-use) which influence action situations, patterns of interactions, and outcomes [35,36]. Our analysis revealed that most of the FLW solutions reported in the VNRs and VLRs focused on technical solutions while socially based solutions were less prominent. This is particularly evident in the 'biophysical conditions' and 'rules-in-use' sections of our IAD framework. The majority of technical solutions discussed in VNRs were rules-in-use based, while the VLRs focused on biophysical solutions. Regarding socially based solutions, the reported education, awareness, and training activities focused mainly on consumers and residents while few initiatives focused on businesses and policymakers. The reports also discussed a wide range of interactions among relevant actors including city-to-city partnerships that span the globe. Outcomes focused reports repeated the same pattern and were also dominated by technically oriented outcomes, while socially oriented outcomes were underexplored. This suggests a need for more integrated approaches that combine technical and social strategies. Building on these results, we highlight key areas for consideration that could strengthen future FLW activities.

First, within the biophysical condition category of our IAD framework, most of the VLRs discussed food waste recycling technologies, while fewer focused on food waste reduction, the most preferred approach according to the widely accepted wasted food scale. This aligns with finding reported by Mourad [86] and Messner et al. [20] who confirm that many existing initiatives concentrate on managing surplus food that has already been produced rather than preventing surplus food in the first place. Very few Reviews (all from Europe including VNRs from Hungary and Spain, and VLRs from cities of Amsterdam and Ghent) discussed the role of shorter food supply chains in preventing FLW, and a small number of VNRs (Egypt, Guyana, Mauritius, and Seychelles) discussed the role of innovative urban farming methods such as aquaponics and hydroponics which reduce water use, water runoff, groundwater pollution, and eutrophication [87]. Further research investigating more waste conscious production techniques in urban communities could further advance more sustainable food systems and reduce losses of highly perishable foods like leafy greens, herbs, and other nutrient dense foods.

Second, concerning the rules in use category of our IAD framework, the overwhelming majority of VNRs and VLRs discussed national and local rules-in-use strategies to address FLW. Our findings align with Shen et al. [73], who state that most current food waste policies are focused on the national strategic level. Some VNRs such as from Norway suggested the importance of legally binding measures to meet the 50% reduction target by 2030, as formal institutions encourage a high degree of compliance. Market interventions such as tax incentives can encourage businesses and individuals to address FLW, yet the small number of reports in this category suggest that these incentive-based strategies may not enjoy wide participation and may fall victim to the free-rider problem, where individuals benefit from the actions of others without contributing [23,24]. As Anderies and Janssen note, larger groups can make it easier for individuals to act as free riders [88], and local and community-based solutions, as well as informal institutions such as unwritten rules and norms are crucial in fostering broad cooperation to achieve desirable outcomes. Informal institutions can therefore influence the adoption of formal policies [24,35]. According to Helmke and Levitsky [89], informal institutions can effectively address limitations in formal institutions and can shape complementary, substituting and accommodating solutions.

https://doi.org/10.3390/su18020833

At the same time, effective interactions between formal and informal instituions can also mitigate tensions resulting from potentially competing instituional aims. O'Hara [24] argues that even in the absence of formal institutional frameworks at the global level, local action can successfully advance desired outcomes. Further research might specifically focus on the role informal institutions can play in shaping behaviors that promote FLW reduction goals not through formal rules and incentives but through conscious lifestyle decisions and mindsets.

Third, the 'attributes of community' category of our IAD framework seems especially important for the evolution of informal institutions. Yet our analysis finds that this category is underexplored in the reviewed VNRs and VLRs. Only a few reports, such as VNRs from Dominica and Iceland and VLRs from Orlando and Sydney, discussed challenges in shifting mindsets to view discarded materials as valuable resources. While different cultures have varying perceptions of what is considered waste, only a small number of VNRs discussed the role of culture and tradition, such as embracing Māori perspectives and the principle of Kaitiakitanga in New Zealand to sustain and restore collective resources; and embracing cultural practices of Indigenous communities in Guyana. This aligns with existing research regarding a lack of studies on FLW and cultural variables [90–92]. Further research will be needed regarding the role of culture and tradition in supporting FLW goals. Several reviews from Europe also discussed the co-benefits of reducing food waste and GHG emissions through lifestyle and dietary changes, such as healthy eating and local food promotion. This supports recent findings from Australia [93] that nutrition-conscious consumers waste less food, partly due to better planning and shopping behaviors. As policymakers play a critical role in developing and implementing FLW program [64], increasing awareness among policymakers may be essential for setting the stage for more effective strategies to promote synergies between formal and informal institutional solutions. Future research should investigate the cultural and social dimensions of FLW, particularly how community attributes and informal institutions shape collective behaviors. Examining local traditions and cultural values could reveal how perceptions of waste vary across contexts. Research should also explore how formal policies can align with these informal, culturally embedded practices to create more effective, integrated FLW strategies. Finally, understanding the behavioral co-benefits of reducing food waste, such as nutrition-conscious consumption and local food promotion, may help identify ways individual actions can reinforce institutional solutions. Interdisciplinary frameworks based on behavioral approaches [94] and food citizenship initiatives [95] can offer useful pathways for advancing community attributes.

Fourth, our findings regarding patterns of interactions reveal that institutional arrangements for FLW solutions are characterized by multiple stakeholders that require coordination and collaboration at the national regional and local levels. The projected surge in urban populations, with 70% of the world's population projected to live in cities by 2050, will further increase the role of local governments creating FLW solutions in cities and metropolitan areas. Strengthening national-local coordination, including strategies for engaging community groups and small-scale FLW entrepreneurial initiatives, will be key to advancing FLW solutions and cross-sector collaboration. This aligns with research regarding the role of start-ups in changing normative and cognitive-cultural institutional pillars [96]. While the reported interactions in the VNRs and VLRs do not necessarily provide information about the level of trust and cooperation they generated, it can be assumed that improved interactions especially among typically segregated types of actors are a step in the right direction in overcoming barriers and reducing the free-rider problem that often stand in the way of effective waste reduction solutions. Further studies on effective collaborations between local governments and non-governmental stakeholders will be important for advancing outcomes and developing effective local and regional FLW reduction strategies. Emerging contemporary debates, such as urban food governance [97], multilevel governance [41,42], food system governance [98], and the political ecology of waste [99] can also make meaningful contributions to advancing effective institutional arrangements to reduce FLW.

Like the reported activities, outcomes were also dominated by technical considerations, such as environmental benefits and the amount of organic waste collected for composting. Our findings highlight the variation in the reported FLW numerical values, including food waste recycling rates, the amount of organic waste collected, the amount of surplus food donated, and food waste per capita. Published research also highlights the challenges associated with measurement issues [6] and much work remains to harmonize FLW definitions and measurements [15]. Environmental impacts were largely focused on GHG emission reductions, with a surprising lack of attention to other co-benefits such as reduced land and water usage. This suggests that research on the water footprint of FLW remains relatively underexplored [100] and further research that studies the indirect resource losses embedded in FLW could particularly focus on the water use and pollution related impacts given that agriculture appropriates approximately 70% of global freshwater use.

Finally, fewer VNRs and VLRs discussed socially oriented outcomes such as improved food security and the benefits of food recovery efforts. This seems surprising since a growing body of research connects FLW and social sustainability in addition to its environmental and economic dimensions [83,86,101]. Other social dimensions, such as the impact of FLW on workers and working conditions are even less explored in the VNRs and VLRs in our sample. In light of new comprehensive assessment tools like Social Life Cycle Assessment, for example, these finding also call for further research that focus on a more systematic and comprehensive evaluation of the social implications of FLW interventions, including their effects on vulnerable populations, labor conditions, and equitable access to recovered food. Such studies would help integrate social sustainability considerations into policy frameworks and local action plans, complementing the existing focus on environmental and economic outcomes.

## 5. Conclusions

This study examined the coverage of FLW solutions in the VNRs and VLRs submitted to the UN during the first ten years since the adoption of the SDGs. Given the key role government policies play in shaping individual and collective FLW solutions, an analysis of VNRs and VLRs submitted by national and local/regional governments is especially instructive in understanding the institutional dimensions that impact the effectiveness of FLW initiatives. A systematic review of 165 VNRs and 116 VLRs submitted between 2016 and 2024 categorized the reports using the IAD framework. The application of the IAD framework is a novel approach that identifies prevalent patterns and research gaps in the FLW solutions reported.

Our analysis found that there are fewer socially oriented solutions than technical solutions in the VNRs and VLRs submitted to date. Similarly, most of the measures used to assess progress toward the FLW goals discussed in the VNRs and VLRs are technical in nature and few reports measured progress in socially oriented objectives. An exception is VNR and VLR reports from Northern Africa, Western Asia and Latin America and the Caribbean. These regions stand out as having a larger number of VNRs that describe socially oriented outcomes and success measures compared to technical ones.

A limitation of this study is the reliance on English-language VNRs and VLRs. This potentially excludes relevant reports published in other UN languages, particularly in Spanish. However, 17% of VNRs and 11% of VLRs in our sample stem from the Latin America and the Caribbean region suggesting a fairly robust representation in our sample.

https://doi.org/10.3390/su18020833

At the same time, activities in the English-speaking countries may be underrepresented given that the United States has yet to submit a VNR and is represented only by the contributions of VLRs from five U.S. cities and one State (Hawai'i).

Our analysis further suggests that progress could be made by giving greater attention to informal institutions (i.e., customs, norms, and shared mindsets) and their interaction with formal institutions (laws, regulations, and policies). The rules-in-use category reported in the VNRs and VLRs focused on activities and strategies rather than regulations to advance FLW solutions. While increased activities through formal institutions such as regulations may be desirable, they often struggle with the free rider problem [23,24]. Informal institutions can play a vital role in reducing free-riding and advancing less formal, voluntary FLW solutions. The IAD category 'attributes of community' is especially important in this regard, and can capture ways to advance the role of informal institutions. This category is underexplored in both VNRs and VLRs suggesting that more attention must be given to the influence of cultural norms and traditions related to food in general and FLW in particular. One solution may be an increased focus on education and social engagement in advancing FLW solutions. SDG Target 12.3 is the explicit expression of global FLW goals. Our findings suggest that a greater focus on systems approaches and cross-boundary collaborations, such as engaging diverse stakeholder views and advancing cooperation models may be needed to guide FLW interventions. Given the range of diverse initiatives evident in the VNRs and VLRs we analyzed, a stronger focus on socially based rather than technical solutions may also advance effective collaborations between local, regional, and global actors that can sustain FLW initiatives beyond the 2030 timeframe of the SDGs.

**Supplementary Materials:** The following supporting information can be downloaded at: https://www.mdpi.com/article/10.3390/su18020833/s1, Table S1: A List of All VNRs Studied; Table S2: A List of All VLRs Studied.

**Author Contributions:** Conceptualization, S.O., E.K. and M.S.; methodology, E.K. and S.O.; formal analysis, E.K. and S.O.; data curation, E.K.; writing—original draft preparation, E.K.; writing—review and editing, E.K., S.O. and M.S.; supervision S.O. All authors have read and agreed to the published version of the manuscript.

**Funding:** This research received no external funding.

**Institutional Review Board Statement:** Not applicable.

**Informed Consent Statement:** Not applicable.

**Data Availability Statement:** All data sources are publicly available and referenced in the text.

**Acknowledgments:** The authors appreciate S.O. for recommending the use of the IAD framework to analyze the complex topic of FLW. The views expressed in this article are those of the authors and do not necessary represent the views of their affiliated organizations.

**Conflicts of Interest:** The authors declare no conflicts of interest.

## References

1. UNEP. Food Waste Index Report 2024. Available online: https://www.unep.org/resources/publication/food-waste-index-report-2024 (accessed on 24 August 2025).
2. Filho, W.L.; Dinis, M.A.P.; Vasconcelos, C.R.P.; Paço, A. Handling the Growing Problem of Offshore Food Waste. *Mar. Policy* **2024**, *163*, 106137. [CrossRef]
3. FAO; IFAD; UNICEF; WFP; WHO. *The State of Food Security and Nutrition in the World 2023. Urbanization, Agrifood Systems Transformation and Healthy Diets Across the Rural–Urban Continuum*; FAO: Rome, Italy, 2023. [CrossRef]
4. UN. Goal 12 Ensure Sustainable Consumption and Production Patterns. Available online: https://sdgs.un.org/goals/goal12 (accessed on 24 August 2025).

https://doi.org/10.3390/su18020833

5. FAO. Sustainability Pathways: Food Wastage Footprint. Available online: https://www.fao.org/home/en/#collapseSearchBox (accessed on 24 August 2025).
6. Cattaneo, A.; Sánchez, M.V.; Torero, M.; Vos, R. Reducing Food Loss and Waste: Five Challenges for Policy and Research. *Food Policy* **2021**, *98*, 101974. [CrossRef]
7. Delgado, L.; Schuster, M.; Torero, M. Food Losses in Agrifood Systems: What We Know. *Annu. Rev. Resour. Econ.* **2023**, *15*, 41–62. [CrossRef]
8. Parfitt, J.; Barthel, M.; Macnaughton, S. Food Waste within Food Supply Chains: Quantification and Potential for Change to 2050. *Phil. Trans. R. Soc. B* **2010**, *365*, 3065–3081. [CrossRef]
9. FAO. *Global Food Losses and Food Waste: Extent, Causes and Prevention*; Food and Agriculture Organization of the United Nations: Rome, Italy, 2011; ISBN 978-92-5-107205-9.
10. UN. The Sustainable Development Goals Report 2024. Available online: https://unstats.un.org/sdgs/report/2024/ (accessed on 24 August 2025).
11. FAO. Tackling Food Loss and Waste: A Triple Win Opportunity. Available online: https://www.fao.org/newsroom/detail/FAO-UNEP-agriculture-environment-food-loss-waste-day-2022/en (accessed on 24 August 2025).
12. Papargyropoulou, E.; Lozano, R.; Steinberger, J.K.; Wright, N.; Ujang, Z.B. The Food Waste Hierarchy as a Framework for the Management of Food Surplus and Food Waste. *J. Clean. Prod.* **2014**, *76*, 106–115. [CrossRef]
13. Magalhães, V.S.M.; Ferreira, L.M.D.F.; Silva, C. Causes and Mitigation Strategies of Food Loss and Waste: A Systematic Literature Review and Framework Development. *Sustain. Prod. Consum.* **2021**, *28*, 1580–1599. [CrossRef]
14. Chauhan, C.; Dhir, A.; Akram, M.U.; Salo, J. Food Loss and Waste in Food Supply Chains. A Systematic Literature Review and Framework Development Approach. *J. Clean. Prod.* **2021**, *295*, 126438. [CrossRef]
15. Boiteau, J.M.; Pingali, P. Can We Agree on a Food Loss and Waste Definition? An Assessment of Definitional Elements for a Globally Applicable Framework. *Glob. Food Secur.* **2023**, *37*, 100677. [CrossRef]
16. Oliveira, M.M.D.; Lago, A.; Dal' Magro, G.P. Food Loss and Waste in the Context of the Circular Economy: A Systematic Review. *J. Clean. Prod.* **2021**, *294*, 126284. [CrossRef]
17. Babbitt, C.W.; Neff, R.A.; Roe, B.E.; Siddiqui, S.; Chavis, C.; Trabold, T.A. Transforming Wasted Food Will Require Systemic and Sustainable Infrastructure Innovations. *Curr. Opin. Environ. Sustain.* **2022**, *54*, 101151. [CrossRef]
18. Croad, T.; Campbell, H.; Mirosa, M. Investigating Systemic and Social Dynamics of Food Loss and Waste: An Application of Waste Regime Theory to Food Production in Aotearoa New Zealand. *Clean. Waste Syst.* **2024**, *7*, 100125. [CrossRef]
19. Gille, Z. From Risk to Waste: Global Food Waste Regimes. *Sociol. Rev.* **2012**, *60*, 27–46. [CrossRef]
20. Messner, R.; Richards, C.; Johnson, H. The "Prevention Paradox": Food Waste Prevention and the Quandary of Systemic Surplus Production. *Agric. Hum. Values* **2020**, *37*, 805–817. [CrossRef]
21. Durkheim, É. *The Rules of Sociological Method*, 1st ed.; Free Press: New York, NY, USA, 1982; ISBN 978-0-02-907930-0.
22. North, D.C. Institutions. *J. Econ. Perspect.* **1991**, *5*, 97–112. [CrossRef]
23. Polski, M.M.; Ostrom, E. An Institutional Framework for Policy Analysis and Design. Available online: https://ostromworkshop.indiana.edu/pdf/teaching/iad-for-policy-applications.pdf (accessed on 24 August 2025).
24. O'Hara, S. Living in the Age of Market Economics: An Analysis of Formal and Informal Institutions and Global Climate Change. *World* **2025**, *6*, 35. [CrossRef]
25. Giddens, A. *Central Problems in Social Theory: Action, Structure, and Contradiction in Social Analysis*; University of California Press: Berkeley, CA, USA, 1979; ISBN 978-0-520-03975-9.
26. Teraji, S. Understanding Coevolution of Mind and Society: Institutions-as-Rules and Institutions-as-Equilibria. *Mind Soc.* **2017**, *16*, 95–112. [CrossRef]
27. Davidson, M.R.; Filatova, T.; Peng, W.; Verbeek, L.; Kucuksayacigil, F. Simulating Institutional Heterogeneity in Sustainability Science. *Proc. Natl. Acad. Sci. USA* **2024**, *121*, e2215674121. [CrossRef]
28. Chua, M.H.; Yau, Y. Institutional Analysis and Development (IAD) Approach for Determining the Effects of the Waste Charging Scheme on Household Food Waste Recycling. *Sustainability* **2022**, *14*, 16120. [CrossRef]
29. Hess, C.; Ostrom, E. (Eds.) *Understanding Knowledge as a Commons: From Theory to Practice*; MIT Press: Cambridge, MA, USA, 2011.
30. National Academies of Sciences, Engineering, and Medicine. *Operationalizing Sustainable Development to Benefit People and the Planet*; The National Academies Press: Washington, DC, USA, 2022. [CrossRef]
31. Smith, W.C.; Susu, A.; Jackaria, I.; Martinez, J.B.; Qu, M.; Niwa, M. Prioritisation of Indicators in SDG 4: Voluntary National Reviews as a Tool of Soft Governance. *Int. Rev. Educ.* **2024**, *70*, 621–649. [CrossRef]
32. Sarwar, M.B.; Nicolai, S. *What Do Analyses of Voluntary National Reviews for Sustainable Development Goals Tell Us About 'Leave No One behind'?* Briefing Note, June; Overseas Development Institute: London, UK, 2018; Available online: https://media.odi.org/documents/What_do_analyses_of_Voluntary_National_Reviews_for_Sustainable_Development_Goa_NWI7pGE.pdf (accessed on 24 August 2025).

33. Fouladvand, J.; Ghorbani, A.; Mouter, N.; Herder, P. Analysing Community-Based Initiatives for Heating and Cooling: A Systematic and Critical Review. *Energy Res. Soc. Sci.* **2022**, *88*, 102507. [CrossRef]
34. Romanelli, J.P.; Boschi, R.S. The Legacy of Elinor Ostrom on Common Forests Research Assessed through Bibliometric Analysis. *CERNE* **2019**, *25*, 332–346. [CrossRef]
35. Ostrom, E. Background on the Institutional Analysis and Development Framework. *Policy Stud. J.* **2011**, *39*, 7–27. [CrossRef]
36. McGinnis, M.D. An Introduction to IAD and the Language of the Ostrom Workshop: A Simple Guide to a Complex Framework. *Policy Stud. J.* **2011**, *39*, 169–183. [CrossRef]
37. Cole, D.H.; Epstein, G.; McGinnis, M.D. The Utility of Combining the IAD and SES Frameworks. *Int. J. Commons* **2019**, *13*, 244–275. [CrossRef]
38. Ostrom, E. A General Framework for Analyzing Sustainability of Social-Ecological Systems. *Science* **2009**, *325*, 419–422. [CrossRef]
39. Meyer, J.W. Reflections on Institutional Theories of Organizations. In *The SAGE Handbook of Organizational Institutionalism*; Greenwood, R., Ed.; Sage: Los Angeles, CA, USA; London, UK, 2017; pp. 788–809.
40. Geels, F.W.; Schot, J. Typology of Sociotechnical Transition Pathways. *Res. Policy* **2007**, *36*, 399–417. [CrossRef]
41. Hooghe, L.; Marks, G. Unraveling the Central State, but How? Types of Multi-Level Governance. *Am. Political Sci. Rev.* **2003**, *97*, 233–243. [CrossRef]
42. Szulecka, J.; Strøm-Andersen, N.; Scordato, L.; Skrivervik, E. Multi-Level Governance of Food Waste: Comparing Norway, Denmark and Sweden. In *From Waste to Value*; Routledge: Abingdon, UK, 2019.
43. Nguyen, T.; Watanabe, T. Autonomous Motivation for the Successful Implementation of Waste Management Policy: An Examination Using an Adapted Institutional Analysis and Development Framework in Thua Thien Hue, Vietnam. *Sustainability* **2020**, *12*, 2724. [CrossRef]
44. Chikowore, N.R.; Kerr, J.M. Waste Governance and Campus Sustainability: Formal and Informal Waste Systems at Football Tailgates in Michigan. *Int. J. Sustain. High. Educ.* **2023**, *24*, 1474–1488. [CrossRef]
45. Oh, J.; Hettiarachchi, H. Collective Action in Waste Management: A Comparative Study of Recycling and Recovery Initiatives from Brazil, Indonesia, and Nigeria Using the Institutional Analysis and Development Framework. *Recycling* **2020**, *5*, 4. [CrossRef]
46. Zhang, Z.; Zhao, L. Voluntary Monitoring of Households in Waste Disposal: An Application of the Institutional Analysis and Development Framework. *Resour. Conserv. Recycl.* **2019**, *143*, 45–59. [CrossRef]
47. De Herde, V. Stronger Together, but How? Lessons from the Walloon Dairy History on the Strategic Relevance of Cooperative Models. *J. Co-Oper. Organ. Manag.* **2023**, *11*, 100198. [CrossRef]
48. Sattler, C.; Barghusen, R.; Bredemeier, B.; Dutilly, C.; Prager, K. Institutional Analysis of Actors Involved in the Governance of Innovative Contracts for Agri-Environmental and Climate Schemes. *Glob. Environ. Change* **2023**, *80*, 102668. [CrossRef] [PubMed]
49. Vidal Merino, M.; Gajjar, S.P.; Subedi, A.; Polgar, A.; Van Den Hoof, C. Resilient Governance Regimes That Support Urban Agriculture in Sub-Saharan Cities: Learning from Local Challenges. *Front. Sustain. Food Syst.* **2021**, *5*, 692167. [CrossRef]
50. Adebayo, A.A.; Lulofs, K.; Heldeweg, M.A. Indicators, Strategies, and Rule Settings for Sustainable Public–Private Infrastructure Partnerships: From Literature Review towards Institutional Designs. *Sustainability* **2023**, *15*, 9422. [CrossRef]
51. Faria, E.; Caldeira-Pires, A.; Barreto, C. Social, Economic, and Institutional Configurations of the Industrial Symbiosis Process: A Comparative Analysis of the Literature and a Proposed Theoretical and Analytical Framework. *Sustainability* **2021**, *13*, 7123. [CrossRef]
52. Qerimi, Q. The Ambitious Modesty of the High-Level Political Forum on Sustainable Development. *Glob. Policy* **2022**, *13*, 281–288. [CrossRef]
53. UN. Voluntary National Reviews. Available online: https://hlpf.un.org/vnrs (accessed on 24 August 2025).
54. UN. SDG Localization and the Voluntary Local Reviews. Available online: https://sdgs.un.org/topics/voluntary-local-reviews (accessed on 24 August 2025).
55. UN. High-Level Political Forum on Sustainable Development. Available online: https://hlpf.un.org/ (accessed on 24 August 2025).
56. Sustainable Development Report 2025. Available online: https://dashboards.sdgindex.org/ (accessed on 24 August 2025).
57. Su, H.; Krol, M.S.; Hogeboom, R.J. The Role of Context in Identifying Linkages between SDG 2 (Food) and SDG 6 (Water). *Sustain. Sci.* **2022**, *17*, 1605–1618. [CrossRef]
58. Smith, W.C. An Exploration of SDG 4 Coverage in Voluntary National Reviews. In *Third International Handbook of Globalisation, Education and Policy Research*; Zajda, J., Ed.; Springer International Publishing: Cham, Switzerland, 2021; pp. 961–980.
59. Ortiz-Moya, F.; Yang, Y. Cities' Review of the Sustainable Development Goals and Insights from Voluntary Local Reviews. *npj Urban Sustain.* **2025**, *5*, 58. [CrossRef]
60. UN Committee of Experts on Public Administration. The Use of Governance Indicators in Voluntary National Review and Voluntary Local Review Processes and What It May Mean for Building Strong Institutions for the SDGs. In Proceedings of the 22nd session of the (CEPA) Committee of Experts on Public Administration, New York, NY, USA, 27–31 March 2023.

61. Liberati, A.; Altman, D.G.; Tetzlaff, J.; Mulrow, C.; Gøtzsche, P.C.; Ioannidis, J.P.A.; Clarke, M.; Devereaux, P.J.; Kleijnen, J.; Moher, D. The PRISMA Statement for Reporting Systematic Reviews and Meta-Analyses of Studies That Evaluate Health Care Interventions: Explanation and Elaboration. *PLoS Med.* **2009**, *6*, e1000100. [CrossRef]
62. Page, M.J.; McKenzie, J.E.; Bossuyt, P.M.; Boutron, I.; Hoffmann, T.C.; Mulrow, C.D.; Shamseer, L.; Tetzlaff, J.M.; Akl, E.A.; Brennan, S.E.; et al. The PRISMA 2020 Statement: An Updated Guideline for Reporting Systematic Reviews. *BMJ* **2021**, *372*, n71. [CrossRef]
63. UN. SDG Indicators: Regional Groupings Used in Report and Statistical Annex. Available online: https://unstats.un.org/sdgs/indicators/regional-groups (accessed on 25 August 2025).
64. Spang, E.S.; Moreno, L.C.; Pace, S.A.; Achmon, Y.; Donis-Gonzalez, I.; Gosliner, W.A.; Jablonski-Sheffield, M.P.; Momin, M.A.; Quested, T.E.; Winans, K.S.; et al. Food Loss and Waste: Measurement, Drivers, and Solutions. *Annu. Rev. Environ. Resour.* **2019**, *44*, 117–156. [CrossRef]
65. US EPA. Wasted Food Scale. Available online: https://www.epa.gov/sustainable-management-food/wasted-food-scale (accessed on 25 August 2025).
66. Nicastro, R.; Carillo, P. Food Loss and Waste Prevention Strategies from Farm to Fork. *Sustainability* **2021**, *13*, 5443. [CrossRef]
67. Fikar, C. A decision support system to investigate food losses in e-grocery deliveries. *Comput. Ind. Eng.* **2018**, *117*, 282–290. [CrossRef]
68. De Korte, M.; Bergman, J.; Van Willigenburg, L.G.; Keesman, K.J. Towards a Zero-Waste Aquaponics-Centered Eco-Industrial Food Park. *J. Clean. Prod.* **2024**, *454*, 142109. [CrossRef]
69. Fattibene, D.; Recanati, F.; Dembska, K.; Antonelli, M. Urban Food Waste: A Framework to Analyse Policies and Initiatives. *Resources* **2020**, *9*, 99. [CrossRef]
70. Petruzzelli, M.; Ihle, R.; Colitti, S.; Vittuari, M. The Role of Short Food Supply Chains in Advancing the Global Agenda for Sustainable Food Systems Transitions. *Cities* **2023**, *141*, 104496. [CrossRef]
71. US EPA. From Field to Bin: The Environmental Impacts of U.S. Food Waste Management Pathways. Available online: https://www.epa.gov/land-research/field-bin-environmental-impacts-us-food-waste-management-pathways (accessed on 25 August 2025).
72. Vittuari, M.; Azzurro, P.; Gaiani, S.; Gheoldus, M.; Burgos, S.; Aramyan, L.; Valeeva, N.; Rogers, D.; Östergren, K.; Timmermans, T.; et al. *Recommendations and Guidelines for a Common European Food Waste Policy Framework*; FUSIONS: Bologna, Italy, 2016. [CrossRef]
73. Shen, G.; Li, Z.; Hong, T.; Ru, X.; Wang, K.; Gu, Y.; Han, J.; Guo, Y. The Status of the Global Food Waste Mitigation Policies: Experience and Inspiration for China. *Environ. Dev. Sustain.* **2024**, *26*, 8329–8357. [CrossRef] [PubMed]
74. Schanes, K.; Dobernig, K.; Gözet, B. Food Waste Matters—A Systematic Review of Household Food Waste Practices and Their Policy Implications. *J. Clean. Prod.* **2018**, *182*, 978–991. [CrossRef]
75. Morris, M.W.; Hong, Y.; Chiu, C.; Liu, Z. Normology: Integrating Insights about Social Norms to Understand Cultural Dynamics. *Organ. Behav. Hum. Decis. Process.* **2015**, *129*, 1–13. [CrossRef]
76. Mariam, N.; Nina, L.; Fabian, B.; Florence, Z.; Simone, A.; Ulf, S.; Daniel, F. The Food Waste Lab: Improving Food Waste Reduction Behavior through Education. *J. Clean. Prod.* **2022**, *370*, 133447. [CrossRef]
77. Li, R.; Roe, B.E. Segmenting U.S. Consumers by Food Waste Attitudes and Behaviors: Opportunities for Targeting Reduction Interventions. *Sustain. Prod. Consum.* **2024**, *45*, 348–358. [CrossRef]
78. Shu, Y.; Booker, A.; Karetny, J.; O'Keefe, K.; Rees, K.; Schroder, L.; Roe, B.E. Evaluation of a Community-Based Food Waste Campaign Using a National Control Group. *Waste Manag.* **2023**, *160*, 101–111. [CrossRef] [PubMed]
79. Ceryes, C.A.; Antonacci, C.C.; Harvey, S.A.; Spiker, M.L.; Bickers, A.; Neff, R.A. "Maybe It's Still Good?" A Qualitative Study of Factors Influencing Food Waste and Application of the E.P.A. Food Recovery Hierarchy in U.S. Supermarkets. *Appetite* **2021**, *161*, 105111. [CrossRef]
80. Ostrom, E. *Governing the Commons: The Evolution of Institutions for Collective Action*; Cambridge University Press: Cambridge, UK, 1990.
81. Ngan, S.L.; Er, A.C.; Yatim, P.; How, B.S.; Lim, C.H.; Ng, W.P.Q.; Chan, Y.H.; Lam, H.L. Social Sustainability of Palm Oil Industry: A Review. *Front. Sustain.* **2022**, *3*, 855551. [CrossRef]
82. Goossens, Y.; Wegner, A.; Schmidt, T. Sustainability Assessment of Food Waste Prevention Measures: Review of Existing Evaluation Practices. *Front. Sustain. Food Syst.* **2019**, *3*, 90. [CrossRef]
83. Sundin, N.; Bartek, L.; Persson Osowski, C.; Strid, I.; Eriksson, M. Sustainability Assessment of Surplus Food Donation: A Transfer System Generating Environmental, Economic, and Social Values. *Sustain. Prod. Consum.* **2023**, *38*, 41–54. [CrossRef]
84. Mayring, P. *Qualitative Content Analysis: Theoretical Foundation, Basic Procedures and Software Solution*; SSOAR: Klagenfurt, Austria, 2014.
85. Morgan, H. Using Triangulation and Crystallization to Make Qualitative Studies Trustworthy and Rigorous. *Qual. Rep.* **2024**, *29*, 1844–1856. [CrossRef]

https://doi.org/10.3390/su18020833

86. Mourad, M. Recycling, Recovering and Preventing "Food Waste": Competing Solutions for Food Systems Sustainability in the United States and France. *J. Clean. Prod.* **2016**, *126*, 461–477. [CrossRef]
87. Stuiver, M.; O'Hara, S.; Roosendaal, L.; Rooij, B.D. Chapter 7: Governance towards Nature-Based City Region Food Systems. In *The Symbiotic City: Voices of Nature in Urban Transformations*; Wageningen Academic Publishers: Wageningen, The Netherlands, 2022; pp. 147–168.
88. Anderies, J.M.; Janssen, M.A. *Sustaining the Commons*; Center for Behavior, Institutions and the Environment, Arizona State University: Tempe, AZ, USA, 2016.
89. Helmke, G.; Levitsky, S. Informal Institutions and Comparative Politics: A Research Agenda. *Perspect. Politics* **2004**, *2*, 725–740. [CrossRef]
90. Elimelech, E.; Segal-Klein, H.; Kaplan Mintz, K.; Katz-Gerro, T.; Ayalon, O. Food Waste Prevention and Reduction: Practices, Cultural and Personal Determinants. *Appetite* **2024**, *200*, 107565. [CrossRef]
91. Long, F.; Aziz, N.A.; Chia, K.W.; Zhang, H. Clear Your Plate! The Impact of Cultural Values and Social Influences on Intention to Reduce Food Waste among Malaysian Consumers. *Cogent Bus. Manag.* **2024**, *11*, 2321796. [CrossRef]
92. Morkunas, M.; Wang, Y.; Borsellino, V. What Really Drives Food Waste in the Restaurant Industry? Evidence from Lithuania. *J. Qual. Assur. Hosp. Tour.* **2025**, 1–38. [CrossRef]
93. Nguyen, T.T.T.; Hetherington, J.B.; O'Connor, P.J.; Malek, L. Sustainable Food Consumption: Sustainability-Conscious Consumers Do Not Reduce Food Waste but Nutrition-Conscious Consumers Do. *Resour. Conserv. Recycl.* **2025**, *219*, 108296. [CrossRef]
94. Danzer, A.; Zeidler, H. Food Waste and Dynamic Inconsistency: A Behavioral Economics Perspective. IZA Discussion Paper No. 17802. Available online: https://ssrn.com/abstract=5196765 (accessed on 9 January 2026).
95. Ben-Othmen, M.A.; Kavouras, J.H. Exploring the Nexus of Food Citizenship and Local Food Consumption in America's Rural-Urban Blend: Insights from Will County (Illinois). *J. Rural. Stud.* **2024**, *109*, 103323. [CrossRef]
96. Närvänen, E.; Mattila, M.; Mesiranta, N. Institutional Work in Food Waste Reduction: Start-Ups' Role in Moving towards a Circular Economy. *Ind. Mark. Manag.* **2021**, *93*, 605–616. [CrossRef]
97. Zerbian, T.; Escario-Chust, A.; Palau-Salvador, G.; Segura-Calero, S. The Multiple and Contested Worlds of Urban Food Governance: The Case of the City of Valencia. *Cities* **2023**, *141*, 104516. [CrossRef]
98. Arthur, H.; Sanderson, D.; Tranter, P.; Thornton, A. A Review of Theoretical Frameworks of Food System Governance, and the Search for Food System Sustainability. *Agroecol. Sustain. Food Syst.* **2022**, *46*, 1277–1300. [CrossRef]
99. Spring, C.; Adams, M.; Hardman, M. Sites of Learning: Exploring Political Ecologies and Visceral Pedagogies of Surplus Food Redistribution in the UK. *Policy Futures Educ.* **2019**, *17*, 844–861. [CrossRef]
100. Marston, L.T.; Read, Q.D.; Brown, S.P.; Muth, M.K. Reducing Water Scarcity by Reducing Food Loss and Waste. *Front. Sustain. Food Syst.* **2021**, *5*, 651476. [CrossRef]
101. Do, Q.; Ramudhin, A.; Colicchia, C.; Creazza, A.; Li, D. A Systematic Review of Research on Food Loss and Waste Prevention and Management for the Circular Economy. *Int. J. Prod. Econ.* **2021**, *239*, 108209. [CrossRef]




https://doi.org/10.3390/su18020833

sustainability

MDPI

Article

# How to Reduce College Students' Food Waste Behavior: From the Perspective of College Canteen Catering Modes

**Amin Wang [1], Xi Luo [2,*], Xiaojun Liu [1] and Yongkai Sun [1]**

[1] School of Management, Xi'an University of Architecture and Technology, Xi'an 710055, China; dln@xauat.edu.cn (A.W.); zxr@xauat.edu.cn (X.L.); syk@xauat.edu.cn (Y.S.)
[2] State Key Laboratory of Green Building, School of Building Services Science and Engineering, Xi'an University of Architecture and Technology, Xi'an 710055, China
* Correspondence: xiluo@xauat.edu.cn

**Abstract:** Reducing consumer food waste plays an important role in achieving the Sustainable Development Goals. Considering the large number of colleges in China, with the largest enrollment in the world, it is especially important to address the issue of food waste among college students. However, the mechanisms underlying the effects that the college canteen catering modes have on the food-saving behavior of college students remain unclear. To fill this gap, an integrated theoretical framework model was constructed from the perspective of "psychological factors–behavioral intention–external environment–actual behavior" based on the theory of planned behavior, the norm activation model, and the attitude–context–behavior theory. Then, 422 valid questionnaires were empirically analyzed by structural equation modeling and hierarchical regression. The main conclusions of this study are as follows: (1) Food-saving intention and herd mentality are the major drivers of college students' food-saving behavior. Personal norms, attitudes, subjective norms, perceived behavior control, and health risk perception are influencing factors on food-saving intention, among which personal norms have the greatest effect. (2) The standard-quantity catering mode has an inhibitory moderating effect, while the large-/small-portion-size and buffet catering modes have promoting moderating effects in the transformation of food-saving intention into actual behavior. Notably, the moderating effects of the buffet catering mode are more pronounced than those of the large-/small-portion-size catering mode. (3) The standard-quantity catering mode has a promoting moderating effect, while the large-/small-portion-size and buffet catering modes have inhibitory moderating effects in the path of the negative impact of herd mentality on food-saving behavior. These conclusions can help colleges recommend strategies to avoid food waste on their campuses from the perspectives of both the individual student and the food provider.

**Keywords:** food waste; food-saving behavior; catering mode; structural equation modeling; hierarchical regression analysis; college canteens; UN Sustainable Development Goals

## 1. Introduction

Food waste, one of the greatest challenges in the context of global climate change, is associated with huge economic losses [1], tremendous resource waste, and severe environmental damage [2,3]. The Sustainable Development Goals (SDGs) were introduced by the United Nations Summit on Sustainable Development on 25 September 2015, in the hope that from 2015 to 2030, people will be able to comprehensively address the three dimensions of development—that is, social, economic, and environmental dimensions—and shift to a path of sustainable development. Item 12.3 of the SDGs directly states that, by 2030, the per capita food waste at the retail and consumption stages should reach a specific target reduction of 50% [4]. According to the latest statistics from the United Nations Environment Program (UNEP), the total amount of food wasted globally has reached 1.03 billion tons per year, about 11% of which occurs at the consumption stage of the food supply chain, while only 2% occurs at the retail stage [5]. Relevant studies have pointed out that intervening in consumers' food waste behaviors at the consumption stage can reduce food waste by about 50% [6]. In summary, reducing food waste in consumers' daily catering consumption activities is the key way to build a sustainable food system and achieve the goal of sustainable development.

In China, the catering industry generates about 17–18 million tons of food waste annually [7], accounting for approximately 3% of China's total annual grain yield. In the 1950s and 1960s, China experienced food shortages, and the food supply was less than the demand. The people born in this period experienced food crises, which resulted in their love of food and their desire to save it, a conclusion verified in the studies of Hou et al. [8] and Wang et al. [9], which also revealed that the youth are the key target of the food waste reduction task. Colleges and universities constitute a vital component of China's social development. Based on a current student scale of 44.3 million in China (2021) [10], the amount of food wasted by Chinese college students each year while dining in canteens is estimated to account for approximately 8% of China's total food waste [11]. This amount is sufficient to feed 10 million people for one year [12]. Consequently, with regard to ensuring the sustainable development of colleges and universities, food waste in college canteens poses an urgent problem.

Both weak food-saving intentions and barriers posed by the external environment tend to increase food waste by college students [13,14]. In response to the weak food-saving intentions among students, Chinese colleges and universities have taken a series of publicity and education measures. Prominent examples include strengthening public food-saving messaging and holding regular themed lectures and class meetings, as well as designing food-saving signs, slogans, and posters for placement in the entrance hall, dining table, wall, and food-recycling areas of university canteens, with the aims of enhancing the food-saving intention of students and reducing food waste in colleges and universities [15,16]. While these measures have improved the food-saving intentions of students to a certain extent, there is still a problem of negativity in their actual food-saving behavior [17]. This is consistent with the research findings of Ellison et al. [18], who showed that educational activities do not affect students' food-saving behavior, except for changing their opinions on saving food. In this sense, simply improving the food-saving intentions of college students through education is not sufficient to improve the status quo of food waste in colleges and universities; the effects of external environmental factors also need to be considered. At present, the main external barriers that prevent college students from saving food are the mismatch between the homogeneous and inflexible food supply modes in college canteens and the differentiated food-saving needs of college students [19]. However, it still remains unclear how the existing catering modes in college canteens affect college students' food-saving behavior and decision-making processes. In other words, the role that each catering mode plays in the transformation of college students' food-saving intentions into actual food-saving behavior has not yet been clarified.

## 2. Literature Review

The existing research related to consumer food-saving behavior focuses on exploring the psychological driving mechanisms behind it with the help of the theory of planned behavior (TPB) and the normative activation model (NAM). For example, Gokarn et al. [20] and Wang et al. [21] constructed a structural equation model with the help of the TPB and pointed out through empirical analyses that subjective norms, attitudes, and perceived behavioral control significantly affect consumers' food-saving intentions. Academics have also pointed out that the TPB has good extensibility and has attained a good explanation rate for explaining the psychological formation mechanism of individual pro-environmental behaviors [22,23]; however, some scholars have criticized the TPB for overemphasizing the influence of rational factors on individual pro-environmental behaviors and ignoring the influence of the individual's irrationality and the ethical motives for their behaviors [24,25]. Thus, Shin et al. [26] questioned the validity of the TPB in explaining pro-environmental behavior.

The NAM suggests that the performance of a behavior by an individual depends on the sense of moral obligation formed in the individual toward the behavior, stressing the explanation of individual pro-environmental behavior from the perspective of moral motives [27]. Wang et al. [28] constructed a model of the moral psychological driving mechanism of consumer food-saving behavior with the help of the NAM and conducted an empirical analysis with the help of structural equation modeling, pointing out that consumers' sense of moral obligation to save food positively affects their food-saving intentions, achieving a good explanation rate. However, some scholars have criticized the NAM for overemphasizing the moral motivation of individual pro-environmental behaviors and neglecting to explain individual environmental behaviors from the perspective of rational motivation [27]. As a result, the validity of the NAM in explaining the formation mechanism of individual pro-environmental behaviors has been questioned [26,29].

With the increasing depth of research, the combined use of the TPB and NAM is considered critical for analyzing the psychological driving mechanisms of consumer food-saving behaviors and developing effective guidance strategies. As Shin et al. [26] and Jia et al. [29] pointed out by constructing the TPB-NAM integrated theoretical model and using questionnaire research combined with structural equation modeling for empirical research, the TPB-NAM integrated model pays attention to both the rational and moral motivations of individual influences on behavior and strives to achieve a balance between pro-ego and pro-social in reducing consumers' food waste behaviors. Although the TPB-NAM integrative theoretical model has gained wide recognition at the level of the psychological formation mechanisms of individual pro-environmental behaviors, some experts have criticized the TPB-NAM integrative theoretical model for only focusing on the psychological factors of pro-environmental behaviors while largely ignoring the roles of external environmental factors in pro-environmental behavioral decision-making [30,31].

Catering mode has been shown to be an important external environmental factor in reducing consumer food waste behavior. Werkman et al. [32] identified the food portion size as an important influencing factor in consumer food waste, suggesting that most consumers with food-saving tendencies end up wasting food because the food portion size exceeds their personal appetites. Li et al. [33] uncovered that, when dining in cafeterias that grant them the freedom to choose both the food category and food portion size instead of a traditional restaurant, consumers often show strong food-saving tendencies. As academics focus on the issue of food waste among college students, research on food waste in colleges has clearly shown that food satisfaction [34] and the food portion size [35] are significant factors that influence the food-saving behaviors of college students; that is, the larger the food portion size, the lower the food satisfaction, and the more likely students are to waste food [35]. According to Zhao and Manning [36], the food category and food plate size both significantly affect students' food waste. Richardson et al. [37] proved that adjusting the size and shape of food plates helps to reduce food waste in college canteens. Based on a field survey of college canteens, Fan et al. [38] pointed out that college students' food-saving behaviors vary with the specific catering modes in college canteens. Moreover, students can save an average of 13.30 g of food per meal when dining at buffet buttery hatches (buttery hatch: a small window for passing food in university canteens) instead of standard-quantity buttery hatches. In summary, the catering modes in college canteens constitute a key factor influencing college students' food-saving behaviors.

The above studies have shown that food portion size and food category are important reasons affecting the food waste of college students at home and abroad; however, different types of catering modes in college canteens have different degrees of restriction on the students' selection of food portion size and food category, which, in turn, have different degrees of influence on the students' food-saving behavior. This study categorized the existing catering modes into the following three types based on the way that students choose their food categories and food portion sizes: the standard-quantity catering mode (fixed portion sizes), the large-/small-portion-size catering mode (portion sizes are optional), and the buffet catering mode (arbitrary choice of food categories and portion sizes). The above research on the catering-mode factors makes up for the lack of external environmental factors in the existing research on food-saving behavior; however, it was limited to identifying the influences of the college canteen catering modes on the food-saving behaviors of college students through regression analysis or field measurements and did not conduct an in-depth analysis of the paths through which the different types of catering modes influence the food-saving behaviors. That is, the mechanisms underlying the effects of the college canteen catering modes on college students' food-saving behavior are still unclear, which is an unresolved issue in the existing research.

The factors that influence consumer food-saving behavior are multi-dimensional [11]. This means that, in order to explore the mechanisms underlying the influences of the external environmental college canteen catering modes on college students' food-saving behaviors, it is necessary to consider the psychological decision-making mechanisms underlying how the college canteen catering modes influence individual food-saving behaviors by combining internal and external factors. The attitude–context– behavior (ABC) theory suggests that external environmental factors have moderating effects on the transformation of a psychological tendency into an actual behavior [39]. Although the ABC theory has achieved wide academic recognition in explaining the mechanisms by which external environmental factors influence individual behavior, some scholars have pointed out that the ABC theoretical model largely ignores the psychological formation mechanisms of individual behaviors, which makes its findings one-sided [40]; therefore, it is usually used in conjunction with theories that explore psychological factors [41]. Therefore, in this study, we investigated the mechanisms of the influences of the college canteen catering modes on the food-saving behavior of college students with the help of three theories: the TPB, NAM, and ABC theory.

When exploring the psychological decision-making mechanisms of the food-saving behavior of college students, it is not only necessary to determine whether the psychological variables selected for the study are the influencing factors of the behavior but also to identify the path through which each psychological factor influences the behavior. The existing studies mostly use structural equation modeling (SEM) to solve this kind of research problem [20,28,31]. SEM is a statistical method that is used to analyze the relationships between the variables based on their covariance matrix. Compared with the traditional regression analysis method, SEM can determine the path of influence of each factor on the behavior while performing a factor analysis, which is the advantage of this method and the reason that it was selected for this study [42,43]. While exploring the moderating effects of the external environmental factors (the college canteen catering modes) on college students' food-saving behavior with the help of the ABC theory, hierarchical regression analysis is the most commonly used and effective method to solve this type of problem in the existing research, and the advantage of choosing the hierarchical regression analysis method for this study is reflected in its ability to exclude other irrelevant variables, preventing them from interfering with the moderating variables [44].

To sum up, in this study, we took college students as the research object, constructed a research framework from the perspective of "psychological factors–behavioral intention–external environment–actual behavior" with the help of the TPB, NAM, and ABC theories, and used structural equation modeling through questionnaire research and stratified regression analysis for empirical testing to explore the influence mechanisms of the college canteen catering modes on college students' food-saving behavior. The results of this study provide a reference for China and other Asian countries for the development of effective strategies to reduce food waste in college canteens.

## 3. Theoretical Basis and Research Hypotheses

### 3.1. TPB and Research Hypotheses

The TPB maintains that the decision to perform a certain behavior involves a rational thinking process. According to the TPB, an individual's intention to behave is influenced by three factors: attitudes, subjective norms, and perceived behavioral control, and the intention to behave is a key predictor of behavior [24]. Attitude refers to a consumer's positive or negative evaluation of food-saving behaviors. In general, a consumer with a more positive evaluation of food-saving behavior is more likely to possess strong food-saving intentions [20]. Subjective norms refer to the social pressure consumers experience when deciding whether to perform food-saving behavior. Generally speaking, when influenced by social conformity norms, an individual under greater social pressure has a stronger tendency to align with the expectations of those who are important to them. If a consumer experiences positive expectations from others (such as parents, teachers, and friends) and recognizes their food-saving behavior, the possibility that they will perform food-saving behavior will increase [21]. Perceived behavioral control refers to consumers' perception of the ease or difficulty of performing food-saving behavior, which is hindered by external environmental factors (such as the absence of (or change in) a purchase plan [45], an unwillingness to consume leftovers, and the inability to control food portion sizes). Perceived behavioral control is also influenced by judgments about the values created by the performance of food-saving behavior. This means that consumers who can better overcome the hindrance of external factors and who regard food-saving behavior as easier to perform will have strong food-saving intentions [21]. According to the TPB, the intention to behave is often considered a strong internal stimulus and an antecedent of behavior [46]. In situations in which it is difficult to measure individual pro-environmental behavior, scholars tend to replace actual behavior with the intention to behave [47]. It has been pointed out that consumer food-saving intentions positively affect their food-saving behaviors [21,46]. To summarize the above, the following hypotheses are proposed:

**H1:** *Students' attitudes have a significant positive effect on their food-saving intentions.*

**H2:** *Students' subjective norms have a significant positive effect on their food- saving intentions.*

**H3:** *Students' perceived behavioral control has a significant positive effect on their food-saving intentions.*

**H4:** *Students' food-saving intentions have a significant positive effect on their food-saving intentions.*

Recently, certain scholars have incorporated herd mentality and risk perception into their research on individual pro-environmental behavior and have identified them as important predictors in the TPB [44,48]. Therefore, in this paper, the above two factors are introduced into the basic model of the TPB, and the extended TPB model is used to explain the factors influencing college students' food-saving behavior at the level of individual rationality. Herd mentality refers to the psychological tendency of individuals to change their behavior under the influence of a particular behavior of the group around them, with the goal of aligning their behavior with that of the people around them [49]. Research has shown that Chinese consumers are more susceptible to the influence of the behavior of other groups than U.S. consumers [50]. When studying the catering consumption behavior of residents, scholars have found that consumers have a herd mentality toward food waste behavior, mainly because of their tendencies to save face and follow trends [22]. College students are the main consumers of college canteen catering activities, and their catering consumption behavior may also change with that of their classmates around them. Moreover, when college students go out for meals, most of them will refer to the treatment of leftovers by other students around them before making their own decisions as to whether to pack leftovers or not [51]. Considering the above-mentioned content, depending on its specific definition, herd mentality affects consumer food-saving intentions and behavior in different manners, either positively or negatively. As this study is aimed at solving the problem of food waste in colleges, herd mentality is defined as the tendency of college students to exhibit a consistent behavior or response when others generate food waste. It was speculated that, under the influence of an environment that tolerates food waste, college students may relax the requirements for themselves in terms of their food-saving intentions and behavior. Accordingly, the following hypotheses are presented:

**H5:** *Students' herd mentality has a significant negative impact on their food-saving intentions.*

**H6:** *Students' herd mentality has a significant negative impact on their food-saving behavior.*

Risk perception is an intuitive judgment about the characteristics and severity of a specific, objective risk [52]. Property loss, environmental deterioration, and health threats are all risks posed by food waste. With the development of society and the improvement in people's quality of life, health risks are receiving increasingly close attention. For the purpose of this study, health risk perception refers to the individual's perception of the health risks posed by stale food and edible leftovers. Research on hotel and home catering consumption behaviors shows that consumers have weakened food-saving intentions when they have concerns about the health and hygiene status of packed leftovers [23], as well as about the potential harms of consuming expired food or leftovers [22,53]. Similarly, the authors suspect that students who perceive more severe health threats from stale food and edible leftovers are more likely to generate food waste. To this end, the following hypothesis is proposed:

**H7:** *Students' health risk perceptions have a significant negative impact on their food-saving intentions.*

### 3.2. NAM and Research Hypotheses

The NAM explains the pro-environmental behavior of an individual from the perspective of the moral norms formed by the individual toward such behavior. The NAM includes three core latent variables: awareness of consequences, ascription of responsibility, and personal norms [28]. Personal norms refer to the sense of moral obligation formed by an individual toward a certain behavior. Ascription of responsibility involves an individual's perception of responsibility for the adverse consequences of not performing a certain behavior. Awareness of consequences relates to an individual's perception of the potential impact that a certain behavior may have on society and others [28]. To a certain extent, consumers' moral norms can influence their daily consumption behavior [27]. The NAM suggests that the source of power that drives individuals to perform food-saving behavior is their sense of moral obligation (i.e., their personal norms). Based on a TPB-NAM integrated model, scholars have pointed out that positive personal norms play an important role in the formation of food-saving intentions [30]. A consumer with a stronger perception of the ascription of responsibility toward food-saving behavior is more likely to self-condemn and feel guilty for the serious adverse consequences caused by food waste; this further stimulates personal norms to promote the performance of food-saving behavior [54]. The stronger the consumers' perception of the adverse consequences of food waste, the more likely it will be that their personal norms will be activated to urge their performance of food-saving behavior [55]. Generally speaking, when individuals become aware of the serious adverse consequences of not performing pro-environmental behaviors, they develop an effective sense of an ascription of responsibility toward such consequences and engage in behaviors that comply with personal norms [56]. In summary, the following hypotheses are proposed:

**H8:** *Students' personal norms have a significant positive effect on their food-saving intentions.*

**H9:** *Students' ascription of responsibility has a significant positive effect on their personal norms.*

**H10:** *Students' awareness of consequences has a significant positive effect on their personal norms.*

**H11:** *Students' awareness of consequences has a significant positive effect on their ascription of responsibility.*

Relying on the extended TPB model, scholars have identified subjective norms as an antecedent of personal norms, suggesting that subjective norms can not only be used to test the social correctness of individual behavior but can also be perceived and internalized as personal norms by individuals [57]. Therefore, if individuals perceive that food-saving behavior is socially accepted, they will feel obligated to perform such behavior. Related studies have also confirmed the connection between individual subjective norms and personal norms [58]. In general, under the combined pressure from society, schools, parents, and teachers, individual students tend to enhance their sense of moral obligation toward pro-environmental behaviors. Hence, the stronger a college student's perception of subjective norms toward food-saving behavior, the stronger their perception of responsibility, and the higher their level of personal norms. In summary, the following hypothesis is proposed:

**H12:** *Students' subjective norms have a significant positive effect on their personal norms.*

### 3.3. ABC and Research Hypotheses

According to the ABC theory, the psychological tendency of an individual when deciding whether to perform a behavior is subject to interference from the external environment [39]. In a favorable external environment, if an individual has a positive psychological tendency toward a certain behavior, then that individual will likely perform the behavior. On the contrary, in an unfavorable external environment, if an individual has a negative psychological tendency toward a certain behavior, such an external environment will hinder that individual from performing the behavior. In this study, the catering modes (external environmental factors) are considered to potentially play moderating roles, as they directly affect consumers' ordering behavior and further influence their food-saving behavior. As a case in point, Li [59] substituted students' food-saving intentions for their actual food-saving behaviors. Using the PRECEDE-PROCEED model, Li pointed out that enabling factors (including catering modes) can moderate the relationship between an individual's food-saving intention and its antecedents.

Currently, there are three main catering modes offered in college canteens: (1) the standard-quantity catering mode, wherein the food portion size served by the canteen staff is basically the same for each student and only two or three dishes are offered with a set meal of rice; (2) the large-/small-portion-size catering mode, wherein a large portion size and a small portion size, allocated by the canteen staff, are available for selection by students as needed; and (3) the buffet catering mode, wherein students can freely choose the category and portion size of food that they like, and the food is then weighed and charged accordingly. At present, the standard-quantity catering mode is the dominant catering mode across college canteens. However, due to the varied food needs of students, even those with strong food-saving tendencies may end up wasting food, as they have no control over the category or portion size of their food. Specifically, students with large appetites may find the standardized portion size insufficient and purchase several portion sizes, while students with small appetites (or those who are picky) may find the standardized portion size excessive. In both cases, food is wasted, and the performance of food-saving behavior is hindered. In addition, when students have a low level of control over the category and portion size of their food, they may relax the requirements for themselves in terms of food-saving behavior under the influence of others who produce food waste, causing food waste. Conversely, when students have a high level of control over the category and portion size of their food, the effect of their food-saving intention on their behavior will be enhanced. At the same time, as they are able to control the category and portion size of their food, they tend to take only according to their needs, in which case the degree of food waste due to the herd mentality influence is reduced. Summing up the above, the following hypotheses are proposed:

**H13:** *The standard-quantity catering mode plays a moderating role between food-saving intention and behavior.*

**H14:** *The large-/small-portion-size catering mode plays a moderating role between food-saving intention and behavior.*

**H15:** *The buffet catering mode plays a moderating role between food-saving intention and behavior.*

**H16:** *The standard-quantity catering mode plays a moderating role between herd mentality and food-saving behavior.*

**H17:** *The large-/small-portion-size catering mode plays a moderating role between herd mentality and food-saving behavior.*

**H18:** *The buffet catering mode plays a moderating role between herd mentality and food-saving behavior.*

Moreover, Guo et al. [60] and Ma et al. [61] concluded that individual-attribute factors, such as gender, education, household registration, and economic status (comprising monthly living expenses and average daily spending on meals), can all affect college students' food-saving behavior. Scholars have also pointed out that the situation of food waste is more serious for female students than male students, as well as more serious for undergraduate students than graduate students [62] and for students with better family economic conditions than those with poorer family economic conditions [11]. To summarize the above, the following hypothesis is proposed:

**H19:** *Individual-attribute factors have significant effects on food-saving behavior.*

In summary, in this study, we integrated the above three classical behavioral theoretical models and established a hypothetical model for extending the TPB-NAM-ABC integrated theoretical framework, as shown in Figure 1. This model can be used to examine the mechanisms underlying the formation of college students' food-saving behavior from two perspectives: internal psychological factors and external environmental factors. To identify the psychological driving mechanisms underlying the food-saving behavior of college students, an extended TPB-NAM model was constructed. To identify the mechanisms by which the catering modes affect college students' food-saving behavior, the ABC model was integrated with the extended TPB-NAM model for a more detailed explanation.

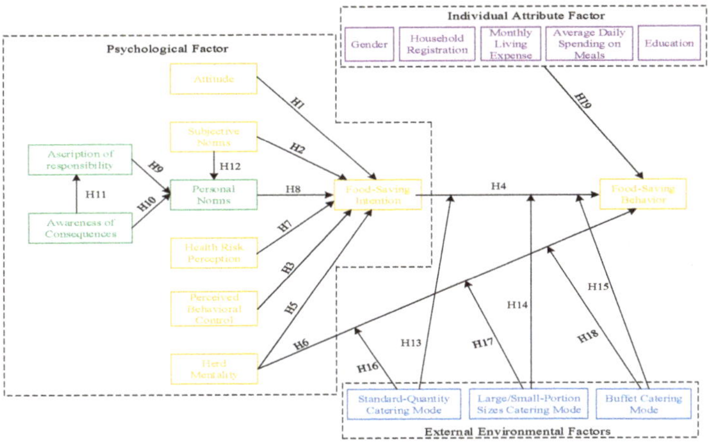


**Figure 1.** Diagram of the hypothetical model for extending the theory of planned behavior (TPB)/norm activation model (NAM)/attitude–context–behavior (ABC) (TPB-NAM-ABC) integrated theoretical framework. Note: The yellow boxes contain the related factors of the extended TPB; the green boxes contain the related factors of the NAM; the blue boxes contain the external environmental factors related to the ABC model; and the purple boxes contain the factors related to the individual attributes.

## 4. Research Methodology

### *4.1. Questionnaire Design*

To quantify the 13 selected latent variables that affect college students' food-saving behavior, a questionnaire was designed, drawing on mature scales from existing studies in behavioral science fields. The questionnaire consisted of four parts: Part 1 sampled the college students' basic information, including their gender, education, monthly living expenses, average daily spending on meals, household registration, and the meal most often wasted in a day, as well as the specific causes of the food waste. Part 2 and Part 3 sampled the observations of individual psychological factors and external environmental factors, respectively. The individual psychological factors involved the nine latent variables: attitudes, subjective norms, perceived behavioral control, food-saving intention, herd mentality, health risk perception, personal norms, ascription of responsibility, and awareness of consequences. The external environmental factors included the three latent variables: the standard-quantity catering mode, the large-/small-portion-size catering mode, and the buffet catering mode. A seven-point Likert scale, with 1, 2, 3, 4, 5, 6, and 7 denoting "strongly oppose", "oppose", "comparatively oppose", "general", "comparatively agree", "agree", and "strongly agree", respectively, was used to measure the aforementioned variables. Part 4 sampled the observation of the college students' food-saving behavior, wherein pertinent measurement methods developed by scholars such as Visschers et al. [63] and Soorani and Ahmadvand [64] were used for reference. In this regard, the college students' food-saving behavior was observed through reverse transformation by gathering their self-reported food waste levels after they dined in the canteens during the previous week. The measurements were conducted using a seven-point Likert scale, with 1, 2, 3, 4, 5, 6, and 7 denoting "no waste", "0–10% waste", "10–20% waste", "20–30% waste", "30–50% waste", "50–70% waste", and "greater than 70% waste", respectively. The design of the observation question items for the latent variables and their sources are shown in Table 1.

**Table 1.** Design of observation question items for latent variables and their sources.

| Latent Variable | Observed Variable | Definition | Reference Scale |
|---|---|---|---|
| Food-saving intention (FSI) | FSI1 | I am willing to try my best to reduce food waste in the daily dining process. | Soorani and Ahmadvand [64] Liao et al. [27] Viccaro et al. [58] |
| | FSI2 | I try to eat all the food I have bought. | |
| | FSI3 | In the future, when I eat in the canteens, I will save food by ordering in moderation. | |
| | FSI4 | In the future, I will take the "Clean Plate Campaign" when I eat in the canteens. | |
| Attitude (AT) | AT1 | I think college students should save food. | Visschers et al. [63] Soorani and Ahmadvand [64] |
| | AT2 | I think it makes sense for college students to save food | |
| | AT3 | I think saving food for college students is worth popularizing. | |
| | AT4 | I think it is immoral for college students to waste food. | |
| Subjective norms (SNs) | SN1 | My parents and relatives approve of my food saving behavior. | Stefan et al. [45] Van der Werf et al. [65] |
| | SN2 | My friends and classmates around me support my food saving behavior. | |
| | SN3 | People who are important to me will support my food saving behavior. | |
| Perceived behavioral control (PBC) | PBC1 | I think I can save food in the daily dining process. | Lorenz et al. [66] Van der Werf et al. [65] |
| | PBC2 | It is entirely up to me to reduce food waste when eating in the canteens. | |
| | PBC3 | I always can share the leftover food with others or pack it and eat it again. | |
| | PBC4 | I am very confident that I will reduce food waste when eating in the canteens in the future. | |
| Herd mentality (HM) | HM1 | When eating in the canteens, I will give priority to the delicious food recommended by everyone. | Ding [22] |
| | HM2 | If my classmates don't pack leftovers, I won't pack them either. | |
| | HM3 | When eating in the canteens, I try to be consistent with my classmates around me. | |
| Health risk perception (HRP) | HRP1 | I am worried that eating leftovers will be bad for my health. | Schmidt [53] Visschers et al. [63] |
| | HRP2 | I'm worried that the leftovers packed back are unsanitary. | |
| | HRP3 | For the sake of health, the leftover food in college canteens should be thrown away. | |
| | HRP4 | Food that is not fresh can jeopardize the health of college students. | |

**Table 1.** *Cont.*

| Latent Variable | Observed Variable | Definition | Reference Scale |
|---|---|---|---|
| Personal norms (PNs) | PN1 | Participating in the "Clean Plate Campaign" is in line with my values. | Obuobi et al. [54] Loh et al. [65] |
| | PN2 | I feel guilty when I produce food waste. | |
| | PN3 | Regardless of what anyone else does, I think it is important to minimize food waste at meals. | |
| | PN4 | I am obligated to take part in the "Clean Plate Campaign" through moral reasoning. | |
| Ascription of responsibility (AR) | AR1 | I should be responsible for global warming caused by food waste. | |
| | AR2 | I should be responsible for the natural resources consumed in grain production. | |
| | AR3 | I am responsible for the environmental pollution caused by food waste. | |
| Awareness of consequences (AC) | AC1 | Reducing food waste by college students can reduce environmental pollution. | |
| | AC2 | Reducing food waste for college students can reduce resource consumption. | |
| | AC3 | Reducing food waste by college students can reduce the food crisis. | |
| Standard-quantity catering mode (SQ) | SQ1 | When I eat at the standard-quantity buttery hatch, the portion size of food is determined by the canteens staff, which is not conducive to saving food for me. | Self-developed |
| | SQ2 | It is limited for me to choose the choice of food when I eat at the standard-quantity buttery hatch, which is not conducive to saving food for me. | |
| | SQ3 | When I eat in the canteens, it is not conducive to save food when I am limited in my choice of food category and the degree of control over food portion size. | |
| Large-/small-portion-size catering mode (LS) | LS1 | Adding the large/small-portion size buttery hatch in college canteens will help me to save food. | Self-developed |
| | LS2 | When eating at the large/small-portion size buttery hatch, I can choose the portion size of food, which is conducive to saving food. | |
| | LS3 | Being able to control the portion size of food to a certain extent helps me to save food. | |
| Buffet catering mode (BF) | BF1 | Adding the buffet buttery hatch in college canteens will help me to save food. | Self-developed |
| | BF2 | When I eat at the buffet buttery hatch, I can pick my favorite food category and control the portion size of food, which is good for me to save food. | |
| | BF3 | When I eat in a canteens, the more freedom I have to choose food categories and decide on portion sizes, the better it is for me to save food. | |
| Food-saving behavior (FSB) | FSB1 | How much of the food you ordered was wasted when you ate in the canteens in the last week? | Stefan et al. [45] Lorenz et al. [66] |
| | FSB2 | How many staple foods have you wasted in the canteens in the last week? | |
| | FSB3 | How many vegetables have you wasted in dining in the canteens in the last week? | |
| | FSB4 | How much meat did you waste when eating in the canteens in the last week? | |
| | FSB5 | How many soups have you wasted in the dining hall in the last week? | |

### *4.2. Ethical Statement*

The content of this study strictly adhered to the Declaration of Helsinki, and because this study did not involve animal or human clinical trials, we applied for an ethical review waiver statement from the Institutional Review Board of the School of Management of Xi'an University of Architecture and Technology prior to the start of the research activities. During the implementation of the research activities, each participant was informed in advance of the following five points: (1) the specific purpose of the research and the type of questions; (2) that all their responses would be anonymous and confidential; (3) that the results of the survey are for academic research use only and will not affect them negatively in any way; (4) that they would each receive a pack of garbage bags as a prize upon completion of the research; (5) that the decision as to whether or not to participate in this research activity would be voluntary; the research questionnaire was distributed to participants for completion after obtaining their signed informed consent. The researcher did not identify individual participants during or after data collection because all the participants were anonymous for this research activity. The researcher inquired about the age of the participants before the questionnaires were distributed to ensure that each participant was 18 years of age or older and that no minors participated in this study.

### 4.3. Questionnaire Distribution and Collection

#### 4.3.1. Questionnaire Distribution

In China, college canteen catering modes can be categorized into three types based on the ways in which students select their food categories and food portion sizes, namely, standard-quantity catering (fixed portion sizes), large-/small-portion-size catering (portion sizes are optional), and self-service catering (arbitrary choices of food categories and portion sizes). In this regard, Xi'an University of Architecture and Technology is not an exception. In this study, we used only Xi'an University of Architecture and Technology as an example to reveal the influence mechanisms of the college canteen catering modes on college students' food-saving behavior, but the results are applicable to all colleges in China. In order to ensure the readability and comprehensibility of the questionnaire design, the researchers first randomly selected 35 students in the target institutions to carry out the pre-survey and, in the process of the pre-survey, recorded the questions that the participants did not understand or found ambiguous. Their findings were then combined with the views of the relevant experts. The questionnaire was repeatedly adjusted to form the final version. The formal research activity was conducted from 1 March to 1 April 2023. In order to ensure that the questionnaire distribution is representative, and the random-sampling method was adopted to collect data by conducting offline questionnaire research at Xi'an University of Architecture and Technology. The questionnaires were distributed in the apartment buildings of male and female students of Xi'an University of Architecture and Technology. The number of questionnaires distributed was 35 times the number of latent variables, that is, 450, of which 225 were for male and 225 for female students. Our method of questionnaire distribution simultaneously satisfies all the characteristics of the simple random sampling method: limitation, personality-by-personality, non-returnability, and equal likelihood. Thus, it also ensured that the questionnaire distribution in this study was representative to an objective degree.

#### 4.3.2. Questionnaire Collection

In order to increase the effective recovery rate of the questionnaire and enhance the representativeness of the research data, we took the following four measures in the process of the questionnaire distribution: (1) to avoid a loss of data due to the mistakes of researchers, the researchers were trained before the start of the research to ensure that they were clear about the purpose of the research and the related precautions; (2) to reduce the loss of questionnaire numbers due to missing questions, once the participants completed the questionnaire, the investigator checked the levels of completion and directed them to fill in any omissions; (3) to fully motivate the participants to fill out the questionnaire seriously, before the questionnaire research, the participants were informed that they would receive the corresponding gift for filling out the questionnaire; (4) to avoid the distortion of the questionnaire data due to the participants' lack of understanding of the questions, they could ask questions about topics they did not understand at any time during the questionnaire completion process, and the researcher described the meaning of the text of the questions rather than inducing responses. In this study, 450 questionnaires were distributed to students, and, after excluding 28 invalid questionnaires, 422 valid questionnaires were obtained. The validity rate of the questionnaires was 93.8%. Referring to Jackson et al. [67], for the sample size requirement when obtaining research data based on questionnaires, the sample size needs to be greater than or equal to more than 10 times the number of latent variables in the study. There were 13 latent variables in this study; therefore, the sample size met the requirement and could be analyzed in the next step. The meals most often wasted in a day by the canteen-dining college students were lunch (60.70%) and dinner (36.70%), and the reasons for the food waste were mainly unreasonable catering modes (61.37%) and the unsatisfactory taste of the food (56.40%). Again, this result demonstrates the necessity of exploring the mechanisms underlying the effects of college canteen catering modes on college students' food-saving behavior. The results of the descriptive statistical analysis of the specific sample are shown in Table 2.

**Table 2.** Descriptive statistical analysis of individual-attribute variables.

| Variable | Categories | Frequency | Percent (%) | Variable | Categories | Frequency | Percent (%) |
|---|---|---|---|---|---|---|---|
| Gender | Male | 224 | 53.1% | Average daily spending on meals | ≤CNY 20 | 93 | 22% |
| | Female | 198 | 46.9% | | CNY 21–30 | 189 | 44.8% |
| Education | Undergraduate | 220 | 52.1% | | ≥CNY 31 | 140 | 33.2% |
| | Postgraduate | 159 | 37.7% | The meal most often wasted in a day | Breakfast | 12 | 2.80% |
| | Doctoral student | 43 | 10.2 | | Lunch | 256 | 60.70% |
| Household registration | Rural areas | 223 | 52.8% | | Dinner | 155 | 36.7% |
| | Urban areas | 199 | 47.2% | Reasons for food waste | Unreasonable catering modes | 259 | 61.37% |
| Monthly living expenses | ≤CNY 1000 | 75 | 17.8% | | Unsatisfactory taste of food | 238 | 56.40% |
| | CNY 1001–1500 | 110 | 26.1% | | Weight loss goal | 44 | 10.43% |
| | CNY 1501–2000 | 135 | 32% | | Dining time is short | 42 | 9.95% |
| | ≥CNY 2001 | 102 | 24.2% | | Poor sanitation and others | 32 | 7.58% |

### 4.4. Data Analysis Methods

A data analysis was performed mainly using SPSS 24.0 and AMOS 24.0. SPSS 24.0 has the following objectives: (1) to conduct a descriptive statistical analysis on the respondents' individual-attribute information; (2) to determine whether the respondents' food-saving behavior differed in terms of the individual-attribute factors by performing independent-sample *t*-tests and one-way ANOVA; (3) to ensure the reasonable design and representativeness of the questionnaire and the acquisition of reliable data through reliability testing, validity testing, and confirmatory factor analysis; and (4) to test the validity of the hypothesis on the moderating effects of external environmental factors through hierarchical regression analysis. AMOS 24.0 was adopted for the following tasks: (1) the construction of a structural equation model; (2) the testing of the model fitness index; (3) path analysis of the latent variables of each psychological factor; and (4) a bootstrap mediating-effect test, which was mainly conducted to test the presence of a mediating effect in the hypothetical model. The research conclusions were drawn by comprehensively utilizing the aforementioned methods.

## 5. Results and Analysis

### 5.1. Difference Test of Individual-Attribute Factors

In this study, difference tests were conducted on the binary variables of gender and household registration in the individual attributes of the college students using independent-sample *t*-tests. Difference tests were also conducted on the multi-level variables in the individual attributes, such as education, monthly living expenses, and average daily spending on meals, using one-way ANOVA. The test results are shown in Table 3.

The data in Table 3 show that the self-reported food-saving behavior of female students was lower than that of male students, and that students with rural household registration were more inclined to save food than those with urban household registration. There was no significant difference in the self-reported food-saving behavior among the college students in terms of education. The self-reported food-saving behavior of the college students decreased with increasing individual monthly living expenses and average daily spending on meals. Thus, H19 is partially valid.

**Table 3.** Independent-sample *t*-test results and one-way ANOVA results for individual-attribute variables.

| Individual Attribute | Categories | Frequency | Mean Self-Reported Food-Saving Behavior | Standard Deviation | $F$ | $p$ |
|---|---|---|---|---|---|---|
| Gender | Male | 224 | 6.161 | 0.974 | 4.512 | 0.000 |
| | Female | 198 | 5.747 | 0.897 | | |

**Table 3.** *Cont.*

| Individual Attribute | Categories | Frequency | Mean Self-Reported Food-Saving Behavior | Standard Deviation | *F* | *p* |
|---|---|---|---|---|---|---|
| Household registration | Rural areas<br>Urban areas | 223<br>199 | 6.068<br>5.853 | 0.849<br>1.062 | 2.306 | 0.022 |
| Education | Undergraduate<br>Postgraduate<br>Doctoral student | 220<br>159<br>43 | 5.978<br>5.928<br>6.051 | 0.869<br>1.004<br>1.224 | 0.308 | 0.735 |
| Monthly living expenses | ≤CNY 1000<br>CNY 1001–1500<br>CNY 1501–2000<br>≥CNY 2000 | 75<br>110<br>135<br>102 | 6.144<br>6.056<br>6.021<br>5.669 | 1.022<br>1.030<br>0.783<br>0.047 | 4.711 | 0.003 |
| Average daily spending on meals | ≤CNY 20<br>CNY 21–30<br>≥CNY 31 | 93<br>189<br>140 | 6.159<br>6.002<br>5.791 | 1.019<br>0.842<br>0.088 | 4.397 | 0.013 |

### 5.2. Reliability and Validity Tests

The overall reliability of the sample data was tested using SPSS 24.0, and the Cronbach's $\alpha$ coefficient (0.906) was greater than 0.8. According to Table 4, the Cronbach's $\alpha$ of each latent variable was greater than 0.8, which testified to the high reliability of the sample data.

**Table 4.** Results of descriptive statistics and reliability and validity tests for each latent variable.

| Latent Variable | Observed Variable | Mean | Standard Deviation | Standardized Factor Loading | Cronbach's $\alpha$ | C.R. | AVE |
|---|---|---|---|---|---|---|---|
| Food-saving behavior (FSB) | FSB1<br>FSB2<br>FSB3<br>FSB4<br>FSB5 | 5.967 | 1.147 | 0.785<br>0.832<br>0.872<br>0.775<br>0.680 | 0.898 | 0.893 | 0.626 |
| Food-saving intention (FSI) | FSI1<br>FSI2<br>FSI3<br>FSI4 | 6.361 | 1.109 | 0.809<br>0.791<br>0.864<br>0.917 | 0.912 | 0.909 | 0.717 |
| Attitude (AT) | AT1<br>AT2<br>AT3<br>AT4 | 6.247 | 1.433 | 0.965<br>0.896<br>0.793<br>0.786 | 0.917 | 0.921 | 0.745 |
| Subjective norms (SNs) | SN1<br>SN2<br>SN3 | 6.332 | 1.098 | 0.936<br>0.926<br>0.971 | 0.961 | 0.961 | 0.892 |
| Perceived behavioral control (PBC) | PBC1<br>PBC2<br>PBC3<br>PBC4 | 5.000 | 1.412 | 0.896<br>0.768<br>0.690<br>0.878 | 0.880 | 0.885 | 0.660 |
| Herd mentality (HM) | HM1<br>HM2<br>HM3 | 4.543 | 1.816 | 0.819<br>0.856<br>0.795 | 0.862 | 0.863 | 0.678 |
| Health risk perception (HRP) | HRP1<br>HRP2<br>HRP3<br>HRP4 | 4.845 | 1.445 | 0.771<br>0.739<br>0.800<br>0.862 | 0.872 | 0.872 | 0.631 |

**Table 4.** *Cont.*

| Latent Variable | Observed Variable | Mean | Standard Deviation | Standardized Factor Loading | Cronbach's α | C.R. | AVE |
|---|---|---|---|---|---|---|---|
| Personal norms (PNs) | PN1<br>PN2<br>PN3<br>PN4 | 6.068 | 1.260 | 0.817<br>0.811<br>0.807<br>0.902 | 0.901 | 0.902 | 0.697 |
| Ascription of responsibility (AR) | AR1<br>AR2<br>AR3 | 5.167 | 1.512 | 0.920<br>0.872<br>0.750 | 0.882 | 0.886 | 0.723 |
| Awareness of consequences (AC) | AC1<br>AC2<br>AC3 | 5.620 | 1.536 | 0.845<br>0.918<br>0.885 | 0.915 | 0.914 | 0.780 |
| Standard-quantity catering mode (SQ) | SQ1<br>SQ2<br>SQ3 | 2.611 | 1.544 | 0.901<br>0.904<br>0.804 | 0.880 | 0.904 | 0.758 |
| Large-/small-portion-size catering mode (LS) | LS1<br>LS2<br>LS3 | 5.573 | 1.313 | 0.885<br>0.894<br>0.835 | 0.894 | 0.905 | 0.760 |
| Buffet catering mode (BF) | BF1<br>BF2<br>BF3 | 6.108 | 1.121 | 0.802<br>0.837<br>0.838 | 0.837 | 0.865 | 0.682 |

A cConfirmatory factor analysis was performed using AMOS 24.0. Table 4 displays the test results. As shown in Table 4, the standardized-factor-loading value of each observed variable exceeded the recommended cutoff of 0.5. The CR value of each latent variable was greater than the recommended cutoff of 0.7. The average-variance-extracted (AVE) value of each latent variable was also greater than the recommended threshold of 0.5. Therefore, all the latent variables involved in this study had good convergent validity. Table 5 shows that the correlation coefficients between the latent variables (the values below the diagonal cells) were all smaller than the specified upper limits (i.e., the square roots of the corresponding AVE values, formatted in bold in the diagonal columns). In this study, the values formatted in bold in the diagonal cells were significantly greater than all the values below the columns in which they were located. Therefore, the discriminant validity was good for all the latent variables involved in this study. By combining Tables 4 and 5, it can be concluded that the questionnaire designed for this study has high validity and reliability.

**Table 5.** Latent variable correlation coefficients and square roots of average variances extracted.

| Latent Variable | FSB | FSI | AT | SN | PBC | HM | HRP | PN | AR | AC | SQ | LS | BF |
|---|---|---|---|---|---|---|---|---|---|---|---|---|---|
| FSB | **0.791** | | | | | | | | | | | | |
| FSI | 0.465 | **0.847** | | | | | | | | | | | |
| AT | 0.370 | 0.682 | **0.863** | | | | | | | | | | |
| SN | 0.320 | 0.729 | 0.594 | **0.944** | | | | | | | | | |
| PBC | 0.379 | 0.587 | 0.568 | 0.518 | **0.812** | | | | | | | | |
| HM | −0.184 | −0.179 | −0.176 | −0.132 | −0.152 | **0.824** | | | | | | | |
| HRP | −0.068 | −0.042 | −0.066 | 0.065 | 0.043 | 0.158 | **0.794** | | | | | | |
| PN | 0.391 | 0.737 | 0.555 | 0.665 | 0.560 | −0.151 | 0.024 | **0.835** | | | | | |
| AR | 0.161 | 0.414 | 0.287 | 0.423 | 0.370 | −0.042 | 0.090 | 0.523 | **0.850** | | | | |
| AC | 0.152 | 0.421 | 0.337 | 0.423 | 0.427 | −0.094 | 0.103 | 0.608 | 0.547 | **0.883** | | | |
| SQ | −0.093 | −0.011 | 0.049 | −0.068 | −0.003 | 0.049 | −0.023 | −0.046 | −0.084 | −0.071 | **0.871** | | |
| LS | 0.198 | 0.189 | 0.202 | 0.204 | 0.164 | 0.037 | −0.163 | 0.263 | 0.146 | 0.208 | −0.352 | **0.872** | |
| BF | 0.299 | 0.186 | 0.118 | 0.227 | 0.146 | −0.085 | −0.056 | 0.236 | 0.156 | 0.191 | −0.409 | 0.478 | **0.826** |

Note: The values formatted in bold represent the square roots of the AVE values.

## *5.3. Tests of Fitness and Validity of Structural Equation Modeling*

### 5.3.1. Tests of Fitness and Validity of Structural Equation Modeling

The model's fitness reflects the degree of consistency between the hypothetical theoretical model and the actual situation; that is, it reflects the correctness of the proposed hypothetical structural equation model. The fitting results of the fitness index of the extended TPB-NAM model are shown in Table 6. The RMSEA value of the model was 0.040, which is lower than the recommended threshold of 0.08. The $X^2/df$ value was 1.689, which is lower than the recommended threshold of 3. The GFI value was 0.888, the NFI value was 0.923, the IFI value was 0.967, the TLI value was 0.963, the CFI value was 0.967, and the AGFI value was 0.869. All these values were greater than the recommended threshold of 0.8; the closer the fitting value to 1, the better the fitness of the model. To sum up the above analysis, the fitting values of the various indices of the structural equation model (i.e., the extended TPB-NAM model) established in this study all met the relevant requirements, indicating the satisfactory fitting results of the model.

**Table 6.** College student food-saving behavior SEM fitting index..

| Index | $X^2/df$ | $p$ | RMSEA | NFI | CFI | GFI | TLI | IFI | AGFI |
|---|---|---|---|---|---|---|---|---|---|
| Recommended threshold | <3 | <0.5 | <0.08 | >0.8 | >0.8 | >0.8 | >0.8 | >0.8 | >0.8 |
| Fitness index value | 1.689 | 0.000 | 0.040 | 0.923 | 0.967 | 0.888 | 0.963 | 0.967 | 0.869 |

### 5.3.2. Comparison of the Extended TPB-NAM Model with Existing Related Research Models

By building an extended TPB-NAM model based on prior studies, we examined the psychological driving mechanisms underlying college students' food-saving behaviors. The extended TPB-NAM model was compared with other related models to test its validity. Specifically, the comparison focused on the model explanation rate (the squared multiple correlation) of each model for the two latent variables of food-saving intention and food-saving behavior. The comparison results are displayed in Table 7. As shown in Table 7, the explanation rates of the extended TPB-NAM model on food-saving intention and food-saving behavior were 79.8% and 28.8%, respectively, which are both higher than those of the existing models. This shows that the extended TPB-NAM model was superior to the models in the existing studies, which confirmed its validity.

**Table 7.** Explanation rates of food-saving intention and food-saving behavior in different models.

| Model | TPB | NAM | Extended TPB | TPB-NAM | Extended TPB-NAM |
|---|---|---|---|---|---|
| FSI (SMC) | 0.739 | 0.647 | 0.750 | 0.789 | 0.798 |
| FSI (SMC) | 0.269 | 0.263 | 0.280 | 0.269 | 0.288 |

## *5.4. Results of Effect Paths of Psychological Factors*

Table 8 presents the empirical test results of the effect paths between the latent variables for psychological factors in the extended TPB-NAM-ABC model. As shown in Table 8, in the extended TPB model, the effects of both attitudes and subjective norms on the food-saving intention were significant at the statistical level of 0.1%. The effect of perceived behavioral control on the food-saving intention was significant at the statistical level of 5%, and that of the health risk perception was also significant at the statistical level of 1%. These results indicate that the students' food-saving intentions were significantly affected by their intuitive evaluation of food-saving behavior, their perception of the social pressure related to saving food, their perception of the ease or difficulty associated with performing food-saving behavior, and their perception of the health threats emanating from stale food and edible leftovers. Thus, H1, H2, H3, and H7 are all valid. The effects of food-saving intention and herd mentality on food-saving behavior are significant at the statistical levels of 0.1% and 5%, respectively. This result shows that the students' food-saving behavior was directly affected by their individual behavioral intentions toward food saving and by the food waste behavior of the classmates around them. Thus, H4 and H6 are valid. The effect of herd mentality on the food-saving intention was non-significant, but it exerted a significant negative impact on the food-saving behavior. This clarifies that even students with strong food-saving intentions might waste food under the influence of classmates, friends, and the luxurious and wasteful environment around them [22]. Thus, H5 is invalid.

**Table 8.** Results of effect paths of psychological factors in SEM.

| Hypothesis | Path | | | Standardized Coefficient | S.E. | C.R. | $p$ | Result |
|---|---|---|---|---|---|---|---|---|
| H1 | AT | → | FSI | 0.303 | 0.039 | 6.714 | *** | Valid |
| H2 | SN | → | FSI | 0.273 | 0.047 | 5.888 | *** | Valid |
| H3 | PBC | → | FSI | 0.094 | 0.04 | 2.168 | 0.03 | Valid |
| H4 | FSI | → | FSB | 0.489 | 0.042 | 9.521 | *** | Valid |
| H5 | HM | → | FSI | −0.007 | 0.022 | −0.239 | 0.811 | Invalid |
| H6 | HM | → | FSB | −0.129 | 0.03 | −2.519 | 0.012 | Valid |
| H7 | HRP | → | FSI | −0.089 | 0.024 | −2.95 | 0.003 | Valid |
| H8 | PN | → | FSI | 0.374 | 0.054 | 7.87 | *** | Valid |
| H9 | AR | → | PN | 0.135 | 0.038 | 2.862 | 0.004 | Valid |
| H10 | AC | → | PN | 0.371 | 0.033 | 7.457 | *** | Valid |
| H11 | AC | → | AR | 0.608 | 0.044 | 11.347 | *** | Valid |
| H13 | SN | → | PN | 0.493 | 0.038 | 11.392 | *** | Valid |

Note: *** $p < 0.001$.

In the NAM, the effect of personal norms on the food-saving intention, the effect of the awareness of consequences on personal norms, and the effect of the awareness of consequences on the ascription of responsibility were all significant at the statistical level of 0.1%. The effect of the ascription of responsibility on personal norms was significant at the statistical level of 1%. Thus, H8, H9, H10, and H11 are all valid. The above results validated the chain mediation between awareness of consequences, ascription of responsibility, and personal norms on the one hand and individual behavioral intention on the other [57]. That is, college students with a higher awareness of consequences are more likely to perceive the negative impacts of massive food waste (such as resource consumption and environmental pollution) at a psychological level [68]. This enables them to realize that saving food helps reduce the consumption of resources and energy and eases environmental crises. They internalize food saving as a sense of responsibility, thereby activating personal norms to enhance their food-saving intentions.

In addition, in the extended TPB-NAM model, the effects of subjective norms on personal norms also passed the significance test at the statistical level of 0.1%. This result suggests that individuals under higher pressure from parents, society, and schools in terms of food-saving behavior could better strengthen their sense of moral obligation toward saving food [69], thereby enhancing their food-saving intentions. Thus, H12 is valid.

The extended TPB-NAM-ABC model developed in this study includes a partial mediation model in its framework. The presence of the above mediating effects was tested using the bootstrap method. According to this method, the first step is to judge whether the total effect is significant (when the 95% bootstrap CI does not include 0, it is significant; otherwise, it is non-significant). When the total effect is significant, the next step is to determine whether the direct effect is also significant. When the direct effect is significant, there is a partial mediating effect; when the direct effect is non-significant, there is a full mediating effect. Otherwise, no mediating effect exists. According to the results of the mediating-effect tests presented in Table 9, ascription of responsibility (AR) played a partial mediating role in the "AC → PN" path, accounting for 18.10% of the total effect. Personal norms (PN) played a partial mediating role in the "SN → FSI" path, accounting for 40.26% of the total effect.

**Table 9.** Results of the mediating-effect test for ascription of responsibility (AR) and personal norms (PN).

| Intermediary Path | Index | Effect Value | Boot SE | Bias-Corrected 95% CI | | Proportion in Total Effect (%) |
|---|---|---|---|---|---|---|
| | | | | Lower | Upper | |
| AC→AR→PN | Total effect | 0.453 | 0.063 | 0.336 | 0.581 | |
| | Direct effect | 0.371 | 0.067 | 0.244 | 0.508 | 81.90% |
| | Indirect effect | 0.082 | 0.030 | 0.025 | 0.145 | 18.10% |
| SN→PN→FSI | Total effect | 0.457 | 0.067 | 0.324 | 0.584 | |
| | Direct effect | 0.273 | 0.076 | 0.131 | 0.425 | 59.74% |
| | Indirect effect | 0.184 | 0.054 | 0.097 | 0.308 | 40.26% |

### 5.5. Moderating Effects of Three Catering Modes

In this study, a hierarchical regression analysis was performed to validate the moderating effects of the three catering modes in the extended TPB-NAM-ABC model. The specific test steps were as follows: Step 1: The construction of Model 1. In this step, a regression analysis was performed on both the independent variable and the dependent variable. This step constitutes the establishment of the moderating effects, wherein the key lies in ensuring that the regression coefficient of the independent variable on the dependent variable is significant. Step 2: The construction of Model 2. A regression analysis of the dependent variable was performed based on Model 1 by adding a moderator variable. This step helps to ascertain whether the moderator variable is an influence factor on the dependent variable. Step 3: The construction of Model 3. Based on Model 2, the interaction term between the independent variable and the moderator variable is added to perform regression analysis on the dependent variable. When the regression results are significant, the moderator variable plays a moderating role in the effect path of the independent variable on the dependent variable. The test results are shown in Tables 10 and 11.

According to Table 10, the results of Model 1 indicate that food-saving intention had a significant positive promoting effect on food-saving behavior. The results of Model 2 show that the standard-quantity catering mode had a significant negative inhibitory effect on food-saving behavior. That is, most students believed that the standard-quantity catering mode, which awards them with a low level of autonomy over the category and portion size of their food, hinders their performance of food-saving behavior. The results of Model 3 clarified that the standard-quantity catering mode played a negative moderating role in the "FSI → FSB" path ($\beta = -0.122$; $p < 0.001$). That is, the more that canteen buttery hatches use the standard-quantity catering mode, the more difficult it is for students to transform their food-saving intentions into food-saving behavior. Thus, H13 is valid.

**Table 10.** Moderating-effect tests for three catering modes in the "FSI → FSB" path.

| Variable | Model 1 | Model 2 | Model 3 | Model 2′ | Model 3′ | Model 2″ | Model 3″ |
|---|---|---|---|---|---|---|---|
| FSI | 0.454 *** | 0.453 *** | 0.449 *** | 0.433 *** | 0.469 *** | 0.414 *** | 0.445 *** |
| SQ | | −0.061 * | −0.036 | | | | |
| LS | | | | 0.092 ** | 0.078 * | | |
| BF | | | | | | 0.217 *** | 0.212 *** |
| FSI * SQ | | | −0.122 *** | | | | |
| FSI * LS | | | | | 0.089 * | | |
| FSI * BF | | | | | | | 0.105 *** |
| $R^2$ | 0.217 | 0.224 | 0.251 | 0.229 | 0.237 | 0.263 | 0.275 |
| Adjusted $R^2$ | 0.215 | 0.221 | 0.245 | 0.225 | 0.231 | 0.260 | 0.270 |
| *F* value | 116.172 *** | 60.598 *** | 46.608 *** | 62.282 *** | 43.246 *** | 74.833 *** | 52.876 *** |

Note: * $p < 0.05$; ** $p < 0.01$; *** $p < 0.001$.

The results of Models 2′ and 2″ in Table 10 indicate that both the large-/small-portion-size catering mode and the buffet catering mode had significant promoting effects on the students' food-saving behavior. That is, most students believed that a certain level of autonomy over the category and portion size of their food facilitates their food-saving behavior performance. The results of Models 3′ and 3″ show that the large-/small-portion-size catering mode and the buffet catering mode both had positive moderating effects in the "FSI → FSB" path (β = 0.089, $p$ = 0.041 < 0.05; and β = 0.105, $p$ < 0.001). That is, increasing large-/small-portion-size and buffet buttery hatches in the canteens would make it easier for students to transform their food-saving intentions into actual food-saving behavior. Thus, H14 and H15 are valid.

**Table 11.** Moderating-effect tests on three catering modes in the "HM → FSB" path.

| Variable | Model 1 | Model 2 | Model 3 | Model 2′ | Model 3′ | Model 2″ | Model 3″ |
|---|---|---|---|---|---|---|---|
| HM | −0.113 *** | −0.111 *** | −0.116 *** | −0.118 *** | −0.122 *** | −0.098 *** | −0.108 *** |
| SQ | | −0.058 | −0.060 | | | | |
| LS | | | | 0.165 *** | 0.173 *** | | |
| BF | | | | | | 0.281 *** | 0.282 *** |
| HM * SQ | | | −0.039 * | | | | |
| HM * LS | | | | | 0.057 * | | |
| HM * BF | | | | | | | 0.068 * |
| $R^2$ | 0.034 | 0.041 | 0.050 | 0.076 | 0.089 | 0.115 | 0.128 |
| Adjusted $R^2$ | 0.032 | 0.036 | 0.044 | 0.071 | 0.082 | 0.110 | 0.122 |
| *F* value | 14.745 *** | 8.945 *** | 7.383 *** | 17.177 *** | 13.554 *** | 27.105 *** | 20.422 *** |

Note: * $p$ < 0.05; *** $p$ < 0.001.

To better understand the moderating effects of the three existing canteen catering modes in the "FSI → FSB" path, the simple-slope analysis proposed by Robinson et al. [70] was used. The different types of catering modes were divided into two levels (high and low) according to their proportions in the total number of buttery hatches in college canteens, and the food-saving intention psychological factors were also divided into high and low levels. Detailed graphs were drawn, as shown in Figure 2. Figure 2a shows that the higher the proportion of standard-quantity buttery hatches in the total number of buttery hatches in college canteens, the more difficult it is for college students to transform their food-saving intentions into actual food-saving behavior. Figure 2b,c show that the higher the proportions of large-/small-portion-size and buffet buttery hatches, the easier it is for college students to transform their food-saving intentions into actual food-saving behavior.

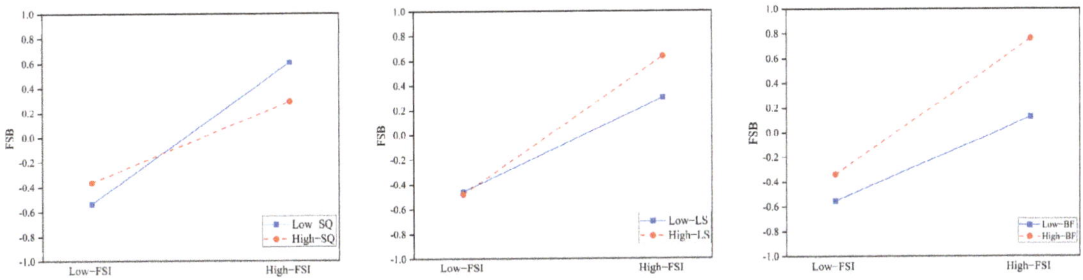


(a) (b) (c)

**Figure 2.** (**a**) Moderating effect of SQ on "FSI → FSB" path; (**b**) moderating effect of LS on "FSI → FSB" path; (**c**) moderating effect of BF on "FSI → FSB" path.

The results of Model 1 in Table 11 indicate that herd mentality had a significant negative impact on food-saving behavior. The results of Model 3 show that the standard-quantity catering mode had a significant promoting moderating effect in the path of the negative impact of herd mentality on food-saving behavior ($\beta = -0.039$, $p = 0.043 < 0.05$). That is, the more canteens offering standard-quantity buttery hatches, the higher the degree of food waste because of the influence of herd mentality. Thus, H16 is valid. The results of Models 3′ and 3″ in Table 11 clarify that both the large-/small-portion-size catering mode and the buffet catering mode had significant inhibitory moderating effects in the path of the negative impact of herd mentality on food-saving behavior ($\beta = 0.057$, $p = 0.016 < 0.05$; $\beta = 0.068$, $p = 0.012 < 0.05$). That is, the higher the proportions of large-/small-portion-size and buffet buttery hatches, the weaker the negative impact of the students' herd mentality on their food-saving behavior. Thus, H17 and H18 are valid. To better understand the moderating effects of the three existing canteen catering modes in the "HM → FSB" path, we drew more detailed graphs in this study, which are presented in Figure 3.

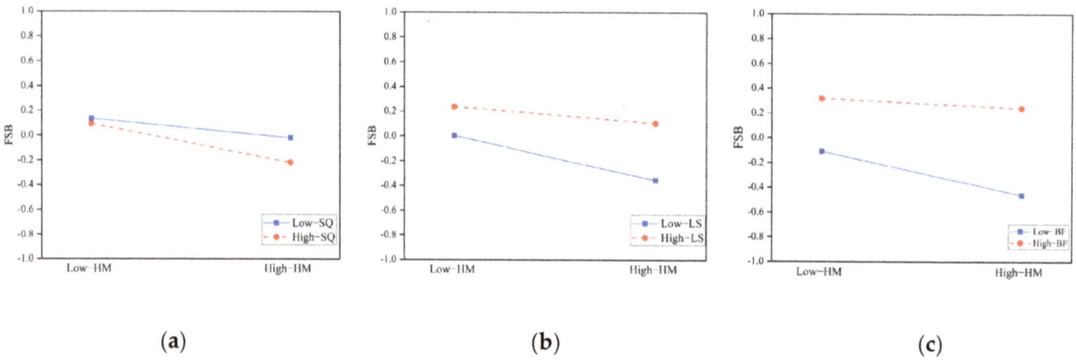


**Figure 3.** (**a**) Moderating effect of SQ on "HM → FSB" path; (**b**) moderating effect of LS on "HM → FSB" path; (**c**) moderating effect of BF on "HM → FSB" path.

According to Figure 3a, the higher the proportion of standard-quantity buttery hatches in the total number of buttery hatches in college canteens, the more susceptible college students are to imitating the food waste behavior of the other classmates around them, and the more serious their food waste behavior. By contrast, according to Figure 3b,c, the higher the proportions of large-/small-portion-size and buffet buttery hatches, the easier it is for college students to overcome the herd mentality and enhance their own food-saving behavior.

To compare the moderating effects of the large-/small-portion-size catering mode and the buffet catering mode, first, the frequencies of the latent variable sample scores of the three catering modes were counted, as shown in Figure 4a. Students generally gave high scores for the large-/small-portion-size catering mode and buffet catering mode. In other words, they generally believed that the buffet catering mode and large-/small-portion-size catering mode facilitate their food-saving behavior. Secondly, the matched-sample $t$-test was conducted on the averages of the latent variable samples of the three catering modes, as shown in Figure 4b. The results showed that the mean score of the buffet catering mode was significantly higher than that of the large-/small-portion-size catering mode. Therefore, the buffet catering mode had a more significant moderating effect than the large-/small-portion-size catering mode in both the "FSI → FSB" path and the "HM → FSB" path.

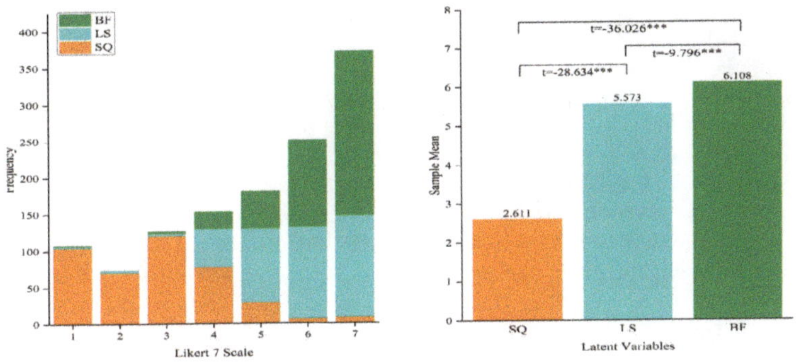


(a) (b)

**Figure 4.** (**a**) Frequency statistics of latent variable scores of three catering modes; (**b**) latent variable-mean paired-sample *t*-test for three catering modes. Note: For (**a**), a latent variable score > N but ≤N + 1 (N = 1, 2, . . ., 6) is counted as N + 1. To be specific, the score of a latent variable is calculated via the average score of the observed variables. For (**b**), *** $p < 0.001$.

## 6. Discussion

### 6.1. Differences in College Students' Food-Saving Behavior in Terms of Individual-Attribute Factors

The difference test results showed that while there was no significant difference in the college students' food-saving behavior in terms of their education levels, significant differences were found in terms of gender, household registration, monthly living expenses, and average daily spending on meals. The food-saving behavior of male students was more pronounced than that of female students, which is consistent with the findings of Painter et al. [71]. This is because the standard-quantity catering mode is the dominant catering mode in college canteens, which means that both male and female students are often given roughly the same portion sizes of food. In this case, female students are more likely to waste food than male students because male students generally have larger appetites [72]. Students with rural household registration have more pronounced food-saving behavior than those with urban household registration, which is consistent with the findings of Guo et al. [60]. The reason for this result is that students with rural household registration have farming experience and have witnessed the hard work that farmers put into grain production, which is why they are more inclined to save food while dining in the canteens. College students' food-saving behavior declines as their economic statuses (comprising monthly living expenses and average daily spending on meals) increase, which is consistent with the findings of Xu et al. [7] on household food waste behavior and those of Hafiz et al. [73] on reducing food waste behavior among college students (i.e., when consumers have a higher monthly income or college students have sufficient monthly living expenses, they tend to spend more on food purchases, which, in turn, leads to a lot of food waste). The finding that there was no significant difference in the college students' food-saving behavior in terms of education is inconsistent with the findings of Wu et al. [62]. A possible explanation is that there is no significant age difference among college students, and that the students in this age bracket have not witnessed periods of extreme food scarcity [74]. As a result, the differential effect of education on the college students' food-saving behavior was non-significant. In summary, college administrators should develop targeted guidance strategies for various student groups to reduce food waste in the canteens.

### 6.2. Psychological Driving Mechanisms Underlying College Students' Food-Saving Behavior

According to the results of a comparison of different structural equation models in terms of the explanation rates for food-saving intention and food-saving behavior, the extended TPB-NAM model has better explanatory power for the psychological driving mechanisms underlying college students' food-saving behavior. The following can be distilled from the SEM results:

First, of the two direct drivers of food-saving behavior, food-saving intention had a more significant effect on food-saving behavior among the college students (β = 0.489) than herd mentality (β = −0.129). This result is similar to the findings of existing studies in that it is feasible to predict an individual's food-saving behavior based on their food-saving intention. For example, a study on household food waste noted that food-saving intention is an antecedent of food-saving behavior [65]. This study also proved that in addition to food-saving intention, the herd mentality also has a direct impact on the food-saving behavior of college students, and the results of this study are consistent with the findings of Hafiz et al. [73]; that is, in the interview process, college students said that they desire to be recognized by society, and that they desire to be included in the surrounding group rather than excluded, and, under this pressure, they behave in the same way as the surrounding group; however, existing studies in colleges and universities only found a direct relationship between food-saving intention and behavior [19,21,64]. To summarize the aforementioned analysis, in situations in which it is difficult to directly observe college students' food-saving behavior, their food-saving intentions can be observed as a substitute.

Second, of the five latent variables that are directly associated with the food-saving intention of college students, personal norms (β = 0.374) had the most significant effect, followed in descending order by attitudes (β = 0.303), subjective norms (β = 0.273), perceived behavioral control (β = 0.094), and perceived health risks (β = −0.089). The above findings do not contradict the results of the existing studies on pro-environmental behaviors, such as college students' food-saving behavior. Wang et al. [21], in a study devoted to college students reduction food waste behaviors, also noted that attitudes, subjective norms, and perceived behavioral control can significantly influence college students' food-saving intentions. Elliott et al. [75] also pointed out that the more positive attitudes that students have toward food-saving behavior, the greater their willingness to reduce food waste; in contrast, negative attitudes are a deterrent to food-saving intentions. The finding that subjective norms are related to college students' food-saving intentions is consistent with the findings of Wu et al. [62]. Thus, if students perceive higher food-saving expectations from people who are important to them, they will be more inclined to save food. Perceived behavioral control refers to students' perceptions of the level of difficulty or ease with which they implement food-saving behaviors in a particular setting, based on the food skills, resources, and knowledge they possess, which, in turn, influences their food-saving intentions. This finding is in agreement with the findings of Truman et al. [76]. That is, if students have fewer relevant food skills, resources, and knowledge, they will judge that the implementation of reduced food waste behaviors in that particular context will be hindered, and thus the food-saving behavior will be lower. Health risk perception negatively influences college students' food-saving intentions. This finding is consistent with the findings of Williams et al. [77] on the influencing factors of food waste in households. Therefore, the higher the health threat of stale food and leftover re-edible meals, the more likely students are to produce food waste. In addition, a study on energy-saving behavior in college student dormitories found that personal norms had the most substantial impact on the students' energy-saving intentions [78]. This conclusion demonstrates that the pro-environmental intentions of highly educated people can be more easily stimulated by elevating their moral levels rather than making them rationally cognizant of such behavior, thereby encouraging their pro-environmental behavior.

Third, the chain mediation model connecting the intention to behave, personal norms, ascription of responsibility, and awareness of consequences was verified. This finding is consistent with the findings of the existing studies. For example, a study that analyzed household food waste behavior from a moral perspective noted that a household's awareness of the consequences of food waste behavior has a positive effect on its ascription of responsibility, that they both positively affect personal norms, and that personal norms can positively affect the food-saving intentions of the household [54]. This study also validates the research conclusions of Viccaro et al. [68] and Wu et al. [69], namely, that personal norms mediate the relationship between subjective norms and the intention to behave. The above findings suggest that two measures can be taken by college administrators to enhance students' ethical norms regarding pro-environmental behaviors: (1) the implementation of pertinent ethical education activities; and (2) cooperation with students' families to compel the pro-environment behavior of college students from both inside and outside the campus.

### 6.3. Moderating Effects of Catering Modes on College Students' Food-Saving Behavior

According to the statistics of the sample data, the latent variables for the food-saving intention had a mean value of 6.361, suggesting that college students are highly willing to save food. However, only 30.33% of the 422 questionnaire samples lived up to the criteria of the "empty plate" campaign in the week preceding the survey, suggesting that there is still a substantial gap between college students' food-saving intentions and their actual behavior. These findings also indicate that students regard unreasonable catering modes in college canteens as the primary cause of food waste, which means that finding a reasonable catering mode may help to narrow this gap.

The results of the moderating-effect test in this study showed that the standard-quantity catering mode had a significant inhibitory moderating effect in the pathway of food-saving intention and food-saving behavior, but it had a significant promoting moderating effect in the pathway of the negative influence of herd mentality on food-saving behavior. This result suggests that the standard-quantity catering mode is not conducive to students' implementation of food-saving behavior, which is consistent with the findings of Fan et al. [38] and Hafiz et al. [73]. That is, if students eat a fixed amount of food every day, it will cause them to produce more food waste. One possible explanation for this result is that the standard-quantity catering mode offers a comparatively fixed portion size to college students while ignoring their varying and unstable appetite levels. As a result, the implementation of food-saving behavior is hindered [38].

The moderating effect test results also showed that both the large-/small-portion-size catering mode and the buffet catering mode had significant promoting moderating effects in the path of food-saving intention and food-saving behavior, while they had a significant inhibitory moderating effect in the path of the negative impact of herd mentality on food-saving behavior. This suggests that the large-/small-portion-size catering mode and the buffet catering mode are both conducive to the implementation of food-saving behavior among students. As stated by Elliott and Truman [79], external environmental factors such as food portion sizes, food choices, and individual food preferences are important factors that hinder students' food-saving behaviors, and if students are provided with larger-portion-size food or food categories that do not meet their dietary preferences during canteen meals, their food-saving behaviors are hindered, which results in the creation of food waste. Conversely, if students are provided with food portion sizes and food types that match their food needs, the greater the likelihood that they will implement reduced food waste behaviors while eating in the canteens. This result justifies the strategic recommendations given in related studies that advocate for the introduction of half-portion-size food, small-portion-size food, and other catering modes in college canteens [37,38]. Based on the results of this study, it can be seen that when college students are in an environment that makes saving food easy (i.e., when the portion size and category of food can be controlled by individuals to a certain extent), they tend to order food according to their dietary preferences and appetite demands. Such an environment not only promotes the transformation of positive food-saving intentions into actual behavior but also alleviates the situation of food waste caused by the influence of herd mentality.

In summary, in this study, we confirmed that college canteen managers should adjust most of the existing standard-quantity buttery hatches in college canteens into large-/small-portion-size and buffet buttery hatches. This will provide a convenient external environment that encourages food-saving behavior in college students, which will help them to transform their knowledge of food saving into actual behaviors, thereby narrowing the gap between food-saving intention and actual food-saving behavior.

## 7. Conclusions and Suggestions

### *7.1. Conclusions*

Reducing food waste in college students' catering activities and guiding them toward the implementation of food-saving behavior contribute to the achievement of the Sustainable Development Goals (SDGs) proposed by the United Nations World Summit on Sustainable Development, and in particular to the achievement of SDG 12.3, "Reduce food waste at the consumption stage". In this study, relying on data from 422 valid questionnaires retrieved from a survey administered at Xi'an University of Architecture and Technology, China, we built an extended TPB-NAM-ABC theoretical framework model. Using this model as a basis, we explored the mechanisms underlying the effects of external environmental factors (i.e., catering modes) on college students' food-saving behavior. This exploration was based on a comprehensive consideration of both internal and external factors using SEM and hierarchical regression analysis. The main conclusions of this study are as follows:

(1) The food-saving degree is higher in male students than in female students, higher in students of rural household registration than in those of urban household registration, and higher in students with better economic status (comprising monthly living expenses and average daily spending on meals).

(2) Food-saving intention and herd mentality are major drivers of college students' food-saving behavior. Personal norms, attitudes, subjective norms, perceived behavior control, and health risk perception are the influencing factors of food-saving intention, among which, personal norms have the greatest effect. Personal norms can be activated either directly through the awareness of consequences or indirectly through the ascription of responsibility, thereby affecting the food-saving intention. Moreover, subjective norms can also activate personal norms.

(3) The standard-quantity catering mode has an inhibitory moderating effect in the transformation of food-saving intention into actual food-saving behavior, but it has a promoting moderating effect in the path through which herd mentality negatively impacts food-saving behavior. Both the large-/small-portion-size catering mode and buffet catering mode have promoting moderating effects in the transformation of food-saving intention into actual food-saving behavior, but they have an inhibitory moderating effect on the path through which herd mentality negatively impacts food-saving behavior. Notably, the moderating effects of the buffet catering mode are more significant than those of the large-/small-portion-size catering mode.

### *7.2. Suggestions*

Sustainable food systems play a crucial role in achieving sustainable development, particularly in eradicating hunger, ensuring food security, and promoting sustainable agricultural development. The key to creating a culture of sustainability and frugality on campus is to urge college students to take responsibility for themselves, others, society, and the environment. Based on the conclusions of this study, the following two suggestions are provided:

Firstly, food waste should be reduced from the perspective of the individual student. Food-saving education and publicity activities should be organized to raise students' awareness of saving food. Positive ethical norms and attitudes regarding food saving can be fostered in students by actively developing food-saving educational activities in colleges and universities (e.g., food-saving lectures, class meetings, and knowledge contests). To increase students' awareness of the detrimental effects of food wastage and encourage them to take actions to avoid it, colleges and universities should also post the total amount of food and resources wasted over the course of the previous week or month on the display screens in their canteens. In addition, as college students must strictly adhere to teaching and training programs, colleges and universities may also consider making environmental courses or environmental practices a mandatory part of their education, which is conducive to raising students' awareness of environmental issues and boosting their food-saving behavior.

Secondly, food waste should be reduced from the perspective of the food supply side. Efforts should be made to meet the diverse food-saving needs of individual students and to optimize the catering modes in college canteens. College canteens should adjust the existing catering modes by increasing the number of large-/small-portion-size and buffet buttery hatches, thereby ensuring that students with different individual attributes can select the most suitable catering mode according to their dining habits. These measures can help transform students' food-saving intentions into actual food-saving behavior.

### 7.3. Limitations and Future Research

In this study, we examined the mechanisms underlying the effects of catering mode on college students' food-saving behavior in conjunction with the related internal psychological driving mechanisms. Valuable conclusions were obtained. However, certain limitations persist, which open avenues for future research. First, in this study, when exploring the influencing mechanism of the catering mode on the food-saving behavior of college students, the types of college canteen catering modes were classified based on the students' methods of selecting their categories and portion sizes of food, and the conclusions of this study will not be applicable if the basis of the catering-mode classification is different (e.g., food prices, cooking methods, etc.). In the future, scholars can try to consider the type of catering mode according to different classification criteria and conduct research with the help of this study's methodology so as to enhance the generalizability of the results of this study to college students in China and other Asian countries or to other consumer groups in society. Second, in this study, the college students' food-saving behavior was measured using self-reported behavior rather than actual experiments, which may not reflect reality, as there is generally a gap between self-reported behavior and actual behavior. In future related research, the food-saving behavior of research subjects can be observed more accurately through field measurement experiments, which will also help to improve the accuracy of the results. Third, the focus of this study was on exploring the influence of different types of catering modes in college canteens on the process of transforming students' food-saving awareness into actual behaviors. Therefore, the psychological driving mechanisms of the food-saving behaviors of college students were not sufficiently considered. However, some studies have pointed out that the religiosity and family upbringing of individuals and their (grand)parents are important factors in the formation of college students' food-saving intentions. Therefore, in the future, researchers should consider exploring the mechanisms of their influences on food-saving intentions from the perspectives of the religiosity and family upbringing of individuals and their (grand)parents in order to fill in the gaps in the existing literature.

**Author Contributions:** Conceptualization, X.L. (Xi Luo) and Y.S.; methodology, A.W.; software, A.W.; validation, A.W., X.L. (Xi Luo) and Y.S.; formal analysis, A.W.; investigation, A.W.; resources, X.L. (Xiaojun Liu); data curation, A.W.; writing—original draft preparation, A.W.; writing—review and editing, X.L. (Xi Luo); visualization, A.W.; supervision, X.L. (Xi Luo) and X.L. (Xiaojun Liu); project administration, X.L. (Xiaojun Liu); funding acquisition, X.L. (Xi Luo) and X.L. (Xiaojun Liu). All authors have read and agreed to the published version of the manuscript.

**Funding:** This research was funded by the National Natural Science Foundation of China (52008328 and 72141018), the Department of Science and Technology of Shaanxi Province (2023KJXX-043), and the Shaanxi Association for Science and Technology (20220425).

**Institutional Review Board Statement:** Ethical review and approval were waived for this study, due to respondents being ensured of confidentiality and anonymity, and all participation was voluntary.

**Informed Consent Statement:** Informed consent was obtained from all subjects involved in the study.

**Data Availability Statement:** The data used in the current study are available from the corresponding author on reasonable request.

**Conflicts of Interest:** The authors declare no conflicts of interest.

## References

1. Wang, R.; Lu, S.; Zhou, L.; Yang, Z.; Tang, Z.; Zhao, M.; Cheng, G. Assessing nutritional and economic aspects of food loss and waste in China. *Sustain. Prod. Consum.* **2023**, *42*, 95–105. [CrossRef]
2. Jeswani, H.K.; Figueroa-Torres, G.; Azapagic, A. The extent of food waste generation in the UK and its environmental impacts. *Sustain. Prod. Consum.* **2021**, *26*, 532–547. [CrossRef]
3. Zhu, J.; Luo, Z.; Sun, T.; Li, W.; Zhou, W.; Wang, X.; Fei, X.; Tong, H.; Yin, K. Cradle-to-grave emissions from food loss and waste represent half of total greenhouse gas emissions from food systems. *Nat. Food* **2023**, *4*, 247–256. [CrossRef]
4. Lee, B.X.; Kjaerulf, F.; Turner, S.; Cohen, L.; Donnelly, P.D.; Muggah, R.; Davis, R.; Realini, A.; Kieselbach, B.; MacGregor, L.S.; et al. Transforming our world: Implementing the 2030 agenda through sustainable development goal indicators. *J. Public Health Policy* **2016**, *37*, 13–31. [CrossRef]
5. Forbes, H.; Quested, T.; O'Connor, C. *Food Waste Index Report 2021*; United Nations Environment Programme: Nairobi, Kenya, 2021. Available online: https://catalogue.unccd.int/1679_FoodWaste.pdf (accessed on 14 September 2023).
6. Xue, L.; Liu, X.J.; Lu, S.J.; Cheng, G.Y. China's food loss and waste embodies increasing environmental impacts. *Nat. Food* **2021**, *2*, 519–528. [CrossRef] [PubMed]
7. Xu, Z.; Zhang, Z.; Liu, H.; Zhong, F.; Bai, J.; Cheng, S. Food-away-from-home plate waste in China: Preference for variety and quantity. *Food Policy.* **2020**, *97*, 101918. [CrossRef]
8. Hou, C.; Zhang, M.; Zhao, X.; Zhang, M.; Guo, X.; Fu, H. The neural mechanism of food waste behavior in Chinese households from the perspective of gender differences. *J. Nat. Resour.* **2022**, *37*, 2531–2543. [CrossRef]
9. Wang, Y.; Xu, S.; Li, Z.; Yu, W.; Gao, L. An empirical stydy on the food waste influence factors of urban residents' repast. *Chin. J. Agric. Resour. Reg. Plan.* **2018**, *39*, 199–204.
10. Ministry of Education of China. Statistical Communique on National Education Development in 2021. 2022. Available online: http://www.moe.gov.cn/jyb_sjzl/sjzl_fztjgb/202209/t20220914_660850.html (accessed on 21 September 2023).
11. Qian, L.; Li, F.; Cao, B.; Wang, L.; Jin, S. Determinants of food waste generation in Chinese university canteens: Evidence from 9192 university students. *Resour. Conserv. Recycl.* **2021**, *167*, 105410. [CrossRef]
12. National Bureau of Statistics. China Energy Statistical Yearbook 2021. 2022. Available online: https://cnki.nbsti.net/CSYDMirror/Trade/yearbook/single/N2022060061?z=Z024 (accessed on 25 September 2023).
13. Aschemann-Witzel, J.; Hooge, I.E.; Rohm, H.; Normann, A.; Bosley, M.B.; Gronhoj, A.; Oostindjer, M. Key characteristics and success factors of supply chain initiatives tackling consumer-related food waste: A multiple case study. *J. Clean. Prod.* **2016**, *155*, 33–45. [CrossRef]
14. Pinto, R.S.; Pinto, R.M.D.S.; Melo, F.F.S.; Campos, S.S.; Cordovil, C.M.-S. A simple awareness campaign to promote food waste reduction in a University canteen. *Waste Manag.* **2018**, *76*, 28–38. [CrossRef] [PubMed]
15. Cai, H.; Biesbroek, S.; Wen, X.; Fan, S.; van't Veer, P.; Talsma, E.F. Environmental footprints of Chinese foods and beverages: Literature-based construction of a LCA database. *Data Brief.* **2022**, *42*, 108244. [CrossRef]
16. Malefors, C.; Sundin, N.; Tromp, M.; Eriksson, M. Testing interventions to reduce food waste in school catering. *Resour. Conserv. Recycl.* **2022**, *177*, 105997. [CrossRef]
17. Qian, L.; Zhao, X.; Liu, G. The association between the awareness campaign and food waste among university students in China. *Resour. Conserv. Recycl.* **2024**, *193*, 107361. [CrossRef]
18. Ellison, B.; Savchenko, O.; Nikolaus, C.J.; Duff, B.R. Every plate counts: Evaluation of a food waste reduction campaign in a university dining hall. *Resour. Conserv. Recycl.* **2019**, *144*, 276–284. [CrossRef]
19. Musicus, A.A.; Amsler Challamel, G.C.; Mckenzie, R.; Eric, B.; Rimm, E.B.; Stacy, A.; Blondin, S.A. Food Waste Management Practices and Barriers to Progress in US University Foodservice. *Int. J. Environ. Res. Public Health* **2022**, *19*, 6512. [CrossRef]
20. Gokarn, S.; Kushwah, S.; Khaba, S.; Choudhary, A.; Rosaline, S. Young consumers' food waste reduction behaviour in a developing nation: Extending the theory of planned behaviour. *J. Consum. Prot. Food Saf.* **2023**, *18*, 291–302. [CrossRef]
21. Wang, H.; Ma, B. What influences students' food waste behavior in campus canteens? *Br. Food J.* **2023**, *125*, 381–395. [CrossRef]
22. Ding, C.Q. Study on Psychological Inducing Mechanism and Simulation of Urban Residents' Catering Waste Behavior. Master's Thesis, China University of Mining and Technology, Xuzhou, China, 2020.
23. Kasavan, S.; Mohamed, A.F.; Halim, S.A. Drivers of food waste generation: Case study of island based hotels in Langkawi, Malaysia. *Waste Manag.* **2019**, *91*, 72–79. [CrossRef]
24. Fishbein, M.; Ajzen, I. Belief, Attitude, Intention and Behavior: An Introduction to Theory and Research. *Contemp. Sociol.* **1977**, *6*, 244–245.
25. Rezaei, R.; Safa, L.; Damalas, C.A.; Ganjkhanloo, M.M. Drivers of farmers' intention to use integrated pest management: Integrating theory of planned behavior and norm activation model. *J. Environ. Manag.* **2019**, *236*, 328–339. [CrossRef]
26. Shin, Y.H.; Im, J.; Jung, S.E.; Severt, K. The theory of planned behavior and the norm activation model approach to consumer behavior regarding organic menus. *Int. J. Hosp. Manag.* **2018**, *69*, 21–29. [CrossRef]

27. Liao, C.; Hong, J.; Zhao, D.; Zhang, S.; Chen, C. Confucian culture as determinants of consumer' food leftover generation: Evidence from Chengdu, China. *Environ. Sci. Pollut. Res.* **2018**, *25*, 14919–14933. [CrossRef] [PubMed]
28. Wang, J.; Li, M.; Li, S.; Chen, K. Understanding Consumers' Food Waste Reduction Behavior-A Study Based on Extended Norm Activation Theory. *Int. J. Environ. Res. Public Health* **2022**, *19*, 4187. [CrossRef]
29. Jia, L.; Qiao, G. Determinants of university student's food waste prevention behaviour—Ananalysis based on the theory of planned behavior and norm activation model. *J. Arid. Land Resour. Environ.* **2022**, *37*, 38–46. [CrossRef]
30. Xuan, V.N.; Loan, L.T.; Hoa, N.M.; Dao, N.T. Using a Unified Model of TPB, NAM, and SOBC to Investigate the Energy- Saving Behavior of Urban Residents in Vietnam: Moderation Role of Cultural Values. *Sustainability* **2023**, *15*, 2225. [CrossRef]
31. Si, H.; Shen, L.; Liu, W.; Wu, G. Uncovering people's mask-saving intentions and behaviors in the post-COVID19 period: Evidence from China. *Sustain. Cities Soc.* **2021**, *65*, 102626. [CrossRef] [PubMed]
32. Werkman, A.; Van, D.J.; Van, I.K. Are you being served? Managing waist and waste via serving size, unit size, and self-serving. *Food Qual. Prefer.* **2022**, *99*, 104568. [CrossRef]
33. Li, H.; Li, M.X.; Meng, F.H.; Yu, Z.C.; Hao, Y.; Hou, J.Q. Investigation of food wastage in different types of restaurants in China and analysis of its influencing factors. *J. Environ. Eng. Technol.* **2021**, *11*, 898–907. [CrossRef]
34. Betz, A.; Buchli, J.; Gobel, C.; Müller, C. Food waste in the Swiss food service industry—Magnitude and potential for reduction. *Waste Manag.* **2015**, *35*, 218–226. [CrossRef] [PubMed]
35. Qian, L.; Li, F.; Liu, H.; Wang, L.; McCarthy, B.; Jin, S. Rice vs. Wheat: Does staple food consumption pattern affect food waste in Chinese university canteens? *Resour. Conserv. Recycl.* **2020**, *176*, 105902. [CrossRef]
36. Zhao, X.; Manning, L. Food plate waste: Factors influencing insinuated intention in a university food service setting. *Br. Food J.* **2019**, *121*, 1536–1549. [CrossRef]
37. Richardson, R.; Prescott, M.P.; Ellison, B. Impact of plate shape and size on individual food waste in a university dining hall. *Resour. Conserv. Recycl.* **2021**, *168*, 105293. [CrossRef]
38. Fan, Q.; Liu, M.; Li, S. Empirical study on the factors of food waste in college canteen in Hubei province. *Sci. Technol. Cereals Oils Foods* **2016**, *24*, 109–113. [CrossRef]
39. Guagnano, G.A.; Stern, P.C.; Dietz, T. Influences on attitude-behavior relationships: A natural experiment with curbside recycling. *Environ. Behav.* **1995**, *27*, 699–718. [CrossRef]
40. Wang, H.; Mangmeechai, A. Understanding the gap between environmental intention and pro-environmental behavior towards the waste sorting and management policy of China. *Int. J. Environ. Res. Public Health* **2021**, *18*, 757. [CrossRef] [PubMed]
41. Wang, L.; Wang, J.; Shen, P.; Liu, S.; Zhang, S. Low-Carbon Travel Behavior in Daily Residence and Tourism Destination: Based on TPB-ABC Integrated Model. *Sustainability* **2023**, *15*, 14349. [CrossRef]
42. Zhang, Y.; Shao, W.; Zhang, M.; Li, H.; Yin, S.; Xu, Y. Analysis 320 coal mine accidents using structural equation modeling with unsafe conditions of the rules and regulations as exogenous variables. *Accid. Anal. Prev.* **2016**, *92*, 189–201. [CrossRef]
43. Crowley, S.L.; Fan, X. Structural equation modeling: Basic concepts and applications in personality assessment research. *J. Pers. Assess.* **1997**, *68*, 508–531. [CrossRef]
44. Li, L.Y.; Fan, F.M.; Liu, X.D. Determinants of rural household clean energy adoption intention: Evidence from 72 typical villages in ecologically fragile regions of western China. *J. Clean. Prod.* **2022**, *347*, 131296. [CrossRef]
45. Stefan, V.; Van Herpen, E.; Tudoran, A.A.; Lähteenmäki, L. Avoiding food waste by Romanian consumers: The importance of planning and shopping routines. *Food Qual. Prefer.* **2013**, *28*, 375–381. [CrossRef]
46. Sultan, P.; Wong, H.Y.; Azam, M.S. How perceived communication source and food value stimulate purchase intention of organic food: An examination of the stimulus-organism-response (SOR) model. *J. Clean. Prod.* **2021**, *312*, 127807. [CrossRef]
47. Obaidellah, U.H.; Danaee, M.; Mamun, M.; Hasanuzzaman, M.; Rahim, N.A. An application of TPB constructs on energy- saving behavioural intention among university office building occupants: A pilot study in Malaysian tropical climate. *J. Hous. Built Environ.* **2019**, *34*, 533–569. [CrossRef]
48. Delistavrou, A.; Tilikidou, I.; Papaioannou, E. Climate change risk perception and intentions to buy consumer packaged goods with chemicals containing recycled $CO_2$. *J. Clean. Prod.* **2023**, *382*, 135215. [CrossRef]
49. Wang, J.; Xue, Y.; Liu, T. Consumer motivation for organic food consumption: Health consciousness or herd mentality. *Front. Public Health* **2023**, *10*, 1042535. [CrossRef]
50. Juan, L.J.; Su, C. How Face Influences Consumption-A Comparative Study of American and Chinese Consumers. *Int. J. Mark. Res.* **2007**, *49*, 237–256. [CrossRef]
51. Li, Z.; Yang, G.; Yu, X.; Hao, W.; Zou, X.; Liu, C. Investigation and Research on the Influence of Conformity on the Low Carbon Life of College students. *Guide Sci. Educ.* **2018**, *14*, 181–182. [CrossRef]
52. Slovic, P. Perception of risk. *Science* **1987**, *236*, 280–285. [CrossRef]
53. Schmidt, K. Predicting the consumption of expired food by an extended Theory of Planned Behavior. *Food Qual. Prefer.* **2019**, *78*, 103746. [CrossRef]
54. Obuobi, B.; Zhang, Y.; Adu-Gyamfi, G.; Nketiah, E. Households' food waste behavior prediction from a moral perspective: A case of China. *Environ. Dev. Sustain.* **2023**, *26*, 10085–10104. [CrossRef] [PubMed]
55. Loh, C.T.; Krishna, M.; Nirossheni, G.; Chong, S.L.; Darrsini, O.; Ho, J.Y.; Kanageswary, A.; Kirutiga, S. Intention to reduce food waste: A study among Malaysians. *J. Air Waste Manag.* **2021**, *71*, 890–905. [CrossRef]

56. Li, C.Y.; Fang, Y.H. Go Green, Go Social: Exploring the Antecedents of Pro-Environmental Behaviors in Social Networking Sites beyond Norm Activation Theory. *Int. J. Environ. Res. Public Health* **2022**, *19*, 14265. [CrossRef]
57. Teisi, M.F.; Noblet, C.L.; Rubin, J. The psychology of eco-consumption. *J. Agric. Food Ind. Organ.* **2009**, *7*, 1–26. [CrossRef]
58. Viccaro, M.; Coppola, A.; D'Angelo, M.C.; Genovese, F.; Romano, S.; Cozzi, M. Young People Are Not All the Same! The Theory of Planned Behaviour Applied to Food Waste Behaviour across Young Italian Generations. *Sustainability* **2023**, *15*, 14741. [CrossRef]
59. Li, J.P. Research on the Willingness of Undergraduates' to Reduce Food Waste Based on NAM. Master's Thesis, North China University of Water Resources and Electric Power, Zhengzhou, China, 2022.
60. Guo, C.; Chen, J.H.; Zhou, X.L. Empirical Analysis on Influencing Factors of College Students' Food Waste Behavior under the Background of Food Security—Based on the Survey of College Students in Nanchang. *J. Green Sci. Technol.* **2022**, *24*, 276–280. [CrossRef]
61. Ma, X.; Li, J.; Xing, L. Research on the influence mechanism of undergraduates' food waste behavior. *J. Arid Land Resour. Environ.* **2021**, *35*, 25–30. [CrossRef]
62. Wu, Y.; Tian, X.; Li, X.; Yuan, H.; Liu, G. Characteristics, influencing factors, and environmental effects of plate waste at university canteens in Beijing, China. *Resour. Conserv. Recycl.* **2019**, *149*, 151–159. [CrossRef]
63. Visschers, V.H.M.; Wickli, N.; Siegrist, M. Sorting out food waste behavior: A survey on the motivators and barriers of self-reported amounts of food waste in households. *J. Environ. Psychol.* **2016**, *45*, 66–78. [CrossRef]
64. Soorani, F.; Ahmadvand, M. Determinants of consumer' food management behavior: Applying and extending the theory of planned behavior. *Waste Manag.* **2019**, *98*, 151–159. [CrossRef]
65. Van der Werf, P.; Seabrook, J.A.; Gilliland, J.A. Food for thought: Comparing self-reported versus curbside measurements of household food wasting behavior and the predictive capacity of behavioral determinants. *Waste Manag.* **2020**, *95*, 18–27. [CrossRef]
66. Lorenz, B.A.S.; Hartmann, M.; Langen, N. What makes people leave their food? The interaction of personal and situational factors leading to plate leftovers in canteens. *Appetite* **2017**, *116*, 45–56. [CrossRef] [PubMed]
67. Jackson, D.L. Revisiting sample size and number of parameter estimates: Some support for the N:Q hypothesis. *Struct. Equ. Model.* **2003**, *10*, 128–141. [CrossRef]
68. Meier, T.; Borstel, T.V.; Welte, B.; Hogan, B.; Finn, S.M.; Bonaventura, M.; Friedrich, S.; Weber, K.; Teran, T.D. Food waste in healthcare, business and hospitality catering: Composition, environmental impacts and reduction potential on company and national levels. *Sustainability* **2021**, *13*, 3288. [CrossRef]
69. Wu, Y.L.; Kurisu, K.; Phuphisith, S.; Fukushi, K. Household food-waste prevention behaviors in Beijing, Shanghai, and Wuhan in China compared with those in Tokyo and Bangkok. *Resour. Conserv. Recycl.* **2023**, *192*, 106901. [CrossRef]
70. Robinson, C.; Tomek, S.; Schumacker, R. Tests of moderation effects: Difference in simple slopes versus the interaction term. *Mult. Linear Regres. Viewp.* **2013**, *39*, 16–24.
71. Painter, K.; Thondhlana, G.; Kua, H.W. Food waste generation and potential interventions at Rhodes University, South Africa. *Waste Manag.* **2016**, *56*, 491–497. [CrossRef] [PubMed]
72. Qian, L.; Rao, Q.; Liu, H.; McCarthy, B.; Liu, L.X.; Wang, L. Food waste and associated carbon footprint: Evidence from Chinese universities. *Ecosyst. Health Sustain.* **2022**, *8*, 2130094. [CrossRef]
73. Hafiz, A.A.; Gallagher, A.M.; Devine, L.; Hill, A.J. University student practices and perceptions on eating behaviours whilst living away from home. *Int. J. Educ. Res.* **2023**, *117*, 102133. [CrossRef]
74. Quested, T.E.; Marsh, E.; Stunell, D.; Parry, A.D. Spaghetti soup: The complex world of food waste behaviours. *Resour. Conserv. Recycl.* **2013**, *79*, 43–51. [CrossRef]
75. Elliott, C.; Truman, E.; Stephenson, N. Food Marketing and Power: Teen-Identified Indicators of Targeted Food Marketing. *Int. J. Environ. Res. Public Health* **2022**, *19*, 7815. [CrossRef]
76. Truman, E.; Raine, K.; Mrklas, K.; Prowse, R.; Hoed, R.C.D.; Watson-Jarvis, K.; Loewen, J.; Gorham, M.; Ricciardi, C.; Tyminski, S.; et al. Promoting children's health: Toward a consensus statement on food literacy. *Can. J. Public Health* **2017**, *108*, e211–e213. [CrossRef] [PubMed]
77. Williams, H.; Wikström, F.; Otterbring, T.; Löfgren, M.; Gustafsson, A. Reasons for household food waste with special attention to packaging. *J. Clean. Prod.* **2012**, *24*, 141–148. [CrossRef]
78. Du, J.; Pan, W. Examining energy saving behaviors in student dormitories using an expanded theory of planned behavior. *Habitat Int.* **2021**, *107*, 102308. [CrossRef]
79. Elliott, C.; Truman, E. The Power of Packaging: A Scoping Review and Assessment of Child-Targeted Food Packaging. *Nutrients* **2020**, *12*, 958. [CrossRef] [PubMed]

sustainability

MDPI

*Article*

# Participatory Visual Methods and Artificial Intelligence-Driven Analysis for Sustainable Consumption Insights

**Kanwal Gul [1], Syeda Fasih [2], Swapnil Morande [1,*] and Muhammad Ramish [3]**

1 Department of Economics and Management, University of Naples, 80126 Napoli, Italy; kanwal.gul@unina.it
2 Department of Management and HR, Institute of Business Management, Karachi 75190, Pakistan; tayyaba.fasih@iobm.edu.pk
3 Department of Marketing & Entrepreneurship, Institute of Business & Health Management, Karachi 74200, Pakistan; sufyan.ramish@duhs.edu.pk
* Correspondence: swapnil.morande@unina.it

**Abstract:** Sustainable consumption is crucial for mitigating global sustainability challenges. Understanding consumer behaviors and motivations, particularly in developing regions, is essential for designing effective interventions. This study pioneers an innovative methodology integrating participatory visual methods (photovoice) and artificial intelligence analysis to investigate food waste perceptions in an emerging economy context. Twenty-six university students participated in the study, documenting their lived experiences and perspectives on household food waste through photographs and narratives. The key results included 32% of participants expressing shock at the extent of food waste in their daily lives, while 28% showed relative indifference. AI-powered (Artificial Intelligence) computer vision and natural language processing were used to efficiently analyze the large visual and textual dataset. The mixed methods approach generated nuanced, situated insights into consumer attitudes, behaviors, and socio-cultural drivers of wastage. The key themes included low waste consciousness, aesthetic and convenience motivations, social norms, and infrastructural limitations. The participatory process proved effective for raising critical consciousness and uncovering consumption practice dynamics. AI analysis enabled rapid knowledge discovery from the qualitative data while mitigating researcher bias. This innovative integration of participatory methodologies and computational analytics advances sustainable consumption research by empowering marginalized voices and generating contextual insights from unstructured data. With further development, such human-centered AI approaches can transform the study and governance of sustainable consumption.

**Keywords:** photovoice; artificial intelligence; food waste; sustainability; participatory research

## 1. Introduction

Sustainable consumption has emerged as a critical frontier for addressing global sustainability challenges. From climate change to biodiversity loss, the impacts of human consumption threaten planetary boundaries and social foundations [1,2]. Transitioning to sustainable consumption and production forms a core target of the United Nations Sustainable Development Goals. However, progress remains slow, with global material footprints continuing to rise [3]. Behavior change is crucial for driving sustainable transformations, yet consumption practices remain tied to complex social, cultural, institutional, and material factors [4]. Designing effective interventions requires a nuanced understanding of consumption drivers beyond individual choices [5].

Food consumption and wastage form a major sustainability hotspot. One-third of all food produced globally is lost or wasted, with significant environmental, social, and economic repercussions [6]. In developing regions, changing diets, supermarketization, and increasing affluence are altering food practices [7–10]. Much waste occurs at the household level, driven by consumer behaviors [5,11,12]. However, research on consumer food waste perceptions and practices in emerging economies remains limited [8,13,14]. Existing studies rely largely on surveys and macro-level data, overlooking lived experiences [15]. Contextual factors like culture, infrastructure, and social norms remain under-examined. Addressing these research gaps is crucial for tailoring food waste reduction strategies to local realities [16–18].

This study proposes an innovative methodology integrating participatory visual methods and artificial intelligence (AI) analysis to investigate contextualized food consumption and waste behaviors. It employs a photovoice approach, engaging consumers as citizen scientists to document their perspectives through photography and narratives [19]. The qualitative insights generated are rapidly analyzed using AI techniques like computer vision and natural language processing (NLP). This mixed methods approach aims to efficiently uncover nuanced, situated understandings of consumption practices and drivers. This study focuses on household food wastage among urban youth in India, a key demographic in a transitioning and emerging economy. It explores the potential of participatory methodologies and AI analysis for advancing sustainable consumption research.

This paper first reviews the literature on sustainable consumption, consumer food waste, and participatory visual methods [4,17,20]. It then outlines the photovoice methodology employed, with students documenting food waste in their socio-cultural contexts. The AI-powered analytical workflow combining computer vision and NLP is detailed. Key findings are presented, highlighting themes of waste (in)visibility, social drivers, convenience orientation, and food system limitations. Sentiment analysis reveals complex emotions around wastage. Theoretical and practical implications are discussed, along with methodological reflections on participatory AI approaches for sustainability research. This paper concludes with recommendations for policymakers and future research directions.

This study makes several contributions. Firstly, it addresses research gaps on consumer food waste perceptions and behaviors in emerging economy contexts [14]. Secondly, it pioneers a participatory visual methodology for eliciting situated, experiential insights on consumption practices [21]. Thirdly, it leverages AI analysis for efficient knowledge discovery from large qualitative datasets. Finally, it explores the potential of participatory machine learning for empowering marginalized voices and uncovering contextualized sustainability solutions [22]. With further development, such human-centered AI approaches can transform sustainable consumption research and policymaking.

## 2. Literature Review

### 2.1. Sustainable Consumption: Challenges and Drivers

Sustainable consumption has become a core challenge of the Anthropocene. Globally, consumption and production account for over two-thirds of greenhouse gas emissions [23]. Material footprints have grown in lockstep with affluence, straining ecosystems. As highlighted by recent studies [24,25], the environmental impacts of consumption often exceed planetary boundaries, necessitating urgent transformations. Researchers increasingly recognize the need to move beyond technological solutions toward demand-side changes [26]. The Sustainable Development Goals position sustainable consumption as crucial for green, inclusive growth [27]. However, progress remains slow, constrained by socio-technical lock-ins and policy inertia [28,29].

Theories of consumption have evolved to consider the complex interplay of individual, social, and structural factors shaping behavior [30]. Social practice theory highlights how consumption is embedded in everyday routines, norms, and infrastructures [31,32]. Social practice theory provides a valuable perspective for understanding and addressing food waste behavior by focusing on the interconnected elements of practices: materials, competencies, and meanings. Instead of seeing food waste only as a result of individual choices, this perspective takes into account the larger social and cultural surroundings in which food-related actions take place [31]. By studying how daily habits, abilities, and societal standards influence how people buy, cook, eat, and throw away food, we can pinpoint areas where we can intervene [4]. For instance, improving food storage infrastructure (materials), enhancing cooking skills (competencies), and shifting cultural perceptions around food abundance and waste (meanings) can collectively foster more sustainable practices. This holistic understanding helps policymakers, educators, and community organizers design more effective strategies to reduce food waste by transforming the social practices that sustain it [4,20,31]. Changing practices requires reconfiguring their interconnected elements and breaking habits [32]. However, much sustainable consumption research and policy remains dominated by individualistic, rational choice models [33]. There is a growing recognition of the need to examine consumption and food waste from a systemic, socio-cultural perspective [34].

Emerging research emphasizes the importance of understanding food waste contexts and drivers beyond individual attitudes and choices. The most commonly cited prescriptive models are the Theory of Reasoned Action (TRA) and the Theory of Planned Behavior (TPB) [35–38]. These models provide frameworks for understanding the structure of consumer behavior. In contrast, other studies highlight the role of social influence, emotions, identity, environment, and structural conditions in shaping practice [39]. Qualitative, experiential insights are valued for uncovering the situated meanings and dynamics of consumption [39,40]. Participatory and ethnographic approaches, such as photovoice, are seen as powerful tools for eliciting emic perspectives and empowering participants as co-researchers [21,41,42]. However, the scalability of such methods remains a challenge, limiting wider adoption. This study explores the integration of AI analysis to enable rapid, large-scale processing of qualitative visual data, enhancing the reach and impact of participatory sustainable consumption research.

### 2.2. *Consumer Food Waste: An Emerging Sustainability Hotspot*

Food waste presents an urgent sustainable consumption challenge. Globally, one-third of all food is lost or wasted, undermining food security, resource conservation, and climate action [6]. The magnitude of food waste has significant environmental, social, and economic implications, with recent estimates suggesting that it accounts for 8–10% of global greenhouse gas emissions [43]. In developed countries, over half of food waste occurs at the consumption stage, largely in households [11]. Wastage is also rising rapidly in emerging economies, driven by urbanization, changing diets, and the growth of supermarkets [40,44]. Studies indicate that consumer behaviors and attitudes are critical intervention points for reducing waste [45]. However, research on consumer food waste perceptions and practices in developing contexts remains limited and fragmented [12].

Existing studies on consumer food waste behaviors have primarily focused on developed countries, using quantitative survey-based methods. These approaches provide valuable insights into general patterns and predictors of food waste [12,46]. However, they often fail to capture the nuanced, contextual factors shaping food practices in diverse settings [47]. Qualitative, interpretive research is needed to unpack the socio-cultural meanings and drivers of food waste in specific contexts [48]. In emerging economies, rapid transitions in food systems and consumption practices are reshaping waste behaviors [49]. The rise of supermarkets, processed foods, and eating out are altering traditional foodways and creating new waste streams. Changing norms around freshness, aesthetics, and abundance also influence waste generation [50]. However, in-depth, localized studies examining these dynamics remain scarce. There is a need for participatory, context-sensitive research approaches to uncover the lived experiences and perceptions of consumers in transitioning societies [15].

Visual methodologies like photovoice offer promising avenues for eliciting grounded, experiential insights into food consumption and waste practices [51]. By empowering participants to document their own realities, photovoice can surface emic perspectives and reveal the embodied, affective dimensions of waste behaviors [19,51]. However, the scalability of analyzing large visual datasets poses challenges for wider adoption [52]. This study pioneers the integration of AI-powered analysis with photovoice to enable rapid, large-scale processing of participant-generated data, enhancing the reach and impact of contextualized food waste research.

### 2.3. Photovoice: A Participatory Visual Methodology

Photovoice is a participatory action research method that empowers participants to document and reflect on their lived experiences through photography [19]. Grounded in the theoretical traditions of critical consciousness [53,54], feminist research, and documentary photography, photovoice aims to surface marginalized perspectives and catalyze social change [55]. Participants are given cameras to capture images representing their realities and perspectives on a given issue. They then contextualize the photos through verbal or written narratives, often in group discussions [41]. This process of visual storytelling can uncover rich experiential insights, challenge dominant narratives, and raise critical awareness [20].

Photovoice has been increasingly applied in sustainable consumption research to examine contextualized practices and drivers. Studies have engaged participants in documenting food waste behaviors, energy use, waste practices, and health and dietary habits. The method enables a granular examination of consumption contexts, meanings, and materialities [56]. It can reveal tensions between attitudes and practices and pathways for change. By empowering participants as co-researchers, photovoice also democratizes knowledge production and enhances research impact [41].

However, the qualitative visual data generated through photovoice can be time-consuming to analyze, limiting wider adoption [52,57]. Emerging studies explore the potential of integrating AI techniques to automate and scale photovoice analysis. Computer vision algorithms can rapidly identify objects, scenes, and themes in large visual datasets (Sambasivan & Holbrook [58]). Natural language processing can derive insights from associated photo captions and narratives (Garg et al., 2021 [59]). While still an emerging area, AI integration could accelerate photovoice analysis and enhance rigor through computational objectivity.

This study builds on these developments by pioneering a participatory AI approach to photovoice analysis for sustainable consumption research. It engages students in India as citizen scientists to document their food waste perceptions and practices, aligning with photovoice principles of empowerment and contextualization. Computer vision and natural language processing are then applied to efficiently derive insights from the visual data. The approach aims to scale photovoice for a wider application while centering participant interpretations and reflexivity. It explores the potential of human-machine collaboration for uncovering situated drivers of food waste and informing localized interventions.

## 3. Methodology

This study is grounded in a participatory research paradigm that values the co-creation of knowledge with communities. Photovoice, as a visual participatory methodology, aligns with this paradigm by empowering participants as co-researchers to document and reflect on their lived realities [19]. The AI-powered analysis, while grounded in positivist computational approaches, is guided by participatory ethics of transparency and interpretive validation with participants [60]. Reliability is ensured through methodological triangulation of visual and textual data sources, iterative thematic coding, and participant checking of AI-generated insights [61,62].

### 3.1. Photovoice Study Design

The methodology used in this study draws on established protocols and best practices [21,42]. A photovoice study was conducted with undergraduate students at a university in India.

A purposive sample of twenty-six university students ranging in age from 18 to 24 were selected to participate in the study. University students were chosen as a key demographic group for several reasons. Firstly, young adults in urban India are experiencing rapid transitions in food consumption practices, with increasing exposure to globalized diets, processed foods, and changing social norms [63]. These shifts have significant implications for food waste behaviors and sustainability [63]. Secondly, university students represent an important target group for sustainable consumption interventions, given their potential as future leaders and change agents [33,64]. Understanding their perceptions and practices around food waste can inform tailored educational and behavioral strategies [65].

While the purposive sample is not statistically representative of the broader population, it aligns with the qualitative, exploratory aims of the study. Photovoice research prioritizes depth, diversity, and contextual insights over generalizability [42]. The selection of university students from different backgrounds and disciplines aimed to capture a range of perspectives and experiences related to food waste. This diversity is crucial for identifying common patterns and divergences and for generating nuanced, situated understandings. The sample size of twenty-six participants is consistent with established photovoice studies [19,21,66], providing sufficient data for thematic saturation while maintaining feasibility for in-depth analysis [42].

As per Figure 1, after informed consent, participants attended a training workshop on photovoice methods and research ethics. The workshop oriented them to the study purpose of documenting food waste perceptions and practices in their daily lives and communities. Participants were given digital cameras and prompted to capture photos representing their experiences and observations around food wastage over a two-week period. They were encouraged to consider wastage at various stages (e.g., purchasing, cooking, eating out, storing, and disposal) and reflect on drivers and impacts. Ethical guidelines were provided on respecting privacy and obtaining consent for human subjects.

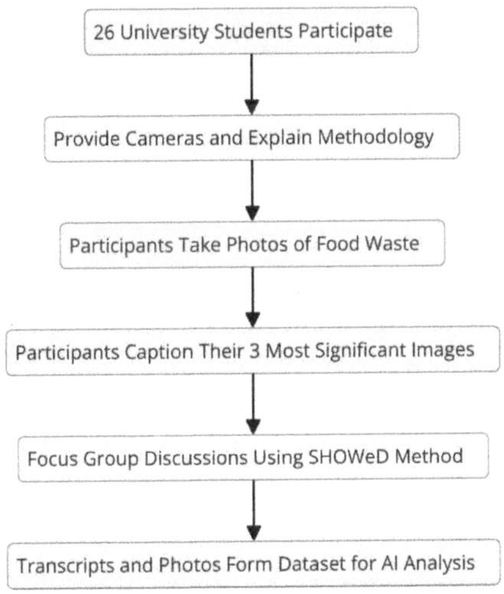


**Figure 1.** Workflow of photovoice activity.

After two weeks, participants returned for a sharing session to discuss their photos. In small groups, they each selected 2–3 of their most meaningful images and wrote accompanying captions explaining the significance. A facilitator then guided a reflection dialogue, prompting participants to share stories behind their photos and build collective interpretations. Discussions were audio-recorded and transcribed with permission. Participants gave informed consent for their anonymized photos, captions, and narratives to be analyzed for research purposes.

### 3.2. AI-Powered Analysis Workflow

The AI-powered analysis workflow used in this study builds on emerging applications of computer vision and natural language processing in qualitative research [67,68]. The Python 3.0 programming language and open-source libraries such as OpenCV, NLTK, and spaCy were used to develop the analysis pipeline. For the visual data, a convolutional neural network (CNN) model was trained to automatically detect and classify objects, scenes, and themes in the participant-generated food waste photographs. The CNN architecture used was MobileNetV2, which has shown high accuracy and efficiency in image classification tasks [69]. The model was pre-trained on the ImageNet dataset and fine-tuned on a manually annotated subset of the study photographs.

For the textual data, a combination of rule-based and machine-learning techniques was applied to analyze the photo captions and participant reflections. Sentiment analysis using the VADER lexicon [70] was conducted to identify emotional valence and arousal in the narratives. Topic modeling using Latent Dirichlet Allocation (LDA) was employed to discover latent themes and concepts related to food waste behaviors and drivers. The CNN model achieved an accuracy of 88% in classifying food waste objects and scenes, comparable to human performance. The research team manually validated the sentiment analysis and topic modeling results to ensure alignment with participant interpretations. Member checking [71,72] was also conducted, with participants reviewing and providing feedback on the AI-generated insights.

While the AI analysis enabled efficient processing of the large qualitative dataset, the interpretive process remained centered on participants' experiences and meanings. The machine learning results served as a complementary lens to surface patterns and prompt further dialogue rather than replace human sense-making. Researcher reflexivity (Finlay, 2002) and contextual awareness were also critically engaged in interrogating and qualifying the AI outputs. The integration of computational and human analysis aimed to leverage the strengths of both approaches for deepened understanding.

### 3.3. Methodological Reflections

The AI-powered photovoice analysis workflow enabled the rapid derivation of insights from the large qualitative dataset. Computational techniques made the visual data analysis feasible at a scale beyond typical manual coding approaches. The unsupervised machine learning algorithms helped identify bottom-up patterns grounded in the data rather than imposing researcher assumptions. Multi-modal analysis integrating visual and textual elements allowed for rich contextualization of themes. However, the AI-generated insights still required careful researcher interpretation and validation. Some nuances of participant narratives were lost in the quantitative topic models. The pre-trained computer vision models also missed culturally specific objects and scenes. Ongoing advancements in few-shot learning and model adaptation could help tailor the AI to context. More broadly, integrating machine and human intelligence requires considered design to preserve the participatory ethics of photovoice [60] (Brown et al., 2021). Reflexive discussion of computer-generated insights with participants could help align machine and emic perspectives. This study provides an initial proof-of-concept for AI-powered photovoice analysis in sustainable consumption research. Further work is needed to refine the techniques and address limitations. Comparative evaluation of manual and AI-assisted analysis could assess efficiency and validity gains. Explainable AI methods could improve algorithmic transparency for participants and researchers. Participatory machine learning, involving participants in training and interpreting AI models, presents an exciting frontier. With mindful development, AI integration could amplify photovoice as a scalable, empowering methodology for eliciting localized sustainability insights.

## 4. Results and Discussion

The participatory photovoice study and AI-powered analysis generated rich insights into urban students' perceptions, practices, and drivers of food waste. The CNN model trained on the participant-generated photographs achieved an overall accuracy of 88% in classifying food waste objects and scenes. The model's precision, recall, and F1 score for each class are presented in Table 1. The high-performance metrics indicate the effectiveness of the MobileNetV2 architecture and transfer learning approach in accurately detecting and categorizing food waste elements from the visual data.

**Table 1.** CNN model performance metrics for food waste classification.

| Class | Precision | Recall | F1 Score |
|---|---|---|---|
| Plate waste | 0.92 | 0.87 | 0.89 |
| Packaging waste | 0.85 | 0.91 | 0.88 |
| Spoiled food | 0.88 | 0.84 | 0.86 |
| Overproduction | 0.86 | 0.90 | 0.88 |
| Miscellaneous | 0.91 | 0.89 | 0.90 |

The model's confusion matrix (Figure 2) further illustrates the classification performance across the different food waste categories. The matrix indicates high diagonal values, signifying accurate predictions with minimal misclassifications between classes. The visualization helps identify specific categories where the model performs exceptionally well (e.g., plate waste) and those where some misclassifications occur (e.g., spoiled food and overproduction).

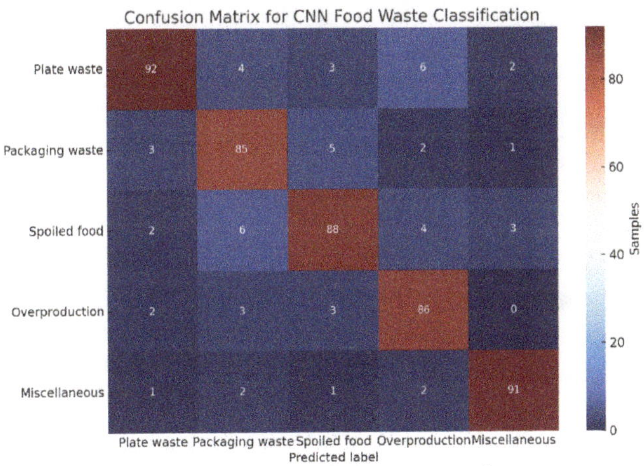


**Figure 2.** Confusion matrix for CNN food waste classification.

The sentiment analysis of the photo captions and reflections revealed a spectrum of emotional responses to food waste. Across the dataset, 32% of text snippets expressed negative sentiments (e.g., shock, guilt, or frustration), 28% conveyed neutral or mixed sentiments, and 24% expressed positive sentiments (e.g., hopefulness or determination). The VADER sentiment analyzer achieved an accuracy of 85% in classifying the sentiments, as validated through manual annotation of a random subset of 100 text samples. The LDA topic modeling identified five dominant themes in the participants' narratives: (1) wasteful consumption habits, (2) food surplus and overproduction, (3) environmental and social consequences, (4) lack of awareness and infrastructure, and (5) opportunities for change. The topic coherence score of 0.62 suggests a reasonable level of semantic consistency within the discovered topics.

Participants' reflections unpacked the social, cultural, and structural drivers of food waste in their contexts. For example, several narratives linked plate waste to changing dietary preferences, time constraints, and the normalization of food abundance. **As one participant shared, "*I often end up throwing away unfinished meals because I'm too busy to eat properly. It's become a habit, even though I feel guilty about it*".

The photovoice process also sparked critical reflection and agency among participants. Many expressed a heightened awareness of their own waste behaviors and a determination to change. As one student reflected, "Participating in this study opened my eyes to how much food I waste daily without realizing it. I'm motivated to start planning my meals better and composting my scraps". Others brainstormed solutions such as food-sharing initiatives, waste-tracking apps, and campus awareness campaigns, demonstrating the potential for bottom-up, community-driven action.

The AI-powered photovoice analysis revealed several key themes around food consumption and waste behaviors among university students. Illustrative photos and quotes are provided to contextualize each theme.

### *4.1. Invisibility of Food Waste*

A dominant theme was the invisibility and normalization of food waste in everyday life. Many photos depicted unfinished meals, buffet leftovers, and discards from food preparation. However, participants often expressed surprise at the cumulative wastage captured, having previously overlooked its prevalence, as previous studies confirmed that lack of knowledge [73] is a determinant of consumer-level food waste. As one student reflected:

> *"I never really noticed how much food goes into the bin each day. It's just become a normal part of cooking and cleaning routine. Seeing all the photos made me realize the scale of waste happening in my own kitchen".*
>
> (Participant 3, female)

The AI analysis corroborated this theme, with "plate waste" and "kitchen scraps" emerging as frequent visual labels. The associated captions and discussions suggested low consciousness of waste volumes among participants. This invisibility poses a barrier to waste reduction, as behaviors remain unexamined [74]. A consumer's lack of knowledge about food waste at the local level hinders the ability to develop place-based solutions that are locally relevant [75]. Photovoice served as a critical reflection tool, making waste visible and prompting a re-evaluation of practices confirming the result of previous research [76].

### *4.2. Socio-Cultural Drivers of Waste*

Photo narratives also highlighted significant socio-cultural drivers of food waste. Images of elaborate family meals, religious feasts, and social gatherings were common. Participants described wastage as an unintended consequence of hospitality norms and aesthetic expectations. Oversized portions, unappetizing buffet leftovers, and guest plate waste were recurrent subjects. As one participant explained:

> *"In our culture, serving excess food is a sign of good hospitality. Hosts feel embarrassed if there isn't more than enough on the table. But this leads to so much waste, especially when catering for large gatherings. The pressure to impress with abundance means a lot gets thrown out".*
>
> (Participant 21, male)

Joint topic modeling of visual and textual data reinforced this link between wastage and socio-cultural expectations. Photos depicting "surplus food" and "leftovers" frequently co-occurred with text themes of "social pressure" and "hospitality". These findings underscore the importance of sociological factors beyond individual attitudes in shaping waste behaviors [74]. Interventions must engage with cultural norms and social dynamics to shift collective practices.

### *4.3. Technological Imaginaries vs. Structural Constraints*

A third prominent theme concerned the infrastructural and systemic barriers to food waste reduction in the context of the developing world. Numerous photos depicted food spoilage due to lack of refrigeration, unreliable electricity, and inadequate packaging. Participant narratives highlighted the challenges of food preservation in humid climates without consistent cold storage. Several noted the compounding effects of long supply chains and lack of processing facilities, leading to high upstream wastage. Food safety concerns and distrust in cold chains also drove over-purchasing and discards.

As one participant shared:

> *"I dream of having a big fridge to keep all my fruits and veggies fresh. But with the constant power cuts in my area, it's not practical. The packaging also feels like too much plastic waste. I wish there were more sustainable and accessible options for food storage".*
>
> (Participant 11, female)

Infrastructural limitations intersected with socio-economic constraints in shaping waste behaviors. Smaller, more frequent purchases were common in low-income households and areas with limited storage space. The informal food sector, including street vendors and wet markets, offered affordable options but with higher perishability risks. Participants recognized the need for systemic solutions beyond individual actions. Improved cold chain infrastructure, local food processing, and packaging innovations were seen as key enablers. Several called for greater public investment and policy support for food waste reduction. These findings align with prior research on the socio-technical barriers to food waste reduction in developing contexts [65]. Soma (2020) similarly found that cold chain limitations and packaging waste concerns hampered the adoption of preservation technologies in Indonesia [77]. Madaan (2024) highlights the need for frugal innovations and circular economy approaches to address infrastructural constraints in developing nations [78]. The present study adds nuance by documenting lived experiences and perceptions of these systemic challenges among urban youth.

The AI-powered analysis revealed patterns of frustration and disempowerment in participants' emotional responses to infrastructural barriers. Sentiment scores were more negative for captions related to systemic issues compared to individual behaviors (0.32 vs. 0.48). Text analysis surfaced terms like "lack", "poor", and "difficult" in relation to infrastructure. These findings echo calls for food waste interventions to address structural drivers and inequities rather than placing the onus solely on consumers [65,77].

The visual data also highlighted innovative coping strategies and adaptations to infrastructural constraints. Photos showed improvised cooling methods like earthen pots and evaporative coolers. Captions described preserving practices like pickling, drying, and composting. Some participants shared photos of community fridges and food-sharing initiatives. These local solutions offer potential models for resilient, low-waste food practices. Participatory AI methods could help surface and scale such community innovations.

### 4.4. *Affective Dimensions of Waste*

Finally, the photovoice data surfaced affective and embodied dimensions of food waste. Sentiment analysis of the photo captions and narratives revealed a mix of emotions, from frustration and guilt to resignation. As per Table 2, images of discarded food were often accompanied by expressions of moral unease, particularly in the context of hunger and food insecurity.

**Table 2.** Sentiment Analysis Chart.

| Sentiment | Percentage |
|---|---|
| Shock | 32% |
| Indifference | 28% |
| Shame | 19% |
| Anger | 14% |
| Concern | 7% |

As one participant reflected:

> *"Every time I scrape perfectly good food into the trash, I feel this pang of guilt. I know there are so many people in my city who don't have enough to eat. But at the moment, convenience just takes over. It's this constant tension between my ideals and actions".*
>
> (Participant 9, female)

The emotional complexity of waste was evident in the AI-generated sentiment clusters. Negative sentiments like "guilt" and "frustration" frequently co-occurred with themes of "discarding food" and "hunger". These effective undercurrents shape waste behaviors in subtle but significant ways. Photovoice provided an avenue for participants to process and articulate these embodied experiences. Waste reduction strategies must engage with these emotional drivers, cultivating positive effects like empowerment and satisfaction.

The findings from this AI-powered photovoice study both align with and extend prior research on consumer food waste drivers. The invisibility of routine waste (Section 4.1) resonates with studies highlighting low consumer awareness of wastage [75,79]. However, our visual methodology uniquely captured the visceral shock of confronting cumulative waste, underscoring photovoice's potential for disrupting normalization. The socio-cultural drivers of waste (Section 4.2), particularly hospitality norms and aesthetic standards, echo findings from other emerging economies, like China [44], India [10,44,80], and Brazil [15]. Our study extends these works by visually documenting the embodied practices and emotional negotiations underlying these drivers. Infrastructural constraints emerged as a novel barrier to waste reduction in the context of the developing world (Section 4.3). While cold chain limitations are recognized in food waste literature [65], our participants' visual accounts viscerally capture the lived experiences of these constraints. The photovoice process also elicited unique socio-technical imaginaries around refrigeration and packaging. AI analysis enabled the distillation of these distinct and cross-cutting themes from the rich visual data. Prior photovoice studies in related domains like energy poverty have relied on manual content analysis. Our integration of participatory methods with computer vision and NLP demonstrates new possibilities for rapid insights from complex qualitative data [71].

## 5. Implications, Limitations, and Future Research

### 5.1. Theoretical Implications

This study advances sustainable consumption scholarship in three key ways. First, it demonstrates the value of participatory visual methodologies in eliciting contextual, embodied accounts of food waste practices. Photovoice can disrupt habitual behaviors, uncover emic meanings, and empower participant agency beyond dominant survey approaches [41,55]. Second, AI analysis enables scalable examination of rich qualitative insights, addressing limitations of manual methods at the expense of some nuance. The computational techniques of computer vision, NLP, and joint topic modeling show promise for rapid theme extraction and data integration [22]. However, they must be guided by participatory research ethics and interpretive validation. Third, the study reveals the socio-technical complexities of food waste behaviors in emerging economies. Waste drivers span socio-cultural norms, material infrastructures, and situated practices in understudied urban contexts like India, Pakistan, and Bangladesh [10].

### 5.2. Practical/Policy Implications

For policymakers and practitioners, this study offers localized insights for tailoring food waste interventions to the realities of the developing world. The low awareness yet high emotionality around waste points to the power of visual messaging and affective appeals in consumer campaigns. Social media is a widely used tool in the developing world, and it is effective in spreading knowledge, emotional messages, and reflection on societal norms [81]. Hence, the content must resonate with the youth (Gen Z) and have a larger impact on their wasteful behavior. Leveraging the normative influence of aspirational groups like university students could shift broader food practices. A survey of 6000 people based in the UK, US, and Canada found that social media platforms inspire more individuals to make sustainable choices rather than TV documentaries or governmental initiatives for behavior change [82]. Social media influencers create a massive impact with a huge number of followers on their pages. In the past five years, governments have hired influencers to spread positive images of political parties, and it has created a massive impact in building the image of the leaders [83,84]. Similarly, the government or authorities can hire higher-paid artists or influencers to create awareness regarding the issue [83,85,86]. The infrastructural barriers highlighted, from cold storage limitations to packaging waste, necessitate systemic solutions. Policies supporting decentralized cold chains, frugal innovations, and circular economy approaches are also needed. Another policy implication that is needed and continuously highlighted by print media is to put regulation and financial penalties on extravagant wedding feasts and Ramzan buffets [10].

### *5.3. Limitations and Future Research*

This study has several limitations that present opportunities for future work. First, the small sample size and university student demographic limit generalizability to broader populations. But this limitation presents a future opportunity. According to a Gallup survey, half of Indian young adults (age 15–24) have mobile and internet access [87,88]. With 58,000 higher education institutes, India has 43.3 million young adult students enrolled in higher education (universities) [88]. Hence, the research or intervention can be replicated to understand its food waste behavior on a larger scale. Second, the two-week photovoice period may not capture longer-term waste behaviors and seasonal variations. Due to limited time and resources, this study could not measure the long-term impact of the intervention. Hence, longitudinal designs could examine the durability of insights and interventions. Third, while the AI analysis enabled rapid theme extraction, the pre-trained models had limitations in recognizing culturally specific objects and practices [89]. Future work should explore participatory machine learning approaches where participants guide model training and interpretation. Fourth, the photography-based methodology may not fully capture the multisensory and embodied dimensions of food practices. Integration with other sensory ethnography methods like smell mapping and taste sampling could enrich insights. Finally, as an exploratory study, the insights generated are more indicative than conclusive. Controlled trials are needed to assess the efficacy of photovoice AI in changing waste behaviors. Participatory implementation research could also examine the real-world impact and community ownership of AI-guided waste reduction strategies. Comparative evaluation with conventional survey and intervention approaches would strengthen the evidence base for participatory AI methodologies in sustainable consumption research. Future research should explore participatory AI applications in diverse cultural contexts and domains of sustainable consumption. A Scopus review on explainable artificial intelligence (XAI) concluded that transdisciplinary collaborations between computer scientists, social scientists, and sustainability practitioners are needed to advance the field and translate insights into contextualized interventions [22]. Hence, methodological studies could compare AI-assisted and manual photovoice analysis to further validate the approach. Consequently, longitudinal research designs could examine the impact of AI-powered photovoice on participants' long-term consumption patterns and change processes.

## 6. Conclusions and Recommendations

This participatory photovoice study, enhanced by AI-powered analysis, aimed to explore urban students' perceptions, practices, and drivers of food waste. The objectives were to (1) visually document and analyze food waste behaviors in context, (2) uncover the social, cultural, and structural factors shaping these behaviors, and (3) generate insights and recommendations for targeted interventions and sustainable consumption strategies. The study's findings and methodological innovations align with and advance these objectives.

Firstly, the photovoice methodology enabled participants to visually document and reflect on their food waste practices in situ. The photographs and narratives provided a rich, contextualized understanding of waste behaviors, from plate waste to food spoilage to overproduction. The AI-driven analysis, particularly the CNN model's accurate classification of food waste scenes and objects, allowed for efficient and comprehensive insights generation from the visual data. This combination of participatory photography and computer vision techniques effectively addressed the first objective of documenting and analyzing food waste behaviors in context. Secondly, the thematic analysis of participants' reflections, augmented by sentiment analysis and topic modeling, revealed the complex social, cultural, and structural drivers of food waste. Factors such as changing dietary preferences, time constraints, aesthetic standards, and consumerism were highlighted as key contributors to wasteful practices. The AI techniques efficiently surfaced these patterns and themes from the large textual dataset, enabling a nuanced understanding of the underlying drivers. By uncovering these contextual factors, the study effectively addressed the second objective and provided a foundation for designing targeted interventions. Thirdly, the photovoice process and AI-powered analysis generated actionable insights and recommendations for food waste reduction and sustainable consumption. Participants' reflections and ideas, such as meal planning, composting, food-sharing initiatives, and awareness campaigns, offer promising avenues for intervention. The granular insights from the CNN model, such as identifying specific wasted food types, can further inform targeted waste reduction strategies. These findings directly align with the third objective of generating insights and recommendations for sustainable consumption.

**Author Contributions:** Conceptualization, K.G.; methodology, S.M.; software, S.M.; validation, K.G.; formal analysis, S.M.; investigation, K.G.; resources, K.G.; data curation, K.G.; writing—original draft preparation, K.G.; writing—review and editing, S.F. and M.R.; visualization, S.F.; supervision, M.R.; project administration, S.F. All authors have read and agreed to the published version of the manuscript.

**Funding:** This research received no external funding.

**Institutional Review Board Statement:** Not applicable.

**Informed Consent Statement:** Informed consent was obtained from all subjects involved in the study.

**Data Availability Statement:** The data presented in this study are available on request from the corresponding author. The data are not publicly available due to privacy restrictions.

**Conflicts of Interest:** The authors declare no conflicts of interest.

## References

1. Richardson, K.; Steffen, W.; Lucht, W.; Bendtsen, J.; Cornell, S.E.; Donges, J.F.; Rockström, J. Earth beyond six of nine planetary boundaries. *Sci. Adv.* **2023**, *9*, eadh2458. [CrossRef] [PubMed]
2. O'Neill, D.W.; Fanning, A.L.; Lamb, W.F.; Steinberger, J.K. A good life for all within planetary boundaries. *Nat. Sustain.* **2018**, *1*, 88–95. [CrossRef]
3. Čuček, L.; Klemeš, J.J.; Kravanja, Z. Overview of environmental footprints. In *Assessing and Measuring Environmental Impact and Sustainability*; Elsevier: Amsterdam, The Netherlands, 2015; pp. 131–193. [CrossRef]
4. Sahakian, M.; Wilhite, H. Making practice theory practicable: Towards more sustainable forms of consumption. *J. Consum. Cult.* **2013**, *14*, 25–44. [CrossRef]
5. Attiq, S.; Danish Habib, M.; Kaur, P.; Junaid Shahid Hasni, M.; Dhir, A. Drivers of food waste reduction behaviour in the household context. *Food Qual. Prefer.* **2021**, *94*, 104300. [CrossRef]
6. FAO. Moving Forward on Food Loss and Waste Reduction. 2019. Available online: https://www.fao.org/policy-support/tools-and-publications/resources-details/en/c/1242090/ (accessed on 15 July 2024).
7. Demmler, K.M.; Klasen, S.; Nzuma, J.M.; Qaim, M. Supermarket purchase contributes to nutrition-related non-communicable diseases in urban Kenya. *PLoS ONE* **2017**, *12*, e0185148. [CrossRef] [PubMed]
8. Thi, N.B.D.; Kumar, G.; Lin, C.Y. An overview of food waste management in developing countries: Current status and future perspective. *J. Environ. Manag.* **2015**, *157*, 220–229. [CrossRef] [PubMed]
9. Wastage at Pakistani Weddings—Daily Times. Available online: https://dailytimes.com.pk/998469/wastage-at-pakistani-weddings/ (accessed on 17 July 2024).

10. Sahoo, A.; Dwivedi, A.; Madheshiya, P.; Kumar, U.; Sharma, R.K.; Tiwari, S. Insights into the management of food waste in developing countries: With special reference to India. *Environ. Sci. Pollut. Res.* **2023**, *31*, 17887–17913. [CrossRef] [PubMed]
11. Parfitt, J.; Barthel, M.; MacNaughton, S. Food waste within food supply chains: Quantification and potential for change to 2050. *Philos. Trans. R. Soc. B Biol. Sci.* **2010**, *365*, 3065–3081. [CrossRef] [PubMed]
12. Aschemann-Witzel, J.; de Hooge, I.; Amani, P.; Bech-Larsen, T.; Oostindjer, M. Consumer-Related Food Waste: Causes and Potential for Action. *Sustainability* **2015**, *7*, 6457–6477. [CrossRef]
13. Aschemann-Witzel, J.; De Hooge, I.E.; Rohm, H.; Normann, A.; Bossle, M.B.; Grønhøj, A.; Oostindjer, M. Key characteristics and success factors of supply chain initiatives tackling consumer-related food waste—A multiple case study. *J. Clean. Prod.* **2017**, *155*, 33–45. [CrossRef]
14. Xue, L.; Liu, G.; Parfitt, J.; Liu, X.; Van Herpen, E.; Stenmarck, Å.; O'Connor, C.; Östergren, K.; Cheng, S. Missing Food, Missing Data? A Critical Review of Global Food Losses and Food Waste Data. *Environ. Sci. Technol.* **2017**, *51*, 6618–6633. [CrossRef] [PubMed]
15. Porpino, G.; Wansink, B.; Parente, J. Wasted Positive Intentions: The Role of Affection and Abundance on Household Food Waste. *J. Food Prod. Mark.* **2016**, *22*, 733–751. [CrossRef]
16. Elimelech, E.; Mintz, K.K.; Katz-Gerro, T.; Segal-Klein, H.; Hussein, L.; Ayalon, O. Between perceptions and practices: The religious and cultural aspects of food wastage in households. *Appetite* **2023**, *180*, 106374. [CrossRef] [PubMed]
17. Interventions to Reduce Food Waste at the Consumer Level—A National Strategy to Reduce Food Waste at the Consumer Level—NCBI Bookshelf. Available online: https://www.ncbi.nlm.nih.gov/books/nbk564040/ (accessed on 18 July 2024).
18. *A National Strategy to Reduce Food Waste at the Consumer Level*; National Academies Press: Washington, DC, USA, 2020. [CrossRef]
19. Wang, C.; Burris, M.A. Photovoice: Concept, Methodology, and Use for Participatory Needs Assessment. *Health Educ. Behav.* **1997**, *24*, 369–387. [CrossRef] [PubMed]
20. Liebenberg, L. Thinking critically about photovoice: Achieving empowerment and social change. *Int. J. Qual. Methods* **2018**, *17*, 1609406918757631. [CrossRef]
21. Budig, K.; Diez, J.; Conde, P.; Sastre, M.; Hernán, M.; Franco, M. Photovoice and empowerment: Evaluating the transformative potential of a participatory action research project. *BMC Public Health* **2018**, *18*, 432. [CrossRef] [PubMed]
22. Johs, A.J.; Agosto, D.E.; Weber, R.O. Explainable artificial intelligence and social science: Further insights for qualitative investigation. *Appl. AI Lett.* **2022**, *3*, e64. [CrossRef]
23. Ivanova, D.; Barrett, J.; Wiedenhofer, D.; Macura, B.; Callaghan, M.; Creutzig, F. Quantifying the potential for climate change mitigation of consumption options. *Environ. Res. Lett.* **2020**, *15*, 093001. [CrossRef]
24. Oswald, Y.; Owen, A.; Steinberger, J.K. Large inequality in international and intranational energy footprints between income groups and across consumption categories. *Nat. Energy* **2020**, *5*, 231–239. [CrossRef]
25. Wiedmann, T.; Lenzen, M.; Keyßer, L.T.; Steinberger, J.K. Scientists' warning on affluence. *Nat. Commun.* **2020**, *11*, 3107. [CrossRef]
26. Creutzig, F.; Roy, J.; Lamb, W.F.; Azevedo, I.M.; Bruine de Bruin, W.; Dalkmann, H.; Edelenbosch, O.Y.; Geels, F.W.; Grubler, A.; Hepburn, C.; et al. Towards demand-side solutions for mitigating climate change. *Nat. Clim. Chang.* **2018**, *8*, 260–263. [CrossRef]
27. GOAL 12: Sustainable Consumption and Production | UNEP—UN Environment Programme. Available online: https://www.unep.org/explore-topics/sustainable-development-goals/why-do-sustainable-development-goals-matter/goal-12 (accessed on 16 July 2024).
28. Newell, P.; Daley, F.; Twena, M. Changing Our Ways: The Report of the Cambridge Sustainability Commission on Scaling Behaviour Change. 2021. Available online: https://www.cambridge.org/engage/coe/article-details/609b92972a6566d64fb44ad9 (accessed on 1 May 2024).
29. Newell, P.; Twena, M.; Daley, F. Scaling behaviour change for a 1.5-degree world: Challenges and opportunities. *Glob. Sustain.* **2021**, *4*, e22. [CrossRef]
30. Welch, D.; Southerton, D. After Paris: Transitions for sustainable consumption. *Sustain. Sci. Pract. Policy* **2019**, *15*, 31–44. [CrossRef]
31. Shove, E.; Pantzar, M.; Watson, M. The dynamics of social practice: Everyday life and how it changes. In *The Dynamics of Social Practice: Everyday Life and How it Changes*; Sage: London, UK, 2012; pp. 1–191. [CrossRef]
32. Klitkou, A.; Bolwig, S.; Huber, A.; Ingeborgrud, L.; Pluciński, P.; Rohracher, H.; Schartinger, D.; Thiene, M.; Żuk, P. The interconnected dynamics of social practices and their implications for transformative change: A review. *Sustain. Prod. Consum.* **2022**, *31*, 603–614. [CrossRef]
33. Fuchs, D.; Di Giulio, A.; Glaab, K.; Lorek, S.; Maniates, M.; Princen, T.; Røpke, I. Power: The missing element in sustainable consumption and absolute reductions research and action. *J. Clean. Prod.* **2016**, *132*, 298–307. [CrossRef]
34. Sahakian, M.; Fuchs, D.; Lorek, S.; Di Giulio, A. Advancing the concept of consumption corridors and exploring its implications. *Sustain. Sci. Pract. Policy* **2021**, *17*, 305–315. [CrossRef]
35. Concari, A.; Kok, G.; Martens, P. A systematic literature review of concepts and factors related to pro-environmental consumer behaviour in relation to waste management through an interdisciplinary approach. *Sustainability* **2020**, *2*, 4452. [CrossRef]
36. Srivastava, S.K.; Mishra, A.; Singh, S.; Jaiswal, D. Household food waste and theory of planned behavior: A systematic review and meta-analysis. *Environ. Sci. Pollut. Res.* **2023**, *30*, 97645–97659. [CrossRef] [PubMed]

37. Schrank, J.; Hanchai, A.; Thongsalab, S.; Sawaddee, N.; Chanrattanagorn, K.; Ketkaew, C. Factors of Food Waste Reduction Underlying the Extended Theory of Planned Behavior: A Study of Consumer Behavior towards the Intention to Reduce Food Waste. *Resources* **2023**, *12*, 93. [CrossRef]
38. Aydin, H.; Aydin, C. Investigating consumers' food waste behaviors: An extended theory of planned behavior of Turkey sample. *Clean. Waste Syst.* **2022**, *3*, 100036. [CrossRef]
39. Bishop, M.; Megicks, P. "Waste Not, Want Not!": Qualitative Insights into Consumer Food Waste Behaviour. *Waste Manag. Environ. IX* **2019**, *231*, 297–308. [CrossRef]
40. Cui, Q.; Lin, G.; Qiu, S.; Duan, T. Identifying the Drivers of Food Waste in University Canteens in China: A Qualitative Approach. *Sage Open* **2023**, *13*, 21582440231198223. [CrossRef]
41. Evans-Agnew, R.A.; Rosemberg, M.A.S. Questioning Photovoice Research: Whose Voice? *Qual. Health Res.* **2016**, *26*, 1019–1030. [CrossRef] [PubMed]
42. Latz, A.O.; Mulvihill, T.M. Photovoice research in education and beyond: A Practical Guide from Theory to Exhibition. In *Photovoice Research in Education and Beyond: A Practical Guide from Theory to Exhibition*; Routledge: New York, NY, USA, 2017; pp. 1–182. [CrossRef]
43. UNEP Food Waste Index Report 2021 | UNEP—UN Environment Programme. Available online: https://www.unep.org/resources/report/unep-food-waste-index-report-2021 (accessed on 16 July 2024).
44. Chang, Y.Y.C. All you can eat or all you can waste? Effects of alternate serving styles and inducements on food waste in buffet restaurants. *Curr. Issues Tour.* **2022**, *25*, 727–744. [CrossRef]
45. Schanes, K.; Dobernig, K.; Gözet, B. Food waste matters—A systematic review of household food waste practices and their policy implications. *J. Clean. Prod.* **2018**, *182*, 978–991. [CrossRef]
46. Stancu, V.; Haugaard, P.; Lähteenmäki, L. Determinants of consumer food waste behaviour: Two routes to food waste. *Appetite* **2016**, *96*, 7–17. [CrossRef] [PubMed]
47. Roodhuyzen, D.M.A.; Luning, P.A.; Fogliano, V.; Steenbekkers, L.P.A. Putting together the puzzle of consumer food waste: Towards an integral perspective. *Trends Food Sci. Technol.* **2017**, *68*, 37–50. [CrossRef]
48. Ganglbauer, E.; Fitzpatrick, G.; Comber, R. Negotiating food waste: Using a practice lens to inform design. *ACM Trans. Comput.-Hum. Interact.* **2013**, *20*, 2463582. [CrossRef]
49. Goss, H.; Tromp, N.; Schifferstein, R. Designing adaptable consumption: A new practice to foster food system transitions. In Proceedings of the IASDR 2023: Life-Changing Design, Milan, Italy, 9–13 October 2023. [CrossRef]
50. Machado-Borges, T. Out of Sight, Out of Mind? Middle-Class Households, Waste, Consumption, and Environmental Awareness in Southeastern Brazil. *J. Lat. Am. Caribb. Anthropol.* **2017**, *22*, 298–319. [CrossRef]
51. Spyreli, E.; Vaughan, E.; McKinley, M.C.; Woodside, J.V.; Hennessy, M.; Kelly, C. Using Online Photovoice to Explore Food Decisions of Families on Low Income: Lessons Learnt During the COVID-19 Pandemic. *Qual. Health Res.* **2024**, *34*, 171–182. [CrossRef] [PubMed]
52. Watson, E.R.; Foster-Fishman, P.G. The Exchange Boundary Framework: Understanding the Evolution of Power within Collaborative Decision-Making Settings. *Am. J. Community Psychol.* **2013**, *51*, 151–163. [CrossRef] [PubMed]
53. Beck, J.; Jenks, C.; Keddie, N.; Young, M.F.D. Pedagogy of the Oppressed. In *Toward a Sociology of Education*; Routledge: New York, NY, USA, 2020; pp. 374–386.
54. Freire, P. *Pedagogy of Indignation*; Routledge: New York, NY, USA, 2015. [CrossRef]
55. Carlson, E.D.; Engebretson, J.; Chamberlain, R.M. Photovoice as a Social Process of Critical Consciousness. *Qual. Health Res.* **2006**, *16*, 836–852. [CrossRef] [PubMed]
56. Johnson, J.L.; Adkins, D.; Chauvin, S. A Review of the Quality Indicators of Rigor in Qualitative Research. *Am. J. Pharm. Educ.* **2020**, *84*, 138–146. [CrossRef] [PubMed]
57. Foster-Fishman, P.; Nowell, B.; Deacon, Z.; Nievar, M.A.; McCann, P. Using methods that matter: The impact of reflection, dialogue, and voice. *Am. J. Community Psychol.* **2005**, *36*, 275–291. [CrossRef] [PubMed]
58. Sambasivan, N.; Holbrook, J. Toward responsible AI for the next billion users. *Interactions* **2018**, *26*, 68–71. [CrossRef]
59. Garg, M.; Wazarkar, S.; Singh, M.; Bojar, O. Multimodality for NLP-centered applications: Resources, advances and frontiers. In Proceedings of the Thirteenth Language Resources and Evaluation Conference, Marseille, France, 20–25 June 2022; pp. 6837–6847.
60. Brown, C.; Chauhan, J.; Grammenos, A.; Han, J.; Hasthanasombat, A.; Spathis, D.; Mascolo, C. Exploring Automatic Diagnosis of COVID-19 from Crowdsourced Respiratory Sound Data. In Proceedings of the ACM SIGKDD International Conference on Knowledge Discovery and Data Mining, Virtual, 6–10 July 2020; pp. 3474–3484. [CrossRef]
61. Creswell, J.D. Mindfulness Interventions. *Annu. Rev. Psychol.* **2017**, *68*, 491–516. [CrossRef] [PubMed]
62. Creswell, J.W. *Research Design: Qualitative, Quantitative, and Mixed Methods Approaches*; Sage: Newcastle upon Tyne, UK, 2009.
63. Usman, A.; Javed, M.; Amjad, A.; Jalil, A.; Shaheen, I.; Amjad, U. Food, City and Young Globavores: Perceptions of Pakistani urban youth about food choices and healthy eating. *Anthropologist* **2017**, *27*, 143–151. [CrossRef]
64. Oinonen, I.; Paloniemi, R. Understanding and measuring young people's sustainability actions. *J. Environ. Psychol.* **2023**, *91*, 102124. [CrossRef]
65. Soma, T.; Li, B.; Maclaren, V. Food Waste Reduction: A Test of Three Consumer Awareness Interventions. *Sustainability* **2020**, *12*, 907. [CrossRef]

66. Saunders, B.; Sim, J.; Kingstone, T.; Baker, S.; Waterfield, J.; Bartlam, B.; Jinks, C. Saturation in qualitative research: Exploring its conceptualization and operationalization. *Qual. Quant.* **2018**, *52*, 1893–1907. [CrossRef]
67. Christou, P.A. The Use of Artificial Intelligence (AI) in Qualitative Research for Theory Development "Philosophies of Hospitality and Tourism: Giving and Receiving" View project The History and Evolution of Tourism View project. *Qual. Rep.* **2023**, *28*, 2739–2755. [CrossRef]
68. Donohue, M.E. A replacement for Justitia's scales: Machine learning's role in sentencing. *Harv. J. Law Technol.* **2018**, *32*, 657–678.
69. Sandler, M.; Howard, A.; Zhu, M.; Zhmoginov, A.; Chen, L.C. MobileNetV2: Inverted Residuals and Linear Bottlenecks. In Proceedings of the IEEE Conference on Computer Vision and Pattern Recognition, Salt Lake City, UT, USA, 18–23 June 2018; pp. 4510–4520.
70. Hutto, C.J.; Gilbert, E. VADER: A Parsimonious Rule-Based Model for Sentiment Analysis of Social Media Text. *Proc. Int. AAAI Conf. Web Soc. Media* **2014**, *8*, 216–225. [CrossRef]
71. Chekijian, S.; Li, H.; Fodeh, S. Emergency care and the patient experience: Using sentiment analysis and topic modeling to understand the impact of the COVID-19 pandemic. *Health Technol.* **2021**, *11*, 1073–1082. [CrossRef] [PubMed]
72. Lincoln, Y.S.; Guba, E.G. But is it rigorous? Trustworthiness and authenticity in naturalistic evaluation. *New Dir. Program Eval.* **1986**, *1986*, 73–84. [CrossRef]
73. Hermanussen, H.; Loy, J.P.; Egamberdiev, B. Determinants of Food Waste from Household Food Consumption: A Case Study from Field Survey in Germany. *Int. J. Environ. Res. Public Health* **2022**, *19*, 14253. [CrossRef] [PubMed]
74. Podjed, D.; Horvat, K.P. The Invisible Life of Food Waste. *Traditiones* **2020**, *49*, 109–124. [CrossRef]
75. Ahmed, S.; Stewart, A.; Smith, E.; Warne, T.; Byker Shanks, C. Consumer Perceptions, Behaviors, and Knowledge of Food Waste in a Rural American State. *Front. Sustain. Food Syst.* **2021**, *5*, 734785. [CrossRef]
76. Díez, J.; Conde, P.; Sandin, M.; Urtasun, M.; López, R.; Carrero, J.L.; Franco, M. Understanding the local food environment: A participatory photovoice project in a low-income area in Madrid, Spain. *Health Place* **2017**, *43*, 95–103. [CrossRef] [PubMed]
77. Soma, T. Space to waste: The influence of income and retail choice on household food consumption and food waste in Indonesia. *Int. Plan. Stud.* **2020**, *25*, 372–392. [CrossRef]
78. Madaan, G.; Singh, A.; Mittal, A.; Shahare, P. Reduce, reuse, recycle: Circular economic principles, sustainability and entrepreneurship in developing ecosystems. *J. Small Bus. Enterp. Dev.* 2024, *ahead-of-print*. [CrossRef]
79. Quested, T.E.; Marsh, E.; Stunell, D.; Parry, A.D. Spaghetti soup: The complex world of food waste behaviours. *Resour. Conserv. Recycl.* **2013**, *79*, 43–51. [CrossRef]
80. Pirani, S. Production HAJ of Cleaner, 2016 undefined. In *Reduction of Food Waste Generation in the Hospitality Industry*; Elsevier: Amsterdam, The Netherlands, 2016. Available online: https://www.sciencedirect.com/science/article/pii/S095965261501077X?casa_token=rrYajvN-BxEAAAAA:R1aXe66XS_xlxO4oTI9p-8tBKNKese0eg_kcyJBTzlUVStw48J-ocsE3YgoMfiE6gK-KCUF2THM (accessed on 29 April 2023).
81. Lim, V.; Funk, M.; Marcenaro, L.; Regazzoni, C.; Rauterberg, M. Designing for action: An evaluation of Social Recipes in reducing food waste. *Int. J. Hum. Comput. Stud.* **2017**, *100*, 18–32. [CrossRef]
82. Can Influencers Solve the World's Food Waste Crisis? | Orderly. Available online: https://orderly.io/blog/can-influencers-solve-the-worlds-food-waste-crisis (accessed on 17 April 2024).
83. Social Media Influencers Are India's New Election Campaigners—The Hindu. Available online: https://www.thehindu.com/sci-tech/technology/social-media-influencers-are-indias-new-election-campaigners/article67996976.ece (accessed on 17 July 2024).
84. Top 15 Sustainability Influencers in India: Eco Warriors Inspiring Green Living—Agency Masala. Available online: https://agencymasala.com/influencer-marketing/list-of-influencers-tbd/top-15-sustainability-influencers-in-india-eco-warriors-inspiring-green-living/ (accessed on 17 April 2024).
85. The Wire: The Wire News India, Latest News, News from India, Politics, External Affairs, Science, Economics, Gender and Culture. Available online: https://thewire.in/politics/how-important-are-social-media-influencers-in-indian-elections (accessed on 17 April 2024).
86. Lok Sabha 2024: The Influencers Driving India's Big Election—BBC News. Available online: https://www.bbc.com/news/world-asia-india-68920953 (accessed on 17 April 2024).
87. India's Youth Dividend: High Hopes for Today and Tomorrow. Available online: https://news.gallup.com/poll/509756/india-youth-dividend-high-hopes-today-tomorrow.aspx (accessed on 17 April 2024).
88. India Releases Updated Higher Education Statistics | British Council. Available online: https://opportunities-insight.britishcouncil.org/news/news/india-releases-updated-higher-education-statistics (accessed on 17 April 2024).
89. Costa, A.P. Qualitative Research Methods: Do digital tools open promising trends? *Rev. Lusofona Educ.* **2023**, *59*, 67–76. [CrossRef]

sustainability

MDPI

*Article*

# Navigating the Digital Divide for Sustainability: Emotional and Cognitive Pathways to Food Waste Reduction Through Social Media Advertising

**Congying Liu [1], Lupin Chen [2,*] and Muhammad Hassan Arshad [3]**

1 School of Tourism, Hospitality and Event Management, Universiti Utara Malaysia, Sintok 06010, Malaysia; liu_congying@gsgsg.uum.edu.my
2 College of International Law and Sociology, Sichuan International Studies University, Chongqing 400015, China
3 School of Business Management, Universiti Utara Malaysia, Sintok 06010, Malaysia; muhammad_hassan_a@oyagsb.uum.edu.my
* Correspondence: 99002315@sisu.ed.cn

**Abstract:** Applying behavioral reasoning theory, this study examines the effect of social media food waste advertising on consumers' intention not to waste food, with customer–brand communal engagement and awareness of food waste consequences acting as mediators, and fast-food customers' altruism as a moderator. Structural equation modeling was applied to data from 422 customers of Malaysian retail fast-food chains. The findings reveal that social media food waste advertising positively affects customers' intentions not to waste food directly and indirectly through two mediators: awareness of food waste consequences and customer–brand communal engagement. The stronger mediator was awareness, and it indicated that cognitive factors play a greater role in pro-environmental behaviors than emotional ones. Additionally, this study demonstrates that customer altruism moderates the cognitive pathway but not the emotional pathway. The results of these findings suggest that social media campaigns should focus on both cognitive and emotional dimensions as well as personal value to maximize effectiveness. The findings have implications for fast-food chains to design more effective social media campaigns that promote food waste reduction as part of broader sustainability goals in the food industry.

**Keywords:** social media advertising; food waste reduction; customer behavior; retail fast-food chains; behavioral reasoning theory

## 1. Introduction

A major global problem related to food security and environmental sustainability is food waste. More than one-third (1.3 billion tons/year) of all food produced is wasted [1], leading to food shortages and exacerbating environmental problems, like greenhouse gas emissions, deforestation and water depletion [2]. Primarily, food waste is essential to relieve pressure on agriculture and to promote conservation [3]. There is a prodigious responsibility that falls on businesses, particularly in the food service sector Smith & Jones, 2019. For example, companies like McDonald's have optimized portion sizes and implemented 'No Waste' campaigns, and Starbucks donates unsold items to local communities [4,5], all of which are actions that brands can take to reduce food waste and enhance their sustainability commitment.

Social media is used by businesses in the digital world to communicate their sustainability initiatives [6–8]. At present, social media platforms that were once used for marketing display corporate social responsibility efforts [9,10], such as food waste reduction [11]. For instance, KFC uses Instagram and Facebook to advertise food donation and portion control, and Tesco tweets food waste tips and charitable partnerships. In addition to helping build brand image, these campaigns also encourage customers to partake in responsible consumption. Yet, the consumers are the key to reducing food waste [12]. A lot of food waste comes from over-ordering, storage errors, and lack of knowledge. To resolve this issue, businesses have been launching campaigns that help to increase awareness and promote mindful consumption practices [13]. However, in addressing the issue of food waste, both businesses and customers are required to share responsibility, and therefore, it is important to understand how businesses influence customer behavior through strategic communication. This study looks at the effect of fast-food brands' social media advertising on customer intention to avoid food waste.

However, human behavior is complex and involves psychological, personal and situational factors that mediate and moderate responses to external influences [14,15]. Following this line of research, this study proposes a dual mediating pathway involving two key variables: customer–brand communal engagement and awareness of food waste consequences. Customer–brand communal engagement refers to the emotional link between customers and a brand based on shared values and perceived care [16]. If customers experience positive emotions such as warmth and attachment, they are likely to be motivated to align their behavior with the brand's sustainability efforts. Awareness of food waste consequences reflects the cognitive understanding of the negative environmental and societal consequences of food waste [17]. Customers recognize these consequences and start to consider the wider consequences of what they do, which leads to an increase in their intention to reduce food waste. Because they address different psychological facets of behavior, emotional and cognitive mediators are crucial to include. Individuals connect emotionally, driven by their support for socially and environmentally responsible brands, with emotional factors. Rational evaluations, i.e., actions carried out based on understanding waste impacts, are prompted by cognitive factors such as awareness of food waste consequences. Combining both dimensions gives a complete picture of how social media advertising affects customer behavior. Moreover, this study introduces altruism (personal human value), which is characterized as a selfless concern for others, as a moderator of prosocial behaviors [18]. Altruistic people are more likely to engage in environmentally friendly initiatives [10,19] and react favorably to messages about reducing waste, making altruism a key moderator in social media advertising. Therefore, altruism highlights the relevance of individual differences in the behavioral reactions to sustainability-oriented advertising.

This study aims to explore the impact of social media advertising on food waste reduction behaviors in the setting of Malaysia's retail fast-food sector. More specifically, this study aims to understand how fast-food chain social media campaigns impact customers' intention not to waste food, through both cognitive and emotional pathways. In order to explore the mediating role between two mediating factors (awareness of food waste consequences, and customer–brand communal engagement) and customers' sustainable behaviors, this research integrates behavioral reasoning theory (BRT). Furthermore, this study examines the moderating effect of altruism in order to more fully understand the psychological mechanisms that shape food waste reduction behavior. This research contributes to filling a gap in the literature by focusing on a developing economy such as Malaysia, where consumption patterns are different, and providing insights for academics and practitioners on how social media can promote sustainable consumption practices.

This study focuses on Malaysia's retail fast-food sector, which has experienced rapid growth over the past decade. The expansion of fast-food chains and dining-out culture has increased food waste, challenging Malaysia's sustainability goals [20]. The Solid Waste Management and Public Cleansing Corporation (SWCorp) reports that the country generates approximately 16,688 tons of food waste daily, with 4080 tons being edible and avoidable waste ending up in landfills [21]. This not only harms the environment but also causes economic losses, undermining resource optimization and food security efforts. Despite government initiatives, food waste remains prevalent, especially in urban areas with high fast-food consumption, necessitating strategies involving both businesses and consumers. While regulatory measures are crucial, promoting behavioral shifts among consumers is equally important. Given the role of fast-food chains in shaping consumption patterns, understanding how these organizations can encourage sustainable practices is essential. This is particularly relevant in Malaysia, where addressing food waste at both the organizational and the consumer level is increasingly recognized.

Despite extensive research, significant gaps remain in understanding the psychological mechanisms influencing customers' intentions to reduce food waste. Most studies examine emotional or cognitive aspects separately, neglecting their combined effect [22,23]. Although emotions like guilt and empathy motivate pro-environmental behaviors [24], while cognitive factors such as awareness inform rational decision-making, this integrated perspective of dual pathways (emotional and cognitive), especially from a food waste standpoint, is missing from the existing literature. This research addresses this gap by introducing a dual mediating model that combines emotional (customer–brand communal engagement) and cognitive (awareness of food waste consequences) factors, offering a more comprehensive understanding of how social media advertising influences customer behavior in Malaysia's fast-food sector.

On a similar note, most existing research is based on developed economies where consumption and waste patterns are substantially different from those of countries like Malaysia [25]. In the developed contexts, studies tend to focus on policies, technological interventions and consumer behavior models that may not be appropriate in light of the distinctive socio-cultural and economic challenges of the emerging markets [26,27]. There is limited empirical evidence from developing economies on how businesses can lead sustainable consumption through advertising strategies such as social media advertising to fight against food waste [28]. This research addresses this critical gap by focusing on Malaysia's retail fast-food sector to understand how social media campaigns can hit the right chords with consumers in a developing economy where affordability, food availability, and sustainability awareness are factors that affect behavior. Informational appeals have been previously studied, but relational constructs such as customer–brand communal engagement have been ignored, without a framework to connect relational factors to cognitive awareness and ultimately promote sustainable behaviors. We introduce customer–brand communal engagement as a mediator to explain the emotional pathway to customers developing responsibility to reduce food waste. Unlike most traditional informational appeals, this approach highlights the value of emotional bonds between customers and brands in encouraging pro-environmental behavior. Social media plays a role in spreading food waste reduction messages, but there are few studies on the effectiveness of social media in actually inducing behavioral change [29]. Previous research has primarily explored its effects on brand awareness and customer engagement but delved little into how targeted social ads on social media affect customer behaviors toward food waste [30]. To fill this gap, our study investigates how retail fast-food chains can utilize social media features like interactivity and broad reach in order to influence consumption patterns and promote sustainable behaviors.

Second, altruism among fast-food customers has been neglected as a moderator. While altruism is a key predictor of environmentally responsible behaviors [18,31,32], little research has been carried out to explore how this can enhance or diminish the impact of social media advertising on shaping customer intentions. Our research aims to investigate how altruism moderates the association between social media food waste advertising and customer intentions to avoid food waste and to present a more nuanced view of the role of individual differences in response to sustainability messaging. The research model proposed includes emotional and cognitive mediators as well as a personal trait moderator, thus representing multiple facets of psychological influence on behavior. By offering a richer understanding of the factors that influence customer intentions, this approach provides clearer advice to businesses and policymakers about how to design effective interventions. Our research fills these gaps and makes considerable contributions to the literature on food waste reduction, and provides new insights that can inform future research and practice in both developed and developing contexts.

## 2. Literature and Theoretical Underpinning

### 2.1. Theoretical Underpinning: Behavioral Reasoning Theory (BRT)

The main theoretical foundation of this study is behavioral reasoning theory (BRT), a rich and comprehensive body of theoretical framework to explain how factors shape individual decision-making and behavior, particularly within the organizational context. Westaby [33] introduces BRT, a theoretical model that extends traditional behavioral theories, such as the theory of planned behavior (TPB), to argue that reasons (for and against a particular behavior) are crucial determinants of both intentions and action. At its core, BRT holds that people act in terms of the reasons they take to be important in mediating between their beliefs and their subsequent actions. These reasons can be both an encouragement and deterrence to behavior depending on how people reason out external cues and internal values [34]. This study applies BRT to explain how fast-food chain social media advertising affects customers' intentions to reduce food waste. In particular, BRT describes two specific pathways through which behavior is shaped, cognitive and emotional. Rational considerations drive cognitive pathways, for instance, awareness of food waste consequences. The focus of this study is on how social media food waste campaigns can raise customer awareness of the environmental and social impacts of their actions. According to BRT, once people realize these reasons, they're more likely to act in ways that conform with the logical motivation to cut down on food waste [11].

On the other hand, BRT stresses the role of affective responses that are also important in influencing behavior. This study examines how customer–brand communal engagement, the emotional bond created through shared values and a feeling of community with the brand, moderates the association between social media food waste advertising and customers' intention not to waste food. BRT says emotional reasons like the feeling of being valued by a socially responsible brand can make a big difference in behavior. In particular, if customers perceive a strong emotional connection to brands of fast-food that market sustainability, they are more likely to engage in pro-environmental behavior, like reducing food waste. In addition, BRT's emphasis on personal values, such as altruism, is especially appropriate for this study. The cognitive and emotional pathways are moderated by altruism (selfless concern for the wellbeing of others) [35]. Social media campaigns aimed at reducing food waste will be more easily engaged with and responded to by altruistic consumers, whose personal values are echoed within the social media campaign. BRT argues that when the reasons for behavior match an individual's values (cognitive or emotional), the reasons become stronger action motivators. Consequently, social media advertising strengthens the effect of customer awareness and emotional engagement with advertising and enhances the purchase intention to reduce food waste. By applying BRT, this study provides a comprehensive understanding of how fast-food chains can effectively use social media advertising to target both rational and emotional dimensions of consumer behavior. It illustrates how awareness of food waste consequences (cognitive pathway), customer–brand communal engagement (emotional pathway), and altruism (moderator) interact to influence food waste reduction behaviors in the fast-food sector

### 2.2. Hypotheses' Development

Social media is one of the best tools to influence consumer behavior because of its broad reach, engaging content and real-time interaction [36]. Duffett [37] argues that research proves social media advertising can impact customer perception, attitude and purchase behavior. Persuasive messaging and visual storytelling can influence consumer behavior [38]. In terms of sustainability and food waste, social media is the ideal channel to educate customers, promote responsible consumption and promote positive behavior change [39]. Not only does it educate customers about sustainable initiatives, but it also makes them accountable as they feel closer to the brand's mission [40]. This study is applied to social media food waste advertising by fast-food chains to influence customer intentions to avoid waste by demonstrating the negative consequences of food waste and the brand's sustainable commitment. Such advertising, in turn, educates customers, informs them, and persuades them to act in ways consistent with environmental values [41]. Emotional appeals such as community engagement in combination with cognitive information about the consequences of food waste increase the effectiveness of promoting pro-environmental behavior [42]. Furthermore, particularly in a country like Malaysia, where government-led sustainability initiatives are becoming more noticeable, corporate involvement through social media for food waste reduction can serve as a complementary means to meet national sustainability targets [43]. Not only are such fast-food chains helping to mitigate environmental harm, but they also benefit their brand image as socially responsible entities. This creates customer loyalty and deepens the brand–consumer relationship, leading to a higher likelihood that consumers will behave in alignment with the brand's sustainability efforts. Consequently, social media is a fundamental platform for designing pro-environmental intentions in Malaysia where consumer awareness of food waste is increasing. BRT argues that reasons that are important, such as environmental sustainability by reducing food waste, play a major role in intentions [44]. In such a context, social media food waste advertising is an external cue that provides the rational reasons and emotional appeals to persuade the customers to avoid food waste. The reason for such an outbreak of noble goals like mitigating environmental harm is what informs the customers' intentions to alter their behavior. Based on this, we propose:

**H1.** *Social media food waste advertising by fast-food chains positively influences customers' intentions to avoid food waste.*

Emotional connections between customers and brands are fostered through social media advertising by presenting shared values and commitments to social and environmental issues [40]. In contrast to traditional advertising, social media allows for real-time engagement and personalized content to strengthen brand-customer relationships [37]. Brands tell compelling stories, show great visuals and interactive features, and communicate messages that stick, leading people to engage with them and feel loyalty and a sense of belonging [38]. This emotional bond influences customer perceptions and behaviors and increases the likelihood that they will support the brand's initiatives and act in line with its values [45]. This study argues that social media food waste advertising provides opportunities to cultivate customer–brand communal engagement by sharing efforts to decrease food waste and increase sustainability. Communal engagement is the emotional bond that customers have with a brand because of shared values and concerns for the community [16]. Brands that communicate their sustainable practices, e.g., donating unsold food or implementing waste reduction, are perceived as socially responsible and caring, which further enhances customers' emotional attachment [46].

Social media advertising in the context of Malaysia's fast-food industry is an ideal platform through which to promote an emotional connection between the brand and the customer, especially on issues such as food waste reduction. With an increasing awareness of environmental concerns, along with the use of social media, the country is a perfect place for fast-food brands to converse with their customers on common ground [47]. Fast-food chains in Malaysia can promote their communal engagement with environmentally conscious consumers by highlighting sustainable practices such as food donation programs or innovative packaging that cuts down on waste. Emotionally resonant campaigns can draw on the societal and cultural value of environmental stewardship in Malaysia to strengthen the emotional tie between the brand and its customers [48]. In addition, Malaysia's socially and economically diverse people are not only after low-cost options but are increasingly attracted to brands that are aligned with their personal and shared values. Fast-food chains have a chance to reach out to their customers and let them know how serious they are about reducing food waste by being leaders in the movement for sustainability. Communal engagement does more than forge emotional attachment between individual consumers and the product; it actually generates emotional attachment among Malaysian consumers to the broader environmental goals. When consumers observe these brand-led sustainability initiatives, they may feel compelled to take waste-reducing behaviors in order to fulfil their moral obligation, and this fits with the emotional pathway that BRT describes. In Malaysia, consumers place great value on social responsibility and community-driven efforts, and this emotional connection is even more important [49]. Fast-food chains can leverage these values to increase customer loyalty and influence certain behaviors which are environmentally sustainable. In the end, this emotional engagement is crucial to cementing the brand's position not as a manufacturer of goods, but as a partner in the journey towards Malaysia's national sustainability goals.

Such a bond motivates customers to support the sustainability commitment of the brand and to adopt behaviors matching its environmental values. This relationship is supported by BRT, as strong affective responses are triggered by emotional reasons (e.g., feeling valued by the brand or noticing the brand's concern for the issues of society) that lead customers to act in a manner consistent with the brand's mission. Reducing food waste may not only be a rational choice for customers, but it may also be a moral obligation driven by the relationship they want with the brand. These reasons are then internalized as they reinforce customers' commitment to behave in a waste-reducing manner and overall commitment to sustainability [44]. As a result, emotional attachment via communal engagement is an important mechanism through which social media advertising impacts intentions to encourage customers to translate positive perceptions of the brand's initiatives into pro-environmental actions. Additionally, emotional reasons stemming from engagement with the community can offer social validation when customers believe they are exposing their support for the brand's values and this reflects positively on how they are perceived by others. This reinforcement actually reinforces their intent to avoid wasting food. Based on this rationale, we propose the following hypotheses:

**H2.** *Social media food waste advertising by retail fast-food chains positively influences customer–brand communal engagement.*

**H3.** *Customer–brand communal engagement positively influences fast-food customers' intentions not to waste food.*

**H4.** *Customer–brand communal engagement mediates the relationship between social media food waste advertising and customers' intentions not to waste food.*

Social media advertising plays a significant role in raising customers' awareness of important social and environmental issues such as food waste [39] because it greatly determines the customers' cognitive perceptions of it. This kind of advertising also helps customers understand the consequences of food waste on the environment and society [17]. By providing factual information, visuals and real-world statistics, customers re-evaluate their consumption habits and their behavior towards responsible consumption, as the level of cognitive recognition of the issue increases [50]. This is because cognitive engagement is important in that it gives customers solid reasons to support sustainability initiatives and conduct themselves responsibly [51]. Greater awareness of the consequences of food waste has a large influence on customer intentions to avoid food waste. According to research, awareness makes a person more likely to have a strong intention to reduce waste [52]. Customers who are aware of the environmental and social consequences of their choices will be more likely to make conscious choices (e.g., eating smaller portions, not over-ordering), which will reduce landfill use and greenhouse gas emissions [53]. Consequently, it is an important driver since it constitutes the cognitive framework for responsible consumption choices. This cognitive pathway is supported by BRT because the person's reasons for behavior are built from beliefs and awareness of the consequences of that behavior [34]. In this respect, customers know the consequences of food waste and have rational reasons to prevent waste considering the whole picture of their activities. Motivational forces that comprise rational reasons are persuasive advertising that leads to responsible consumption behaviors consistent with the customers' knowledge and beliefs [54].

Food waste has become a serious environmental and social issue in Malaysia, where social media campaigns to raise awareness are critical [55]. The Malaysian population is becoming increasingly exposed to the negative consequences of food waste on the environment, greenhouse gas emissions and depletion of natural resources due to the fact that approximately 16,000 tons of food are lost daily [56]. Fast-food chains have an ideal social media platform to execute targeted campaigns to educate consumers on the need for food waste reduction, leveraging real-time data and visual content to prompt customers to reconsider their consumption behaviors. Fast-food chains can use social media to disseminate localized information regarding the specific impact of food waste in Malaysian urban areas and thereby enhance consumers' cognitive engagement with the issue. In addition, these campaigns not only explain to customers what the broader global impacts of the problem are but also tailor the message to reflect Malaysia's context, which makes the problem feel more immediate and relevant. According to BRT, the idea is to give consumers rational reasons to act, such as knowing the environmental damage that accompanies food waste, to increase their consumption intentions for responsible consumption behaviors. In Malaysia, where there are rising government initiatives, societal awareness of sustainability and these rational reasons are even more compelling when these are delivered through platforms that the Malaysian population is used to using. In addition, social media campaigns in Malaysia quite often show brands working together with local communities through campaigns like food banks and reducing landfill waste [57]. Connecting fast-food brands' factual awareness with these tangible local efforts will not only help in creating a greater cognitive impact on fast-food brands' campaigns but also in making consumers internalize the significance of food waste reduction. Social media advertising aligned with local sustainability initiatives offers Malaysian customers stronger rational motivations to change their behavior, consistent with the cognitive pathway described in BRT.

We argue that the relationship between customers' intention not to waste food and social media food waste advertising is mediated by awareness of consequences. Awareness campaigns on the harmful effects of food waste have provided strong reasons for behavior change [17]. These relate the brand's advertising to customers' behavioral intentions and the role of knowledge in advertising to action. We integrate awareness as a mediating mechanism in association with emotional engagement to explain how social media advertising influences behavior. BRT serves as a valuable lens to understand how social media food waste advertising influences customer intentions via cognitive perceptions. Social media serves as a stimulus and makes the customers conscious of reasons to reduce food waste. These reasons are behind very strong intentions not to waste food and mediate the effect of advertising on behavior through its alignment with brand values. This integration highlights the role of cognitive processes in driving customer behavior along with the emotional pathway of customer and brand communal engagement. Based on this rationale, we propose the following hypotheses:

**H5.** *Social media food waste advertising by retail fast-food chains positively influences awareness of food waste consequences.*

**H6.** *Awareness of food waste consequences positively influences fast-food customers' intentions not to waste food.*

**H7.** *Awareness of food waste consequences mediates the relationship between social media food waste advertising and customers' intentions not to waste food.*

Core values underpin attitudes, guide responses to stimuli and determine behavioral intentions [58,59]. Perceptions of desirability and behavior are driven in specific contexts by them [60,61]. Individuals process information more deeply when the messages correspond to personal values and may change their behavior as a result [62]. Altruism (or a selfless concern for others) is a key personal value that greatly influences consumer behavior [63] in the context of sustainability and food waste reduction. Ferguson et al. [18] show that altruistic individuals are more willing to engage in behaviors that benefit society and the environment at a personal cost. They are more responsive to social media messages that discuss social and environmental responsibility. Altruistic customers respond dramatically to social media ads that communicate a brand's dedication to reducing food waste, as it is a way they can contribute to the greater good [64]. Consequently, we hypothesize that altruism will moderate the relationship between social media food waste advertising and customer–brand communal engagement and awareness of the consequences of food waste and will, ultimately, influence customer intentions not to waste food. Because altruism plays a role, it is expected that altruistic customers will react more to social media food waste advertising by creating a stronger emotional connection to the brand. If they observe that the brand shares their values and acts as a socially responsible entity, they are likely to have strengthened emotional bonds. The heightened engagement of customers to support the brand's sustainability initiatives makes them support the brand's sustainability initiatives and increase the effect of social media advertising on their behavioral intentions [42]. It is also expected that altruism will moderate the effect of social media advertising on awareness of food waste consequences. Altruistic customers are more likely to cognitively engage with messages about the negative impacts of food waste, and as a result, reflect on how their behaviors affect society and the environment [65]. Processing these ads more in depth makes them more aware of factual information about food waste, and more receptive to this information, and results in stronger intentions to reduce waste. As a result, altruism is a significant factor in increasing emotional and cognitive engagement with sustainability-oriented social media advertising. This study considers altruism's moderating role to shed light on how personal values interact with external cues, such as social media advertising, and influence behavior. In the context of BRT, altruism becomes a reason to motivate action in accordance with the brand's values, either by emotional affiliation or cognitive understanding [66]. More altruistic customers have stronger emotional and cognitive responses to social media food waste advertising, leading to more robust intentions to cut food waste. The blending of emotional and cognitive pathways to influence pro-environmental behavior underscores the role of personal values in guiding pro-environmental behavior and supplies guidance for the design of effective social media campaigns. According to this rationale, we propose the following hypothesis:

**H8.** *Altruism moderates the indirect effect via customer–brand communal engagement (H8a), and altruism moderates the indirect effect via awareness of food waste consequences (H8b).*

The conceptual model of this study is given in Figure 1.

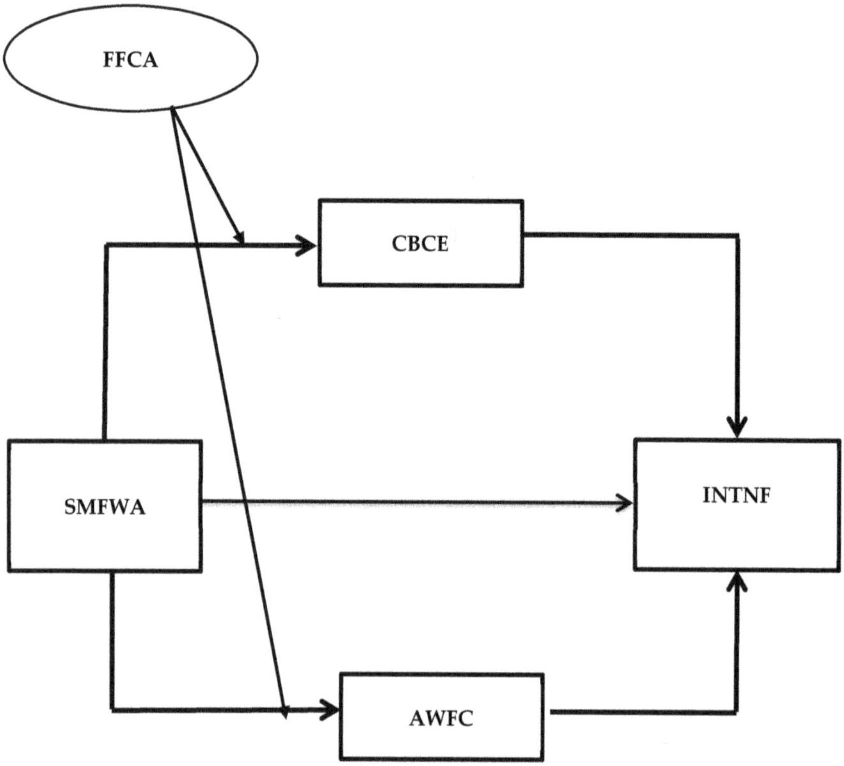


**Figure 1.** Hypothetical model showing the relationships between the study variables. Social media food waste advertising (SMFWA) is the independent variable (IV) and intention not to waste food (INTNF) is the dependent variable (DV). The model includes two mediators: customer–brand communal engagement (CBCE) and awareness of food waste consequences (AWFC). Fast-food customers' altruism (FFCA) is the moderator.

## 3. Material and Methods

### 3.1. Participants and Data Collection Procedure

This study collected data from customers of retail fast-food chains in three major Malaysian cities: Penang, Kuala Lumpur and Johor Bahru. The locations were due to their high concentration of fast-food outlets and diverse demographics, which is representative of Malaysia's urban population. As a hub of commercial activity, and a prime spot for fast food, Kuala Lumpur, the capital city, is a good place to observe the fast-food scene. With a reputation for rich tourism and a food culture that is so lively, Penang has a mix of local and international customers. Due to its proximity to an international market, Johor Bahru, near the Singaporean border, gives us insights into cross-border consumer behavior. In the process of data collection, our study sampled fast-food customers from three major Malaysian cities: Kuala Lumpur, Penang, and Johor Bahru. These cities were chosen for their diverse demographics and significant fast-food consumption. While Johor Bahru is known for its cross-border flow of visitors from Singapore, our study did not explicitly distinguish between local residents and foreign tourists in the analysis. The primary objective was to assess the general consumer behavior towards food waste reduction, regardless of the origin of the customers. This study utilized a convenience sample of fast-food customers. Data were collected using the paper–pencil method, with self-administered questionnaires distributed to respondents at fast-food outlets during non-peak hours. In this study, we did not limit our analysis to a specific social media platform. Instead, we focused on the general use of social media by fast-food chains to engage customers in sustainability initiatives. This approach was chosen because fast-food brands commonly use multiple platforms (e.g., Facebook, Instagram, Twitter) to reach a broad and diverse audience. Each platform offers distinct features that contribute to creating awareness and promoting behavioral change regarding food waste, making a general focus on social media more representative of the industry's practices. Potential respondents were approached by trained research assistants at popular fast-food outlets such as McDonald's, KFC and Subway during non-peak hours in order to minimize disruption. The participants were briefed on the objectives of this study and promised anonymity, and participated voluntarily. Strict ethical considerations were followed, with the participants being informed that their data would be used for academic purposes only and that they were free to withdraw at any time without any consequences. Data collection was approved by the research ethics committee of the lead author's institution. Using Cochran's formula, 400 respondents were considered adequate for this study. Since survey research is likely to result in low response rates, 700 questionnaires were distributed. There were 468 returns out of the 700 questionnaires. After screening for completeness and accuracy, some questionnaires were excluded because of incomplete answers or repetitive response patterns suggesting possible bias. Finally, 422 were retained as valid for data analysis, which was the recommended sample size and guaranteed the reliability and validity of the findings. Table 1 provides an overview of the socio-demographic characteristics of the respondents.

**Table 1.** Socio-demographic details of respondents.

| Variable | Category | Frequency | Percentage (%) |
|---|---|---|---|
| Gender | Male | 215 | 50.9 |
| | Female | 200 | 47.4 |
| | Other/prefer not to say | 7 | 1.7 |
| Age Group | Under 18 | 10 | 2.4 |
| | 18–25 | 132 | 31.3 |
| | 26–35 | 140 | 33.2 |
| | 36–45 | 85 | 20.1 |
| | 46–55 | 45 | 10.7 |
| | 56 and above | 10 | 2.4 |
| Highest Educational Level | Primary school | 5 | 1.2 |
| | Secondary school | 60 | 14.2 |
| | Diploma | 75 | 17.8 |
| | Bachelor's degree | 180 | 42.7 |
| | Master's degree or higher | 102 | 24.2 |
| Frequency of Visits | Rarely | 45 | 10.7 |
| | Occasionally | 110 | 26.1 |
| | Frequently | 157 | 37.2 |
| | Very frequently | 110 | 26.1 |

### 3.2. Measures

Variables were measured using established scales on a five-point Likert scale from 1 (Strongly Disagree) to 5 (Strongly Agree), chosen for high reliability and validated through a pilot test with 50 respondents [67]. The main variable, social media food waste advertising, was measured with four items from the work of Herman et al. [68] to capture how fast-food chains promote food waste reduction (e.g., "The retail fast-food chains I patronize frequently remind customers on social media to minimize food waste"). Customer–brand communal engagement was assessed using seven items based on the work of Song et al. [16] and Aggarwal [69], reflecting emotional bonds with brands engaged in pro-environmental activities (e.g., "This retail fast-food chain cares for me as a valued customer"). Awareness of food waste consequences was measured with four items from Attiq et al. [17] who modified the scale of Davies et al. [70] from the perspective of food waste, assessing customers' understanding of food waste's impact (e.g., "I consider reducing food waste at the retail fast-food chains I visit to be a major way to reduce pollution"). Altruism was evaluated using four items from the work of Ferguson et al. [18] to measure support for environmentally friendly practices (e.g., "By choosing eco-friendly options at retail fast-food chains, I would be fulfilling my duty to society"). Lastly, customers' intentions not to waste food were measured using four items from the work of Bell and Ulhas [71], focusing on efforts to minimize waste at fast-food chains (e.g., "I try not to throw away any food at the retail fast-food chains I patronize").

To address potential common method bias, procedural and statistical techniques were employed. Questionnaire items were presented in random order to prevent pattern detection, reducing artificial covariation. Anonymity and confidentiality were emphasized to encourage honest responses without fear of judgment. To mitigate social desirability bias, the items were worded neutrally. Statistically, Harman's single-factor test was conducted to assess common method variance. The results indicated that the maximum variance explained by a single factor was 32.33%, which is indicative that no single factor accounted for most variance, suggesting common method bias was not a significant concern [72,73]. Although skewness and kurtosis can be useful to assess normality and decrease the risk of common method bias (CMB), we employed other robust techniques to deal with CMB. In particular, we deployed procedural remedies that have been found to reduce the risks of CMB, including respondent anonymity and varying scale formats. We also ran Harman's single factor test and found that CMB was not a concern in this dataset. Thus, while skewness and kurtosis were not included, the measures undertaken were sufficient to address any CMB.

These steps ensured the validity and reliability of the responses, providing a more accurate assessment of the relationships between variables.

## 4. Results

Data analysis was conducted using Structural Equation Modeling (SEM) with Smart-PLS 3.0 software, chosen for its ability to test complex models with multiple constructs and handle latent variables and moderating effects [74]. PLS-SEM is a variance-based method that is particularly suitable for predictive modeling and complex models with latent constructs. This methodology is suitable for studying behavioral intentions [75] that are a function of different cognitive and emotional factors, and, in this way, it allows us to simultaneously estimate the relationships between different exogenous (independent) and endogenous (dependent) variables. In our model, the exogenous variable (independent variable) that predicts other constructs in the model is social media food waste advertising. Variables that are influenced by other constructs, such as customer intentions not to waste food, are endogenous [76,77]. In addition, we include mediator variables (awareness of food waste consequences and customer–brand communal engagement) and a moderator variable (altruism). A mediator variable explains the mechanism through which an independent variable affects a dependent variable, and a moderator variable modifies the nature and strength of that relationship [78,79]. Using a reflective measurement model (i.e., the latent constructs are indicated by their observed measures), we use a dimensional model to express the interrelationships between different latent constructs. The indicators of each latent variable (awareness of food waste consequences; customer–brand communal engagement) reflect the latent variable. By using this approach, the measurement model captures the constructs well [80]. Several model fit indicators were used in order to assess the fit of our PLS-SEM model, such as the Normed Fit Index (NFI) and the Standardized Root Mean Square Residual (SRMR). The NFI is used to compare the model's fit with a null model and values closer to 1 reflect a better fit. The threshold of 0.90 typically recommended in the literature was met with an NFI of 0.91 in this study [81,82]. A good model fit was observed with the SRMR value of 0.065, which is lower than the acceptable threshold of 0.08. Other fit criteria, combined with these indicators, assure us that our model is in good enough shape for the data and that our results are robust.

The results are presented in Table 2, showing factor loadings, composite reliability, average variance extracted (AVE), and R-square values for each construct. Abbreviations used include social media food waste advertising (SMFWA), awareness of food waste consequences (AWFC), customer–brand communal engagement (CBCE), and fast-food customers' intentions not to waste food (FFCINWF).

**Table 2.** Quality statistics of the measurement model.

| Variable | Item Statement | Factor Loading | CR | AVE | R Square |
|---|---|---|---|---|---|
| AWFC | | | 0.912 | 0.722 | 0.733 |
| | I consider reducing food waste at the retail fast-food chains I visit to be a major way to reduce pollution. | 0.858 | | | |
| | I believe that reducing food waste at the retail fast-food chains I patronize creates a better environment. | 0.861 | | | |
| | I believe that reducing food waste at the retail fast-food chains I visit is a major way to reduce landfills. | 0.855 | | | |
| | I consider reducing food waste at the retail fast-food chains I visit to be a major way to conserve resources. | 0.824 | | | |
| CBCE | | | 0.930 | 0.656 | 0.675 |
| | I would have warm feelings whenever I think of this retail fast-food chain. | 0.748 | | | |
| | This retail fast-food chain cares for me as a valued customer. | 0.706 | | | |
| | This retail fast-food chain offers assistance when I have issues with orders. | 0.783 | | | |
| | I would miss this retail fast-food chain if I had to switch to another. | 0.878 | | | |
| | This retail fast-food chain treats me with special attention. | 0.865 | | | |
| | This retail fast-food chain appreciates customers like me. | 0.863 | | | |
| | I care about my relationship with this retail fast-food chain. | 0.812 | | | |
| FFCA | | | 0.954 | 0.839 | |
| | By choosing eco-friendly options at retail fast-food chains, I would be fulfilling my duty to society. | 0.910 | | | |
| | By supporting environmentally conscious fast-food chains, I would be helping others. | 0.922 | | | |
| | By opting for sustainable choices at fast-food chains, I contribute to future generations. | 0.913 | | | |
| | I believe the world would be better if everyone chose sustainable fast-food options. | 0.918 | | | |
| FFCINF | | | 0.982 | 0.915 | 0.676 |
| | I try to eat all the food I order at fast-food chains. | 0.952 | | | |
| | I try not to throw away any food at fast-food chains. | 0.967 | | | |
| | I try to produce minimal food waste when dining at fast-food chains. | 0.964 | | | |
| | I try to leave as little food as possible when dining at fast-food chains. | 0.953 | | | |
| | I intend not to leave any edible food at fast-food chains in the coming weeks. | 0.947 | | | |
| SMFWA | | | 0.936 | 0.787 | |
| | The fast-food chains I patronize remind customers to minimize food waste on social media. | 0.774 | | | |
| | The fast-food chains I patronize clearly communicate messages about reducing food waste on social media. | 0.927 | | | |
| | The fast-food chains I patronize showcase innovative ways to minimize food waste on social media. | 0.936 | | | |
| | The fast-food chains I patronize use effective slogans to promote food waste reduction. | 0.903 | | | |

All items have significant factor loadings above 0.7, confirming strong convergent validity. The composite reliability values exceed 0.7 [83,84], ensuring internal consistency, while the AVE values above 0.5 indicate satisfactory variance explained for each construct [75,85]. The R-square values show the amount of variance explained by independent variables for each endogenous construct. The criteria used for analysis were based on relevant research. In marketing research, an $R^2$ value of 0.50 or higher is considered substantial [86,87]. This threshold was used for evaluating the predictive power of the model. The evaluation criteria, including composite reliability and AVE, follow established standards in the field. The analysis reveals that SMFWA significantly influences both AWFC and CBCE, highlighting the impact of social media campaigns in driving cognitive and emotional responses. SMFWA explains 73.3% of the variance in AWFC and 67.5% in CBCE, indicating substantial explanatory power.

To ensure that each construct in the model is distinct and measures unique aspects of the respondents' perceptions and behaviors, discriminant validity was assessed using both the correlation matrix (Table 3a) and the Heterotrait–Monotrait Ratio of Correlations (HTMT) (Table 3b). First, the Fornell and Larcker [88] criterion was applied, which requires that the square root of each construct's AVE be higher than its correlations with any other construct. Additionally, HTMT values were calculated, with values below 0.85 indicating sufficient discriminant validity. The correlation matrix presents relationships between constructs, with the square root of each construct's AVE shown along the diagonal. Discriminant validity is confirmed when this value is greater than its correlations with other constructs, indicating that each construct is distinct in what it measures [89,90]. For example, if the square root of the AVE for SMFWA exceeds its correlation with AWFC or any other construct, it confirms that SMFWA captures a unique aspect. Table 3 provides more detail.

**Table 3.** (**a**) Fornell–Larcker Criterion Matrix. (**b**) HTMT Matrix.

(a)

| Variable | AWFC | CBCE | FFCA | FFCINF | SMFWA |
|---|---|---|---|---|---|
| AWFC | 0.85 | | | | |
| CBCE | 0.401 | 0.81 | | | |
| FFCA | 0.341 | 0.448 | 0.916 | | |
| FFCINF | 0.311 | 0.423 | 0.382 | 0.956 | |
| SMFWA | 0.437 | 0.417 | 0.392 | 0.422 | 0.887 |

(b)

| Variable | AWFC | CBCE | FFCA | FFCINF | SMFWA |
|---|---|---|---|---|---|
| AWFC | - | | | | |
| CBCE | 0.806 | - | | | |
| FFCA | 0.613 | 0.815 | - | | |
| FFCINF | 0.663 | 0.738 | 0.504 | - | |
| SMFWA | 0.711 | 0.551 | 0.637 | 0.621 | - |

The hypothesis results reveal significant relationships within the proposed model. For H1, SMFWA positively influences FFCINF ($\beta = 0.115$, $t = 2.043$, $p = 0.041$), indicating that targeted social media campaigns can directly impact customer intentions to reduce waste. H2 shows that SMFWA significantly affects customer–brand communal engagement (CBCE) ($\beta = 0.383$, $t = 7.925$, $p < 0.001$), suggesting that social media ads create emotional bonds by demonstrating the brand's commitment to sustainability. H3 confirms that CBCE positively influences FFCINF ($\beta = 0.190$, $t = 3.128$, $p = 0.001$), indicating that stronger emotional attachment motivates customers to align their behaviors with the brand's values and reduce food waste. For H5, SMFWA has a substantial positive effect on AWFC ($\beta = 0.653$, $t = 16.022$, $p < 0.001$), highlighting the role of social media in enhancing customers' understanding of food waste. H6 is supported, as AWFC significantly influences FFCINF ($\beta = 0.562$, $t = 7.715$, $p < 0.001$), demonstrating that customers with greater awareness are more likely to minimize food waste. Regarding mediation hypotheses H4 and H7, both CBCE and AWFC are significant mediators. The indirect effect of SMFWA on FFCINF through CBCE is significant ($\beta = 0.073$, $t = 2.778$, $p = 0.006$), showing that emotional engagement mediates the relationship between social media advertising and customer intentions. Similarly, the indirect effect through AWFC is significant ($\beta = 0.367$, $t = 5.793$, $p < 0.001$), indicating that increased awareness is a key cognitive pathway through which SMFWA influences intentions. Specifically, the analysis of the mediator effect showed that the indirect effect of awareness of food waste consequences (AWFC) ($\beta = 0.367$) was considerably stronger than others'. The finding that increasing awareness of the consequences of food waste affects customers' intention not to waste food indicates that food waste awareness increases. Although social media food waste advertising has a significant strong indirect effect through AWFC rather than a direct effect on customer behavior, the effect is still significant. This is especially pertinent to the indirect effect of AWFC, which emphasizes the need for informational campaigns to engage customers emotionally, but also cognitively increase their understanding of food waste in a broader sense. This finding suggests that in order to achieve more effective behavioral change, fast-food chains need to focus on educational content that would raise awareness among customers in their social media campaigns. Table 4 summarizes these results, while Figure 2 presents the full structural model of this study.

**Table 4.** Hypotheses' validation.

| Hypothesis | Beta Coefficient | S.D | *t*-Statistics | *p*-Values | Bias Corrected Lower CI | Bias Corrected Upper CI | Result |
|---|---|---|---|---|---|---|---|
| **H1**: SMFWA → FFCINF | 0.115 | 0.056 | 2.043 | 0.041 | 0.009 | 0.226 | Supported |
| **H2**: SMFWA → CBCE | 0.383 | 0.048 | 7.925 | 0.000 | 0.288 | 0.485 | Supported |
| **H3**: CBCE → FFCINF | 0.19 | 0.061 | 3.128 | 0.001 | 0.082 | 0.312 | Supported |
| **H5**: SMFWA → AWFC | 0.653 | 0.041 | 16.022 | 0.000 | 0.563 | 0.729 | Supported |
| **H6**: AWFC → FFCINF | 0.562 | 0.073 | 7.715 | 0.000 | 0.403 | 0.684 | Supported |
| **H4**: SMFWA → CBCE → FFCINF | 0.073 | 0.026 | 2.778 | 0.006 | 0.033 | 0.133 | Supported |
| **H7**: SMFWA → AWFC → FFCINF | 0.367 | 0.063 | 5.793 | 0 | 0.24 | 0.479 | Supported |
| **H8a**: Moderating Effect 1 → AWFC → FFCINF | 0.056 | 0.015 | 3.656 | 0 | 0.025 | 0.082 | Supported |
| **H8b**: Moderating Effect 2 → CBCE → FFCINF | 0.015 | 0.008 | 1.897 | 0.059 | −0.004 | 0.032 | Not Supported |

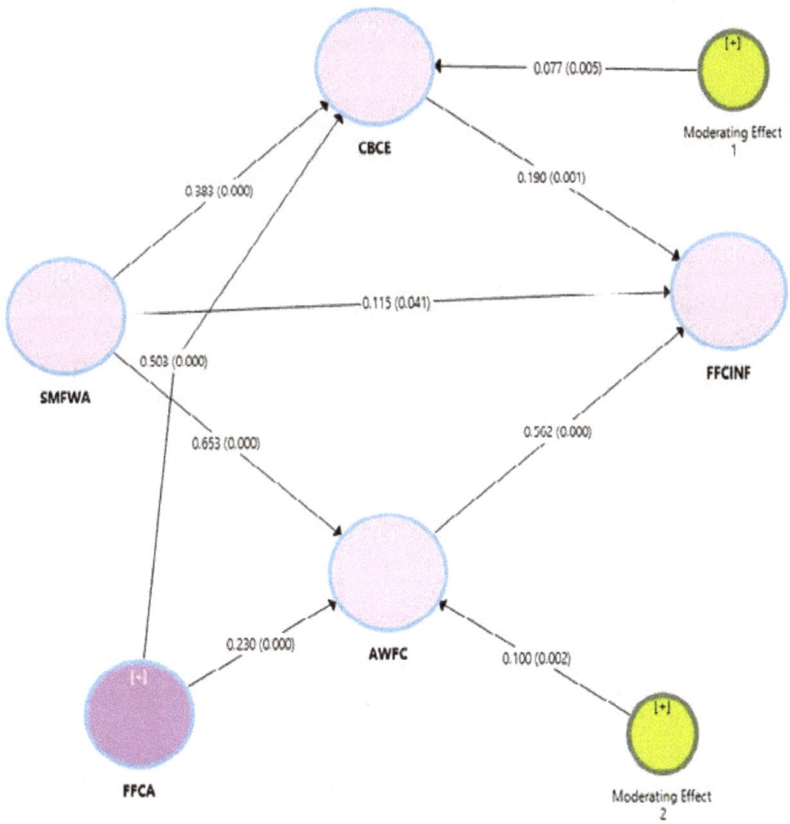


**Figure 2.** The structural model.

## 5. Discussion

Our findings reveal that social media food waste advertising positively impacts fast-food customers' intentions not to waste food ($\beta = 0.115$, $p = 0.042$), demonstrating that social media is an effective channel for promoting sustainable behaviors. This aligns with previous research indicating social media's influence on consumer behavior through compelling messaging [40]. This study finds results which corroborate the previous research which stated that social media campaigns can influence consumer behavior regarding sustainability. Persuasive messaging and interactive content on Facebook and Instagram platforms can dramatically boost consumer awareness of pro-environmental behaviors [37]. Our results also show that social media food waste advertising is a successful means to encourage consumers to adopt responsible consumption behavior. While there are some differences, one important distinction is that in the Malaysian context, awareness campaigns focus on emotional and cognitive engagement through messages, which might be more impactful given the cultural emphasis on food. This reflects a critical difference in terms of psychological factors compared to other contexts, for example, Western countries, where individualism might moderate the effect of such campaigns [91]. Thus, these findings help to explain the ways in which cultural context influences consumer responses to sustainability messaging. However, the modest strength of this direct effect suggests that its impact is significantly enhanced when emotional and cognitive factors are also considered.

The mediation analysis uncovers two key pathways. First, there is a strong positive relationship between social media food waste advertising and awareness of food waste consequences ($\beta = 0.653$, $p < 0.001$), highlighting the campaigns' effectiveness in enhancing customers' understanding of environmental and social impacts. The indirect effect of awareness of food waste consequences on intentions to reduce food waste is highly significant in the Malaysian context, indicating the importance of raising awareness as a strong driver of behavioral change. This result is consistent with research in other Southeast Asian countries where food waste is a growing concern [92], and where campaigns to promote food waste reduction have been effective. Yet, this study extends the literature by showing how social media can be used by fast-food chains to incorporate educational content into their brand strategies. Awareness is apparently more influential in pro-environmental intentions in Malaysia; the psychological factors in Malaysia, including the collectivist nature of the society and the cultural implication of not wasting food, seem to reinforce awareness's influence on pro-environmental intentions.

This cognitive pathway is crucial, as increased awareness leads to stronger intentions to reduce food waste ($\beta = 0.562$, $p < 0.001$). This underscores the importance of educational content and factual information in driving behavioral change, consistent with studies emphasizing awareness in promoting pro-environmental behaviors [46,54]. Second, customer–brand communal engagement significantly mediates the relationship between social media advertising and customers' intentions not to waste food ($\beta = 0.073$, $p = 0.006$). Social media campaigns that showcase a brand's commitment to sustainability foster emotional bonds, encouraging customers to align their behavior with the brand's values. The positive effect on customer–brand communal engagement ($\beta = 0.383$, $p < 0.001$) demonstrates that highlighting shared values strengthens emotional attachment [17]. While both mediators are significant, awareness of food waste consequences has a stronger impact on intentions than communal engagement, emphasizing the critical role of cognitive factors in shaping behavior within food waste reduction.

The moderating effects of customers' altruism offer further insights. Altruism significantly enhances the indirect effect of social media advertising on intentions through awareness of food waste consequences ($\beta = 0.056$, $p < 0.001$), suggesting that altruistic customers are more receptive to cognitive aspects of advertising. However, altruism does not significantly moderate the effect through customer–brand communal engagement ($\beta = 0.015$, $p = 0.059$), indicating limited impact on the emotional pathway. This implies that while altruistic individuals respond more to cognitive messages, their altruism does not notably affect emotional engagement with the brand. The novelty of this study is in its integration of both emotional and cognitive pathways in understanding consumer behavior being taken through the lens of BRT. Cognitive factors like awareness have been studied in prior research, but what remained under-explored is the emotional variables, like communal engagement, that a fast-food chain can utilize to create stronger emotional bonds with their customers. This dual focus is part of a growing body of literature on sustainability and food waste reduction and complements previous research by demonstrating that emotional engagement, through the sense of community, can be as important as cognitive awareness in promoting pro-environmental actions. Our results also matter for fast-food brands in Malaysia. Strong emotional connections induced by customer–brand communal engagement offer a chance for brands to build stronger relationships with consumers by promoting shared values in terms of sustainability. The cultural emphasis on community and collectivism may make this connection even more important in Malaysia. For this reason, emotional and cognitive appeals should continue to be leveraged in brands' sustainability messaging to drive long-term behavioral change.

In sum, these findings align with BRT, which posits that behavior is driven by rational and emotional reasons. The cognitive pathway captures rational reasons for reducing food waste, while the emotional pathway reflects affective motivations. The role of altruism supports BRT, showing that altruistic customers are more influenced by rational considerations (awareness) than emotional ones (communal engagement). Integrating emotional and cognitive pathways, along with personal values, provides a comprehensive understanding of how social media advertising shapes intentions not to waste food. This offers valuable theoretical and practical implications for designing effective sustainability campaigns in the fast-food industry.

### 5.1. Theoretical Implications

This research extends our understanding of how social media advertising can influence customer behavior through dual pathways, cognitive and emotional engagement, offering a sophisticated perspective on the psychological mechanisms underlying pro-environmental behavior. By integrating BRT, this study demonstrates how social media food waste advertising shapes customers' intentions not to waste food through the formation of both rational reasons (awareness of food waste consequences) and emotional reasons (customer–brand communal engagement). This dual-pathway approach advances the existing literature, which often focuses on cognitive or emotional factors in isolation, by illustrating the simultaneous and distinct roles of these dimensions in influencing behavior.

Additionally, the incorporation of fast-food customers' altruism as a moderator enriches the application of BRT by showing how personal values can amplify the cognitive pathway. This finding challenges the notion that personal values uniformly affect all types of reasoning, suggesting instead that altruism has a more pronounced impact on cognitive responses while having a limited effect on emotional engagement. As such, this study provides a more comprehensive framework for understanding the complex interplay between social media advertising, cognitive and emotional factors, and personal values, contributing to the broader fields of consumer behavior and environmental psychology. These insights lay a foundation for future research to explore additional personal and contextual factors that may moderate the effectiveness of sustainability-focused campaigns, thereby expanding the theoretical scope of BRT in the context of sustainable consumer behavior.

From a social media perspective, this research underscores the critical role of digital platforms in shaping consumer perceptions and behaviors towards sustainable consumption, particularly in food waste reduction. This study highlights social media's unique strengths, engaging users through interactive content, facilitating two-way communication, and rapidly disseminating information, making it an ideal medium for promoting pro-environmental behaviors. Our findings show that social media food waste advertising can enhance both cognitive and emotional engagement, providing brands with a powerful tool to not only inform customers about sustainability issues but also establish a sense of communal responsibility. The dual pathways identified in this study suggest that social media content can be strategically designed to address both rational and emotional aspects of customer behavior, thereby making campaigns more impactful. By illustrating how social media advertising can foster awareness and build emotional bonds, this research contributes to the growing body of literature on digital marketing and social media's role in promoting sustainable behaviors.

### 5.2. Practical Implications

The results of this research can help retail fast-food chains that want to use social media advertising to promote sustainable consumption behaviors. We find that social media campaigns can have a strong effect on customers' intentions not to waste food by increasing customers' cognitive understanding and emotional connection to the brand. Fast-food chains should make content that educates customers about the consequences of food waste and conveys shared responsibility in order to have the maximum impact. For example, by combining real-world statistics with emotionally engaging narratives, such as stories of food donation initiatives, it is possible to effectively influence customers' knowledge and emotions. It is this dual approach that makes customers more open to pro-environmental behaviors that are in line with the sustainability goals of the brand. Social media should be the tool fast-food chains use to promote transparency and promote awareness of their sustainability efforts. Customers can be made more aware and intentional about not wasting food by sharing how the brand reduces food waste through operational efficiencies, partnerships with food banks, or innovative packaging. Interactive features such as polls, quizzes, and challenges can help engage audiences in discussions about food waste reduction and make the issue more relatable and easier to understand. Customer–brand communal engagement is emphasized by emphasizing the brand's social and environmental commitments. Campaigns communicating shared values such as reducing environmental footprint or supporting communities strengthen emotional bonds with customers. This engagement, in turn, encourages customers to harmonize their behavior with the brand's sustainable practices. For example, working with environmental groups to demonstrate collaborations can make the brand a sustainability partner, promoting customers to be pro-environmental. The role of altruism should also be considered and fast-food chains should design their social media messages to attract altruistic customers. Content that highlights the societal benefits of reducing food waste, such as reducing food insecurity or pollution, can be very impactful, leading to stronger behavioral intentions. Segmenting campaigns by customer values to tailor messages increases overall effectiveness. So, in short, fast-food chains should develop a complete social media strategy that includes educational and emotionally appealing content to raise awareness about food waste. They can better influence intentions not to waste food by addressing cognitive and emotional dimensions and aligning messages with customers' values. This helps with sustainability goals and reinforces the brand's image as a socially responsible one.

Moreover, this study has important implications for fast-food chains interested in increasing their sustainability efforts through social media advertising. There are existing campaigns, such as McDonald's, that have already shown what can be carried out by using social media to encourage food waste reduction. For instance, McDonald's has run campaigns to urge responsible consumption and promote sustainability. While these efforts are a good start, there are other ways to integrate new and innovative parts into such campaigns. The integration of educational content that displays awareness of food waste consequences is one area for improvement. We found that awareness is a critical driver of behavioral change. To take those efforts further, fast-food chains could build on existing efforts, using interactive tools such as real-time data on food waste reduction, gamification techniques to engage customers, or partnerships with local food banks to illustrate the tangible impacts of waste reduction. Furthermore, campaigns should aim to deepen rooted emotional ties via customer–brand communal engagement by telling stories of the brand's involvement in the community and reaffirming the brand's commitment to sustainability, as our findings indicate that this is important to inform consumer behavior. In addition, fast-food chains in a crowded market can differ from their competitors by incorporating cognitive and emotional appeals that reach both rational awareness and communal values. Existing campaigns have made progress in boosting sustainability, but can do much more by personalizing messages, providing localized content and customer testimonials to enhance emotional engagement and responsibility with consumers.

### 5.3. Limitations and Future Research Directions

Despite its valuable insights, this study has limitations. First, since it was conducted within Malaysia's retail fast-food sector, the findings may not generalize to other cultural or regional settings. Customer perceptions and responses to social media advertising can vary due to cultural, economic, and social differences. Future studies should replicate this research in diverse contexts to test the model's applicability or adapt it to different markets. Second, relying on cross-sectional data limits our ability to establish causal relationships. While integrating emotional and cognitive pathways enhances understanding of how social media food waste advertising influences customer behavior, longitudinal studies could offer deeper insights into how these impacts evolve over time and whether intentions lead to sustained behavioral changes. Experimental designs could also isolate the effects of different types of social media content on engagement and behavior, providing stronger evidence of causality.

Third, this study focuses on specific psychological constructs, awareness of food waste consequences and customer–brand communal engagement, as mediators between social media advertising and behavioral intentions. While these variables provide valuable insights, other constructs like perceived social norms, moral obligations, or environmental concerns may also influence customer behavior. Future research could explore these additional mediators to gain a more comprehensive understanding of the factors influencing pro-environmental behaviors in response to social media advertising. Another limitation is the reliance on self-reported data, which may be susceptible to social desirability bias. Although measures were taken to mitigate this, such as ensuring anonymity and including attention-check questions, future studies could combine self-reported and behavioral data to validate the findings. Observing actual food waste behaviors in a controlled setting could provide more objective evidence of the effectiveness of social media advertising. Lastly, one of the major limitations of this study is the absence of separation between locals and foreigners, especially in Johor Bahru, due to a lot of cross-border tourists. This study contributes valuable information on overall consumer behavior in the fast-food sector in Malaysia, yet further research should segment respondents according to their origin. Consequently, this will enable a more thorough comprehension of how local versus foreign consumers react to social media food waste campaigns and become involved with sustainability undertakings. This distinction could inform more targeted strategies to promote pro-environmental behaviors in diverse consumer groups.

## 6. Conclusions

This paper integrates both cognitive and emotional pathways within a behavioral reasoning framework for a comprehensive examination of how social media food waste advertising can affect fast-food customers' intentions not to waste food. We found that social media food waste advertising has a direct positive effect on customer behavioral intentions, and has significant indirect effects on customer behavioral intentions via increased customer awareness of food waste consequences and customer–brand communal engagement. These dual pathways show that social media campaigns seeking to promote sustainable consumption practices should be designed to address both the rational and emotional aspects of customer behavior. Furthermore, customers' altruism moderates the effectiveness of these advertising efforts, indicating an important role of personal values in influencing the effectiveness of these efforts, particularly by magnifying the cognitive engagement of customers with higher levels of altruism. In general, this study adds to the literature by showing how social media can be used as an effective platform to drive customer behavior toward food waste reduction in the retail fast-food sector. Practical recommendations are provided for fast-food chains to create social media messages that educate and emotionally connect with customers while simultaneously persuading them to engage in pro-environmental behaviors. This research bridges the gap between theory and practice to enhance the theoretical understanding of the effects of social media advertising on customer behavior, and to provide actionable insights for businesses seeking to achieve alignment between marketing strategy and sustainability objectives. Given the world's struggles with food waste, this study demonstrates the potential of social media to contribute to a more sustainable food consumption culture.

**Author Contributions:** All of the authors contributed to the conceptualization, formal analysis, investigation, methodology, and writing and editing of the original draft. All authors have read and agreed to the published version of the manuscript.

**Funding:** This research was funded by the following project: The '14th Five-Year Plan' for Educational Science (Project No.: K24YC3090027).

**Institutional Review Board Statement:** This study received ethical approval from the Research Ethics Committee of Universiti Utara Malaysia. The ethical approval reference number is UUM/REC/Sep/2023/078. All participants provided informed consent prior to their involvement in the study.

**Informed Consent Statement:** Informed consent was obtained from all subjects involved in this study.

**Data Availability Statement:** Data will be made available on a reasonable request by contacting the corresponding author.

**Conflicts of Interest:** The authors declare no conflicts of interest.

## References

1. Biocycle. UN's 2024 Food Waste Index Attributes 60% of Food Waste to Households. Available online: https://www.biocycle.net/2024-food-waste-index/ (accessed on 20 June 2024).
2. Sarker, A.; Ahmmed, R.; Ahsan, S.; Rana, J.; Ghosh, M.K.; Nandi, R. A comprehensive review of food waste valorization for the sustainable management of global food waste. *Sustain. Food Technol.* **2024**, *2*, 48–69. [CrossRef]
3. Liu, C.; Ahmad, N.; Jiang, M.; Arshad, M.Z. Steering the path to safer food: The role of transformational leadership in food services to combat against foodborne illness. *J. Retail. Consum. Serv.* **2024**, *81*, 103958. [CrossRef]
4. McDonald's Corporation. McDonald's Corporation Purpose & Impact Report 2022–2023. Available online: https://corporate.mcdonalds.com/content/dam/sites/corp/nfl/pdf/McDonalds_PurposeImpact_ProgressReport_2022_2023.pdf (accessed on 20 June 2024).
5. STARBUCKS. STARBUCKS Global Impact Report. Available online: https://stories.starbucks.com/uploads/2024/02/2023-Starbucks-Global-Impact-Report.pdf (accessed on 20 June 2024).
6. Sivarajah, U.; Irani, Z.; Gupta, S.; Mahroof, K. Role of big data and social media analytics for business to business sustainability: A participatory web context. *Ind. Mark. Manag.* **2020**, *86*, 163–179. [CrossRef]
7. Zhang, D.; Mahmood, A.; Ariza-Montes, A.; Vega-Muñoz, A.; Ahmad, N.; Han, H.; Sial, M.S. Exploring the impact of corporate social responsibility communication through social media on banking customer e-wom and loyalty in times of crisis. *Int. J. Environ. Res. Public Health* **2021**, *18*, 4739. [CrossRef]
8. Gupta, S.; Nawaz, N.; Alfalah, A.A.; Naveed, R.T.; Muneer, S.; Ahmad, N. The relationship of CSR communication on social media with consumer purchase intention and brand admiration. *J. Theor. Appl. Electron. Commer. Res.* **2021**, *16*, 1217–1230. [CrossRef]
9. Gupta, S.; Nawaz, N.; Tripathi, A.; Muneer, S.; Ahmad, N. Using social media as a medium for CSR communication, to induce consumer–brand relationship in the banking sector of a developing economy. *Sustainability* **2021**, *13*, 3700. [CrossRef]
10. Ahmad, N.; Naveed, R.T.; Scholz, M.; Irfan, M.; Usman, M.; Ahmad, I. CSR communication through social media: A litmus test for banking consumers' loyalty. *Sustainability* **2021**, *13*, 2319. [CrossRef]
11. Jenkins, E.L.; Brennan, L.; Molenaar, A.; McCaffrey, T.A. Exploring the application of social media in food waste campaigns and interventions: A systematic scoping review of the academic and grey literature. *J. Clean. Prod.* **2022**, *360*, 132068. [CrossRef]
12. Rohm, H.; Oostindjer, M.; Aschemann-Witzel, J.; Symmank, C.; Almli, V.L.; De Hooge, I.E.; Normann, A.; Karantininis, K. Consumers in a sustainable food supply chain (COSUS): Understanding consumer behavior to encourage food waste reduction. *Foods* **2017**, *6*, 104. [CrossRef]

13. Parvatiyar, A.; Sheth, J.N. Confronting the deep problem of consumption: Why individual responsibility for mindful consumption matters. *J. Consum. Aff.* **2023**, *57*, 785–820. [CrossRef]
14. Ahmad, N.; Ahmad, A.; Siddique, I. Beyond self-interest: How altruistic values and human emotions drive brand advocacy in hospitality consumers through corporate social responsibility. *Corp. Soc. Responsib. Environ. Manag.* **2024**, *31*, 2439–2453. [CrossRef]
15. Liu, Y.; Ahmad, N.; Lho, L.H.; Han, H. From boardroom to breakroom: Corporate social responsibility, happiness, green self-efficacy, and altruistic values shape sustainable behavior. *Soc. Behav. Personal. Int. J.* **2024**, *52*, 1–14. [CrossRef]
16. Song, B.; Tao, W.; Wen, T.J. Building consumer communal relationships through cause-related marketing: From the perspective of persuasion knowledge. *J. Consum. Behav.* **2021**, *20*, 1388–1402. [CrossRef]
17. Attiq, S.; Habib, M.D.; Kaur, P.; Hasni, M.J.S.; Dhir, A. Drivers of food waste reduction behaviour in the household context. *Food Qual. Prefer.* **2021**, *94*, 104300. [CrossRef]
18. Ferguson, E.; Farrell, K.; Lawrence, C. Blood donation is an act of benevolence rather than altruism. *Health Psychol.* **2008**, *27*, 327. [CrossRef]
19. Guan, X.; Ahmad, N.; Sial, M.S.; Cherian, J.; Han, H. CSR and organizational performance: The role of pro-environmental behavior and personal values. *Corp. Soc. Responsib. Environ. Manag.* **2023**, *30*, 677–694. [CrossRef]
20. Ab Aziz, N.S.a.A.; Izleen Hasmady, N.I.; Shafie, F.A.; Mohd Yatim, S.R.; Azmi, A.; Clark, A. Food Waste and Carbon Footprint Assessment of Eateries in Kelantan, Malaysia. *Malays. J. Med. Health Sci.* **2022**, *18*, 1–8. [CrossRef]
21. SWCorp. When Good Food Goes to Waste. 2024. Available online: https://philanthropynewsdigest.org/news/other-sources/article/?id=14595233&title=When-good-food-goes-to-waste (accessed on 4 November 2024).
22. Russell, S.V.; Young, C.W.; Unsworth, K.L.; Robinson, C. Bringing habits and emotions into food waste behaviour. *Resour. Conserv. Recycl.* **2017**, *125*, 107–114. [CrossRef]
23. Ma, Y.; Al Mamun, A.; Ali, M.H.; Hoque, M.E.; Lili, Z. Modeling the intention and adoption of food waste prevention practices among Chinese households. *Humanit. Soc. Sci. Commun.* **2023**, *10*, 840. [CrossRef]
24. Chen, X.; Cheng, Z.-f.; Yang, H.-j. Empowering pro-environmental behavior in tourists through digital media: The influence of eco-guilt and empathy with nature. *Front. Psychol.* **2024**, *15*, 1387817. [CrossRef]
25. Rahman, T.; deb, N.; Alam, M.Z.; Moniruzzaman, M.; Miah, M.S.; Horaira, M.A.; Kamal, R. Navigating the contemporary landscape of food waste management in developing countries: A comprehensive overview and prospective analysis. *Heliyon* **2024**, *10*, e33218. [CrossRef] [PubMed]
26. Mont, O.; Lehner, M.; Dalhammar, C. Sustainable consumption through policy intervention—A review of research themes. *Front. Sustain.* **2022**, *3*, 921477. [CrossRef]
27. Yu, H.; Shabbir, M.S.; Ahmad, N.; Ariza-Montes, A.; Vega-Muñoz, A.; Han, H.; Scholz, M.; Sial, M.S. A contemporary issue of micro-foundation of CSR, employee pro-environmental behavior, and environmental performance toward energy saving, carbon emission reduction, and recycling. *Int. J. Environ. Res. Public Health* **2021**, *18*, 5380. [CrossRef] [PubMed]
28. Song, B.L. The Effectiveness of Social Media in Promoting Sustainable Food Consumption: Evidence from Developing Countries. *Bus. Res. Proc.* **2023**, *1*, 27–28. [CrossRef]
29. Evans, W.D.; Abroms, L.C.; Broniatowski, D.; Napolitano, M.A.; Arnold, J.; Ichimiya, M.; Agha, S. Digital media for behavior change: Review of an emerging field of study. *Int. J. Environ. Res. Public Health* **2022**, *19*, 9129. [CrossRef]
30. Young, W.; Russell, S.V.; Robinson, C.A.; Barkemeyer, R. Can social media be a tool for reducing consumers' food waste? A behaviour change experiment by a UK retailer. *Resour. Conserv. Recycl.* **2017**, *117*, 195–203. [CrossRef]
31. Deng, Y.; Cherian, J.; Ahmad, N.; Scholz, M.; Samad, S. Conceptualizing the role of target-specific environmental transformational leadership between corporate social responsibility and pro-environmental behaviors of hospital employees. *Int. J. Environ. Res. Public Health* **2022**, *19*, 3565. [CrossRef]
32. Xu, L.; Mohammad, S.J.; Nawaz, N.; Samad, S.; Ahmad, N.; Comite, U. The role of CSR for de-carbonization of hospitality sector through employees: A leadership perspective. *Sustainability* **2022**, *14*, 5365. [CrossRef]
33. Westaby, J.D. Behavioral reasoning theory: Identifying new linkages underlying intentions and behavior. *Organ. Behav. Hum. Decis. Process.* **2005**, *98*, 97–120. [CrossRef]
34. Sahu, A.K.; Padhy, R.; Dhir, A. Envisioning the future of behavioral decision-making: A systematic literature review of behavioral reasoning theory. *Australas. Mark. J.* **2020**, *28*, 145–159. [CrossRef]
35. Mansoor, M.; Paul, J. Consumers' choice behavior: An interactive effect of expected eudaimonic well-being and green altruism. *Bus. Strategy Environ.* **2022**, *31*, 94–109. [CrossRef]
36. Tuten, T.; Mintu-Wimsatt, A. Advancing our understanding of the theory and practice of social media marketing: Introduction to the special issue. *J. Mark. Theory Pr.* **2018**, *26*, 1–3. [CrossRef]
37. Duffett, R.G. Influence of social media marketing communications on young consumers' attitudes. *Young Consum.* **2017**, *18*, 19–39. [CrossRef]
38. Ashley, C.; Tuten, T. Creative strategies in social media marketing: An exploratory study of branded social content and consumer engagement. *Psychol. Mark.* **2015**, *32*, 15–27. [CrossRef]
39. Schmidt, C.G.; Wuttke, D.A.; Ball, G.P.; Heese, H.S. Does social media elevate supply chain importance? An empirical examination of supply chain glitches, Twitter reactions, and stock market returns. *J. Oper. Manag.* **2020**, *66*, 646–669. [CrossRef]

40. Hall, J.A.; Xing, C.; Ross, E.M.; Johnson, R.M. Experimentally manipulating social media abstinence: Results of a four-week diary study. *Media Psychol.* **2021**, *24*, 259–275. [CrossRef]
41. Sutinen, U.-M.; Närvänen, E. Constructing the food waste issue on social media: A discursive social marketing approach. *J. Mark. Manag.* **2022**, *38*, 219–247. [CrossRef]
42. White, K.; Habib, R.; Hardisty, D.J. How to SHIFT consumer behaviors to be more sustainable: A literature review and guiding framework. *J. Mark.* **2019**, *83*, 22–49. [CrossRef]
43. Trade. Malaysia—Country Commercial Guide. Available online: https://www.trade.gov/country-commercial-guides/malaysia-environmental-technology (accessed on 21 October 2024).
44. Tufail, H.S.; Yaqub, R.M.S.; Alsuhaibani, A.M.; Ramzan, S.; Shahid, A.U.; Refat, M.S. Consumers' purchase intention of suboptimal food using behavioral reasoning theory: A food waste reduction strategy. *Sustainability* **2022**, *14*, 8905. [CrossRef]
45. Jabeen, F.; Dhir, A.; Islam, N.; Talwar, S.; Papa, A. Emotions and food waste behavior: Do habit and facilitating conditions matter? *J. Bus. Res.* **2023**, *155*, 113356. [CrossRef]
46. Šagovnović, I.; Pivac, T.; Kovačić, S. Examining antecedents of residents' support for the European Capital of Culture project–event's sustainability perception, emotional solidarity, community attachment and brand trust. *Int. J. Event Festiv. Manag.* **2022**, *13*, 182–202. [CrossRef]
47. Kumar, J.; Konar, R.; Balasubramanian, K. The impact of social media on consumers' purchasing behaviour in Malaysian restaurants. *J. Tour. Sustain. Well-Being* **2020**, *8*, 197–216.
48. Yarar, A. Halal advertising on YouTube: A case study of fast food brands in Malaysia. *Food Res.* **2020**, *4*, 157–163. [CrossRef]
49. Vodus. Malaysian Consumer Shopping Behaviour 2024: Habits & Attitude. Available online: https://vodus.com/article/malaysian-consumer-shopping-behaviour-2024-habits-and-attitude (accessed on 21 October 2024).
50. Huang, Y.; Ma, E.; Wang, D. Message framing strategies, food waste prevention, and diners' repatronage intentions: The mediating role of corporate social responsibility. *J. Sustain. Tour.* **2021**, *29*, 1694–1715. [CrossRef]
51. Moser, D.A.; Doucet, G.E.; Dima, D.; Schumann, G.; Bilder, R.; Frangou, S. An integrated brain–behavior model for working memory. *Mol. Psychiatry* **2018**, *23*, 1974–1980. [CrossRef]
52. Vermeir, I.; Weijters, B.; De Houwer, J.; Geuens, M.; Slabbinck, H.; Spruyt, A.; Van Kerckhove, A.; Van Lippevelde, W.; De Steur, H.; Verbeke, W. Environmentally sustainable food consumption: A review and research agenda from a goal-directed perspective. *Front. Psychol.* **2020**, *11*, 1603.
53. Hartikainen, H.; Riipi, I.; Katajajuuri, J.-M.; Silvennoinen, K. From measurement to management: Food waste in the Finnish food chain. In *Food Waste Management: Solving the Wicked Problem*; Palgrave Macmillan: Cham, Germany, 2020; pp. 415–439.
54. Ajina, A.S.; Ali, S.; Zamil, A.M.A.; Khalid, N.; Bait Ali Sulaiman, M.A. Unleashing the potential of social media celebrities to promote food waste reduction in educational institutions: Developing an extended model based on the value-belief-norm theory. *Br. Food J.* **2024**, *126*, 2787–2808. [CrossRef]
55. Nordin, N.; Kaida, N.; Othman, N.; Akhir, F.; Hara, H. Reducing food waste: Strategies for household waste management to minimize the impact of climate change and contribute to Malaysia's sustainable development. In *Proceedings of the IOP Conference Series: Earth and Environmental Science*; IOP Publishing: Kuala Lumpur, Malaysia, 2020; p. 012035.
56. Ashley, Y. Read, Debate: Engage. Available online: https://www.fairplanet.org/editors-pick/how-malaysians-are-battling-food-waste-with-surplus-food/ (accessed on 22 October 2024).
57. MIDA. Sustainability Agenda. Available online: https://www.mida.gov.my/why-malaysia/sustainability-agenda/ (accessed on 22 October 2024).
58. Schwartz, S.H. Universals in the content and structure of values: Theoretical advances and empirical tests in 20 countries. In *Advances in Experimental Social Psychology*; Elsevier: Amsterdam, The Netherlands, 1992; Volume 25, pp. 1–65.
59. Ahmad, N.; Ullah, Z.; Arshad, M.Z.; waqas Kamran, H.; Scholz, M.; Han, H. Relationship between corporate social responsibility at the micro-level and environmental performance: The mediating role of employee pro-environmental behavior and the moderating role of gender. *Sustain. Prod. Consum.* **2021**, *27*, 1138–1148. [CrossRef]
60. Ball-Rokeach, S.J. Values and violence: A test of the subculture of violence thesis. *Am. Sociol. Rev.* **1973**, *38*, 736–749. [CrossRef]
61. Peng, J.; Samad, S.; Comite, U.; Ahmad, N.; Han, H.; Ariza-Montes, A.; Vega-Muñoz, A. Environmentally specific servant leadership and employees' Energy-specific pro-environmental behavior: Evidence from healthcare sector of a developing economy. *Int. J. Environ. Res. Public Health* **2022**, *19*, 7641. [CrossRef]
62. Verplanken, B.; Holland, R.W. Motivated decision making: Effects of activation and self-centrality of values on choices and behavior. *J. Personal. Soc. Psychol.* **2002**, *82*, 434. [CrossRef] [PubMed]
63. Schwartz, S.H.; Caprara, G.V.; Vecchione, M. Basic personal values, core political values, and voting: A longitudinal analysis. *Political Psychol.* **2010**, *31*, 421–452. [CrossRef]
64. Ahmad, N.; Ahmad, A.; Siddique, I. Responsible tourism and hospitality: The intersection of altruistic values, human emotions, and corporate social responsibility. *Adm. Sci.* **2023**, *13*, 105. [CrossRef]
65. Bautista, R.; Dui, R.; Jeong, L.S.; Paredes, M.P. Does altruism affect purchase intent of green products? A moderated mediation analysis. *Asia-Pac. Soc. Sci. Rev.* **2020**, *20*, 159–170. [CrossRef]
66. Ribbers, D.; van Herpen, E.; Pandelaere, M.; Geuens, M. Morals over money: The relationship between moral motivations and Consumers' food waste. *Sustain. Prod. Consum.* **2024**, *51*, 359–370. [CrossRef]

67. Lewandowska, A.; Berniak-Woźny, J.; Ahmad, N. Competitiveness and innovation of small and medium enter-prises under Industry 4.0 and 5.0 challenges: A comprehensive bibliometric analysis. *Equilibrium. Q. J. Econ. Econ. Policy* **2023**, *18*, 1045–1074. [CrossRef]
68. Herman, L.E.; Udayana, I.B.N.; Farida, N. Young generation and environmental friendly awareness: Does it the impact of green advertising? *Bus. Theory Pract.* **2021**, *22*, 159–166. [CrossRef]
69. Aggarwal, P. The effects of brand relationship norms on consumer attitudes and behavior. *J. Consum. Res.* **2004**, *31*, 87–101. [CrossRef]
70. Davies, J.; Foxall, G.R.; Pallister, J. Beyond the intention–behaviour mythology: An integrated model of recycling. *Mark. Theory* **2002**, *2*, 29–113. [CrossRef]
71. Bell, A.E.; Ulhas, K.R. Working to reduce food waste: Investigating determinants of food waste amongst Taiwanese workers in factory cafeteria settings. *Sustainability* **2020**, *12*, 9669. [CrossRef]
72. Ahmad, N.; Ullah, Z.; Ryu, H.B.; Ariza-Montes, A.; Han, H. From corporate social responsibility to employee well-being: Navigating the pathway to sustainable healthcare. *Psychol. Res. Behav. Manag.* **2023**, *16*, 1079–1095. [CrossRef] [PubMed]
73. Liu, Y.; Cherian, J.; Ahmad, N.; Han, H.; de Vicente-Lama, M.; Ariza-Montes, A. Internal corporate social responsibility and employee burnout: An employee management perspective from the healthcare sector. *Psychol. Res. Behav. Manag.* **2023**, *16*, 283–302. [CrossRef] [PubMed]
74. Hair Jr, J.F.; Matthews, L.M.; Matthews, R.L.; Sarstedt, M. PLS-SEM or CB-SEM: Updated guidelines on which method to use. *Int. J. Multivar. Data Anal.* **2017**, *1*, 107–123. [CrossRef]
75. Ahmad, N.; Samad, S.; Mahmood, S. Sustainable pathways: The intersection of CSR, hospitality and the United Nations' sustainable development goals. *Curr. Issues Tour.* **2024**, *27*, 3924–3943. [CrossRef]
76. Ahmad, N.; Ullah, Z.; AlDhaen, E.; Siddique, I. Promoting the advocacy behavior of customers through corporate social responsibility: The role of brand admiration. *Bus. Soc. Rev.* **2023**, *128*, 367–386. [CrossRef]
77. Han, H.; Al-Ansi, A.; Chua, B.-L.; Ahmad, N.; Kim, J.J.; Radic, A.; Bobby Ryu, H. Reconciling civilizations: Eliciting residents' attitude and behaviours for international Muslim tourism and development. *Curr. Issues Tour.* **2023**, *26*, 1463–1481. [CrossRef]
78. Ahmad, N.; Ullah, Z.; AlDhaen, E.; Han, H.; Scholz, M. A CSR perspective to foster employee creativity in the banking sector: The role of work engagement and psychological safety. *J. Retail. Consum. Serv.* **2022**, *67*, 102968. [CrossRef]
79. Ahmad, N.; Ullah, Z.; AlDhaen, E.; Han, H.; Ariza-Montes, A.; Vega-Muñoz, A. Fostering advocacy behavior of employees: A corporate social responsibility perspective from the hospitality sector. *Front. Psychol.* **2022**, *13*, 865021. [CrossRef]
80. Ahmad, N.; Ullah, Z.; AlDhaen, E.; Han, H.; Araya-Castillo, L.; Ariza-Montes, A. Fostering hotel-employee creativity through micro-level corporate social responsibility: A social identity theory perspective. *Front. Psychol.* **2022**, *13*, 853125. [CrossRef]
81. Ahmad, N.; Scholz, M.; Arshad, M.Z.; Jafri, S.K.A.; Sabir, R.I.; Khan, W.A.; Han, H. The inter-relation of corporate social responsibility at employee level, servant leadership, and innovative work behavior in the time of crisis from the healthcare sector of pakistan. *Int. J. Environ. Res. Public Health* **2021**, *18*, 4608. [CrossRef]
82. Molnár, E.; Mahmood, A.; Ahmad, N.; Ikram, A.; Murtaza, S.A. The interplay between corporate social responsibility at employee level, ethical leadership, quality of work life and employee pro-environmental behavior: The case of healthcare organizations. *Int. J. Environ. Res. Public Health* **2021**, *18*, 4521. [CrossRef] [PubMed]
83. Ahmad, N.; Han, H.; Kim, M. Elevated emotions, elevated ideas: The CSR-employee creativity nexus in hospitality. *J. Serv. Theory Pract.* **2024**, *34*, 891–914. [CrossRef]
84. Ahmad, N.; Ahmad, A.; Lewandowska, A.; Han, H. From screen to service: How corporate social responsibility messages on social media shape hotel consumer advocacy. *J. Hosp. Mark. Manag.* **2024**, *33*, 384–413. [CrossRef]
85. Ahmad, N.; Samad, S.; Han, H. Charting new terrains: How CSR initiatives shape employee creativity and contribute to UN-SDGs in a knowledge-driven world. *J. Innov. Knowl.* **2024**, *9*, 100557. [CrossRef]
86. Hair, J.F.; Ringle, C.M.; Sarstedt, M. PLS-SEM: Indeed a silver bullet. *J. Mark. Theory Pract.* **2011**, *19*, 139–152. [CrossRef]
87. Ahmad, N.; Samad, S.; Han, H. Travel and Tourism Marketing in the age of the conscious tourists: A study on CSR and tourist brand advocacy. *J. Travel Tour. Mark.* **2023**, *40*, 551–567. [CrossRef]
88. Fornell, C.; Larcker, D.F. Evaluating structural equation models with unobservable variables and measurement error. *J. Mark. Res.* **1981**, *18*, 39–50. [CrossRef]
89. Zhou, X.; Ahmad, N.; Lho, L.H.; Han, H. Social ripple: Unraveling the impact of customer relationship management via social media on consumer emotions and behavior. *Soc. Behav. Personal. Int. J.* **2023**, *51*, 1–12. [CrossRef]
90. Fu, B.; Ahmad, N.; Lho, L.H.; Han, H. Triple-E effect: Corporate ethical responsibility, ethical values, and employee emotions in the healthcare sector. *Soc. Behav. Personal. Int. J.* **2023**, *51*, 1–14. [CrossRef]
91. Hofstede, G. Dimensionalizing cultures: The Hofstede model in context. *Online Read. Psychol. Cult.* **2011**, *2*, 8. [CrossRef]
92. Joshi, P.; Visvanathan, C. Sustainable management practices of food waste in Asia: Technological and policy drivers. *J. Environ. Manag.* **2019**, *247*, 538–550. [CrossRef] [PubMed]

*sustainability*

MDPI

*Article*

# Factors Associated with Food Waste Among University Students in Colombia

**Edna Magaly Gamboa-Delgado [1,*], Oscar F. Herrán [1] and Doris Cristina Quintero-Lesmes [2]**

[1] Escuela de Nutrición y Dietética, Universidad Industrial de Santander (UIS), Bucaramanga 680001, Colombia; herran@uis.edu.co

[2] Centro de Investigaciones, Fundación Cardiovascular de Colombia, Floridablanca 681004, Colombia; dorisquintero@fcv.org

* Correspondence: emgamboa@uis.edu.co

**Abstract:** (1) Background: Globally, millions of tons of food are wasted annually, with a significant portion occurring at the household level. This study aimed to quantify food waste generated by university students and identify key factors associated with this behavior. (2) Methods: A cross-sectional study was conducted in a community trial involving 227 Colombian university students. Participants self-reported their daily food waste (grams per person per day), as categorized by food type, which was collected through an online-based questionnaire under prior training. The data were analyzed using a binomial regression model to calculate adjusted prevalence ratios (PRaj) and identify associations between food waste and demographic variables. (3) Results: Of the participants, 63% were women, with a mean age of 20.4 ± 3.8 years. Overall, 65% (n = 148) reported food wastage. The average food waste per person per day was 22.6 ± 15.5 g, increasing to 94.0 ± 5.0 g among those who reported wastage. Significant associations were found between food waste and several factors: Among those with a higher prevalence of food waste were women (PRaj = 1.49; 95% CI: 1.19, 1.89); students from rural areas (PRaj = 1.39; 95% CI: 1.15, 1.67); and those with higher income levels, with adjusted prevalence ratios of 1.38 (95% CI: 1.14, 1.68) for those earning 1–2 Minimum Legal Monthly Wages (LMMW) and 1.49 (95% CI: 1.15, 1.93) for those earning 3+ LMMW. Household size and socioeconomic status were not significantly associated with food waste ($p > 0.05$). (4) Conclusions: On average, the population generates 8.25 kg of food waste per person per year, rising to 34.31 kg among those who report wasting food. Gender, geographic location, and income were key predictors of daily food waste.

**Keywords:** food loss and waste; students; universities; Colombia

## 1. Introduction

Food loss and waste (FWL) refers to a phenomenon that is a product of unsustainable food systems. Specifically, it is the surplus food mass discarded at different stages of the agri-food chain. Thus, FWL is considered a global problem of widespread magnitude and frequency that affects sustainability and food security. This phenomenon goes against Sustainable Development Goals such as Zero Hunger and Responsible Production and Consumption [1]. Additionally, it contrasts with the prevalence of moderate or severe food insecurity of 29.3% in the world and hunger figures that by 2021 reached 828 million people [2].

It is essential to distinguish between food losses and waste. Food losses occur in the production stages up to the retailer, excluding surpluses, and include the production or harvest, post-harvest, storage, and industrial processing phases. On the other hand, waste is generated in later stages, from the retailer to consumption (distribution, retail, and consumption phases), whether domestic or bulk, from public or private gastronomic services. This waste is often a result of behavior, purchasing and consumption habits, and the handling of food [3].

At least one-third of the food produced in the world, equivalent to about 1.3 billion tons destined for human consumption, is lost or wasted annually. In Latin America, per capita food losses are approximately 200 to 225 kg/year, with the production stage being the main scenario due to the lack of the necessary capacity for processing and preserving products that meet each country's food demand [3].

Colombia supplies 28.5 million tons of food, of which 34% is lost or wasted, equivalent to 9.76 million tons [4], enough to feed 8 million people annually. Of the total food lost and wasted, 64% of the losses occur in the production, post-harvest, storage, and industrial processing stages. The remaining 36% is generated in the distribution and household consumption stages [4]. In contrast, it has been estimated that about 40% of waste is produced during agricultural production, and 16% is due to poor management within the household [4]. In the case of Colombia, FWLs represent a monetary value of 15.4 trillion pesos, equivalent to 2% of the Gross Domestic Product (GDP), 30% of the agricultural value, and five times the investment budget of the Colombian Ministry of Agriculture [5].

The scenario of multiple determinants of food wastage is complemented by people's consumption habits and the low nutritional education that generates the purchase of larger quantities of food than necessary, lack of menu planning, extra-large portion sizes, advertising, and food marketing that incites these unnecessary and compulsive purchases that increase the probability of food expiration [5]. FWL generates multiple social impacts, as food systems are disrupted to the point of affecting the global supply of safe, accessible, healthy diets. Furthermore, FWLs contribute to deepening the problem of contemporary food conflict, which consists of the coexistence of overweight, hunger, and inequalities in global food distribution [6]. FWLs also have a negative economic impact, as they affect food systems, resulting in higher food prices and creating difficulties for equitable access to food. All of the above favor food insecurity and poverty while increasing water, land, and carbon footprints [7].

Multiple potential solutions have been attempted to address the problem of FWL, including public policies and actions at the stages of food production, processing, and distribution aimed at impacting consumption, e.g., food guidelines and good manufacturing practices (GMP), food donations to food bank networks, and strengthening the circular economy to promote changes in sustainable production and consumption patterns [8].

Various strategies have been implemented globally to address food waste, aiming to positively impact the environment, the economy, and food security. Notable initiatives include efforts at both governmental and international organizational levels, as well as actions taken by businesses and consumers.

In national policies, many countries have developed action plans against food waste, such as laws prohibiting supermarkets from discarding food in good condition and mandating that they donate it to charitable organizations. There are also laws promoting food waste reduction and encouraging public education and collaboration among different sectors as well as campaigns focused on changing shopping and consumption habits by promoting good practices such as meal planning, utilizing leftovers, and composting organic waste.

At the corporate level, some companies have adopted the methodology of measuring and reporting their food waste levels and implementing technological solutions to optimize the supply chain and reduce surplus food products. Technological platforms are also emerging to connect consumers with businesses to utilize food that would otherwise be discarded, such as through various apps [9].

In Colombia, the legislation that regulates the measures for the prevention of and reduction in food waste is Law 1990 of 2019 [10]. This law aims to establish concrete strategies and actions to reduce food losses and waste throughout the production, distribution, and consumption chain to improve food security and contribute to sustainability. Law 1990 sets measures for reducing food waste at the stages of production, distribution, and consumption; promotes the donation of food fit for human consumption that, for commercial reasons, cannot be sold; and encourages the implementation of educational campaigns and public awareness to promote responsible consumption and reduce household food losses.

This national legislation is complemented by the 2030 Agenda for Sustainable Development, which sets several key objectives related to food waste reduction, such as Sustainable Development Goal (SDG) 12, which refers to ensuring sustainable consumption and production patterns. The target most directly related to food waste (target 12.3) is to halve the per capita global food waste at the retail and consumer levels by 2030.

On the other hand, measuring household food waste is challenging due to the diversity of methods available and variations in consumer behaviors. Several methodologies measure household food wastage, including wastage diaries, direct food weighing, visual plate estimation, image-based tools, waste composition studies, and questionnaires/surveys/forms [11].

Food waste affects all levels of society, including university students. Food waste is a global issue with significant environmental, social, and economic implications, and studying it within the context of university students is particularly relevant due to the characteristics of their lifestyle. University students often live under budgetary constraints and in shared or temporary accommodations, and they frequently manage irregular schedules due to academic and work commitments. Thus, in a study conducted in New Zealand among university students, it was found that the most important factors leading to food waste were the lack of meal and shopping planning as well as the improper separation of food waste from other types of waste, which resulted in the disposal of a considerable amount of avoidable food waste, such as fresh food and leftovers [12]. Meanwhile, a study conducted among university students in Italy and Poland investigated the influence of living conditions and dietary practices on students' perception of food waste. The findings indicated that 39% of respondents in Italy and 31% in Poland were highly aware of food waste. It was also identified that students belonging to Generation Z (ages 18 to 25) are more likely to be influenced by environmental campaigns aimed at reducing food waste compared to Millennials (ages 26 to 35). This highlights the need for educational initiatives targeting this population regarding the importance of sustainable consumption and strengthening university curricula to enhance students' environmental commitment [13]. Measuring food waste among university students is relevant to developing effective awareness, education, and reduction strategies at the school or university level. Thus, this study aimed to estimate food waste derived from consumption by university students and determine its associated factors. The scope of this study is to provide a comprehensive estimation of food waste generated by university students and to explore the factors that contribute to this behavior. By examining key variables such as gender, geographic location, and socioeconomic status, the study aims to shed light on the underlying causes of food waste within this demographic. The findings offer insights into the potential strategies for reducing food waste in academic settings, thereby informing policy development and educational interventions to promote more sustainable consumption habits among university students.

## 2. Material and Methods

A cross-sectional analytical quantitative study was conducted in Bucaramanga, Colombia, during a community trial. The target population of this study was university students. By convenience, two hundred and twenty-seven undergraduate students from a state university were selected. The sample size was calculated in the Open Epi program, considering the following criteria: population size (n = 18,264), significance level of 95%, power (1-beta, % probability of detection) of 80%, hypothetical prevalence of food waste of 18%, risk/prevalence difference of 10, and design effect of 1.

The dependent variable of this study was the total wastage/person/day (g). Other variables were (1) biological and socioeconomic variables: the characteristics of the respondent and his/her household, including sex, age, socioeconomic level, marital status, weight and height, geographic location of the household (urban/rural), career, university to which he/she belongs, and household income; (2) shopping habits and attitudes: who makes the food purchases, the criteria for making and the frequency of food purchases, and money spent per week on food; (3) food habits and attitudes: the frequency of meals at home and outside the home and the preparation of meals; and (4) food waste: the types of food most discarded, the causes of food waste, and the habits and customs that encourage it.

*Data collection*

All the information was collected through an online-based questionnaire, under prior training, to evaluate all the variables of interest. For the questions related to food wastage, the instrument designed by the FAO was used as a basis to measure food wastage in households [11].

The amount of food wasted and the other variables were measured through self-reporting. Senior students of Nutrition and Dietetics explained to the participants, before filling out the survey form, the concepts of food losses and food waste; emphasized the concept of waste, which is the object of this study; and explained in detail how they should fill out the form in terms of food waste for each of the meal times and in terms of the amounts and reasons to report. A pilot test was conducted with 20 university students to validate the level of understanding of each question and the correct completion of the questions related to the amount of food wasted.

The questionnaire used was structured and took approximately 30 min to complete. It aimed to inquire about the levels of food waste produced in the domestic environment in terms of the volume (g), type of food wasted, reasons why food is wasted, consumer behaviors, and attitudes towards the correct use of food. For further details regarding the structure and questions of the surveys, please contact the corresponding author.

*Data analysis*

If a student self-reported wasting at least one meal time, it was considered food wasting. The prevalence of students who wasted and did not waste food for each category of the variables of interest was contrasted to establish differences between them with chi$^2$ tests. Using binomial regression models with the prevalence of wasting as the dependent variable (0/1), we calculated the adjusted prevalence ratios (PRaj) by sex, socioeconomic stratum, geographic area, number of people in the household, and income in terms of legal monthly minimum wages (LMMW). In addition to the estimator for the adjusted variable in the model (PRaj), the 95% confidence interval (95% CI) and the *p*-value (*p*-value) were reported. The grams/person/day of waste estimated for each subject as the dependent variable was transformed to approximate a normal distribution as (grams/person/day)$^{1/3.3}$. The mean and standard error (S.E.) described the back-transformed values. In addition, based on Simple Linear Regression or ANOVA models, the *p*-value was estimated for each variable of interest. Finally, Multiple Linear Regression models were performed with grams/person/day as the dependent variable to estimate the adjusted differences using adjusted *B* coefficients (βaj) in each category of the variables of interest. The absolute adjusted difference was reported with its 95% CI and *p*-value. All data were analyzed using the software Stata v. 15.

*Ethical approval*

The Scientific Research Ethics Committee of the Universidad Industrial de Santander approved this study (Project Code FS202201). All participants signed an informed consent form.

## 3. Results

Sixty-three percent of the participants were women. The mean age was 20.5 $\pm$ 3.5 years (mean $\pm$ S.E.). Food wastage was reported by 65% (n = 148). The wastage/person/day (g) in the total population analyzed was 22.6 $\pm$ 15.5; among students reporting wastage, it was 94.0 $\pm$ 5.0.

The highest frequency of food wastage occurred during lunch (49%), followed by breakfast (33%). The food consumed was purchased in 71% of the cases by the parents and 20% by the participants. Quality (33%), food price (30%), and consumption habits (30%) were the aspects that determined the purchase of food. The frequency of weekly purchases predominated (44%), followed by biweekly (28%). The parents prepared the food in 66% of cases and the participants in 22%. Cereals/tubers/roots were the food group with the highest self-reported wastage (15%), followed by the meat/legumes/egg group (6%). The lowest self-reported wastage was for the milk/dairy group (1%). Among students who wasted food, the main reason for wastage was because the food or preparation was over-purchased or over-prepared (43%) or because the food or preparation deteriorated due to poor preservation or storage (14%).

Table 1 shows the proportion of wastage by type of food. Sex, income, and food preparation frequency determined whether or not a student reported food wasting; *p*-value < 0.05 (Table 2). Table 3 shows the P.R.s for the biological and socioeconomic variables: Being male, living in urban areas, and low socioeconomic income (<1 LMMW) are protective against food wastage. Table 4 shows the grams/person/day of wastage for each category of the variables of interest and the *p*-values for the raw differences. Table 5 presents the adjusted differences for grams/person/day for each of the categories of the variables of interest, sex, income in LMMW units, and frequency of purchase, which remain as the determinant variables of wastage: women, PRaj = 1.49 (95% CI: 1.19, 1.89); students residing in rural areas, PRaj = 1.39 (95% CI: 1.15, 1.67); and students with higher economic income, PRaj = 1.38 (95% CI: 1.14, 1.68) and PRaj = 1.49 (95% CI: 1.15, 1.93) (1–2 Minimum Legal Monthly Wages (LMMW) and 3+ LMMW versus < 1 LMMW, respectively), presented higher risk of wastage/person/day. Household size and socioeconomic level were not associated; $p < 0.05$.

**Table 1.** Number and percentage (%) of students who generated food waste by type of food (n = 227).

| Type of Food | Number (%) |
|---|---|
| Total | 148 (65) |
| Breakfast | 75 (33) |
| Mid-morning | 37 (16) |
| Lunch | 112 (49) |
| Mid-afternoon | 24 (11) |
| Dinner | 63 (28) |

**Table 2.** Number and percentage (%) of students who generated food waste according to biological and socioeconomic variables.

| Variable | Total (%) | Freq [mean], (%) | Yes (%) | *p*-Value |
|---|---|---|---|---|
| Waste—grams * | 227 (100) | 79 [0.0] | 148 [94.0] | <0.0001 |
| Age—year | 227 (100) | 79 [20.3] | 148 [20.5] | 0.751 |
| BMI—kg/m$^2$ | 204 (100) | 70 [22.7] | 134 [22.5] | 0.663 |
| Sex | | | | |
| Female | 143 (63) | 36 (46) | 107 (72) | <0.0001 |
| Male | 84 (37) | 43 (54) | 41 (28) | |
| Socioeconomic status | | | | |
| Very low and low | 148 (65) | 57 (72) | 91 (61) | 0.108 |
| Medium and high | 79 (35) | 22 (28) | 57 (39) | |
| Area | | | | |
| Rural | 21 (9) | 6 (8) | 15 (10) | 0.529 |
| Urban | 206 (91) | 73 (92) | 133 (90) | |

**Table 2.** *Cont.*

| Variable | Total (%) | Freq [mean], (%) | Yes (%) | *p*-Value |
|---|---|---|---|---|
| LMMW ** | | | | |
| <1 | 95 (42) | 42 (53) | 53 (36) | 0.039 |
| 1–2 | 97 (43) | 28 (36) | 69 (47) | |
| ≥3 | 35 (15) | 9 (11) | 26 (17) | |
| Frequency of food purchase | | | | |
| Weekly | 126 (56) | 49 (62) | 77 (52) | 0.059 |
| Biweekly | 62 (27) | 14 (18) | 48 (32) | |
| Monthly | 39 (17) | 16 (20) | 23 (16) | |
| Weekly food expenditure | | | | |
| <50 (12.39/11.38) *** | 26 (11) | 9 (11) | 17 (11) | 0.357 |
| 50 to 100 (24.77/22.75) | 93 (41) | 34 (43) | 59 (40) | |
| >101 (25.02/22.98) | 88 (39) | 26 (33) | 62 (42) | |
| I do not know | 20 (9) | 10 (13) | 10 (7) | |
| Do you prepare food? | | | | |
| Never | 7 (3) | 7 (9) | 0 | 0.002 |
| Rarely | 45 (20) | 19 (24) | 26 (18) | |
| Sometimes | 112 (49) | 35 (44) | 77 (52) | |
| Usually | 37 (16) | 12 (15) | 25 (17) | |
| Always | 26 (12) | 6 (8) | 20 (13) | |
| Household members | | | | |
| <5 | 163 (72) | 54 (68) | 109 (74) | 0.398 |
| ≥5 | 64 (28) | 25 (32) | 39 (26) | |

* Based on transformed values $(g)^{1/3.3}$. ** LMMW: Legal Monthly Minimum Wages. *** COP(USD/EUR): thousands of Colombian pesos/U.S. dollars/euros.

**Table 3.** Adjusted prevalence ratios (PRaj) for food waste generation according to biological and socioeconomic variables.

| Variable | PRaj * | 95% I.C. | *p*-Value |
|---|---|---|---|
| Sex | | | |
| Female | Ref | | <0.0001 |
| Male | 0.67 | 0.53 to 0.84 | |
| Socioeconomic status | | | 0.768 |
| Very low and low | Ref | - | |
| Medium and high | 1.03 | 0.85 to 1.24 | |
| Area | | | |
| Rural | Ref | - | 0.001 |
| Urban | 0.72 | 0.60 to 0.87 | |
| LMMW ** | | | |
| <1 | Ref | - | |
| 1–2 | 1.38 | 1.14 to 1.68 | 0.001 |
| ≥3 | 1.49 | 1.15 to 1.93 | 0.002 |
| Household members | | | |
| <5 | Ref | - | 0.829 |
| ≥5 | 0.98 | 0.82 to 1.17 | |

* PRaj: achieved in a multiple binomial regression model where the dependent variable is food waste generation (0/1). 95% ** LMMW: Legal Monthly Minimum Wages. 95% I.C.: confidence intervals for the PRaj.

**Table 4.** Food waste generation according to biological and socioeconomic variables. Mean (g) and Standard Error by variable category.

| Variable | Total (%) | Freq [mean] * | S.E. | p-Value |
|---|---|---|---|---|
| Waste—grams | 227 | 148 [94.0] | 0.00 | |
| Age—year | 227 | 148 [20.5] | 0.35 | 0.751 |
| BMI—kg/m$^2$ | 204 | 134 [22.5] | 0.27 | 0.663 |
| Sex | | | | |
| Female | 143 (63) | 107 [83.3] | 0.00 | 0.145 |
| Male | 84 (37) | 41 [119.4] | 0.01 | |
| Socioeconomic status | | | | |
| Very low and low | 148 (65) | 91 [112.2] | 0.00 | 0.028 |
| Medium and high | 79 (35) | 57 [67.3] | 0.00 | |
| Area | | | | |
| Rural | 21 (9) | 15 [96.2] | 0.06 | 0.921 |
| Urban | 206 (91) | 133 [92.3] | 0.00 | |
| LMMW ** | | | | |
| <1 | 95 (42) | 53 [115.8] | 0.01 | 0.281 |
| 1–2 | 97 (43 | 69 [84.8] | 0.01 | |
| ≥3 | 35 (15) | 26 [72.4] | 0.01 | |
| Frequency of food purchase | | | | |
| Weekly | 126 (56) | 77 [97.0] | 0.01 | 0.900 |
| Biweekly | 62 (27) | 48 [86.2] | 0.01 | |
| Monthly | 39 (17) | 23 [92.3] | 0.01 | |
| Weekly food expenditure | | | | |
| <50 (12.39/11.38) *** | 26 (11) | 17 [146.3] | 0.06 | 0.198 |
| 50 to 100 (24.77/22.75) | 93 (41) | 59 [104.4] | 0.01 | |
| >101 (25.02/22.98) | 88 (39) | 62 [71.7] | 0.00 | |
| I do not know | 20 (9) | 10 [91.5] | 0.19 | |
| Do you prepare food? | | | | |
| Never | 7 (3) | - | - | 0.026 |
| Rarely | 45 (20) | 26 [52.5] | 0.01 | |
| Sometimes | 112 (49) | 77 [124.1] | 0.01 | |
| Usually | 37 (16) | 25 [67.9] | 0.01 | |
| Always | 26 (12) | 20 [81.9] | 0.03 | |
| Household members | | | | |
| <5 | 163 (72) | 109 [93.8] | 0.00 | 0.819 |
| ≥5 | 64 (28) | 39 [88.5] | 0.01 | |

* Based on transformed values (g)$^{1/3.3}$. S.E.: Standard Error. ** LMMW: Legal Monthly Minimum Wages. *** COP(USD/EUR): thousands of Colombian pesos/U.S. dollars/euros.

**Table 5.** Partially adjusted regression coefficients (βaj)—adjusted differences—for food waste generation according to biological and socioeconomic variables.

| Variable | βaj * | 95% I.C. | p-Value |
|---|---|---|---|
| Sex | | | |
| Female | Ref | - | 0.038 |
| Male | −0.66 | −1.29 to −0.04 | |
| Socioeconomic status | | | |
| Very low and low | Ref | - | 0.671 |
| Medium and high | −0.16 | −0.90 to 0.58 | |

**Table 5.** *Cont.*

| Variable | βaj * | 95% I.C. | *p*-Value |
|---|---|---|---|
| Area | | | |
| Rural | Ref | - | 0.216 |
| Urban | −0.69 | −1.78 to 0.41 | |
| LMMW ** | | | 0.048 |
| <1 | Ref | - | - |
| 1–2 | 0.62 | −0.09 to 1.34 | 0.087 |
| ≥3 | 0.79 | −0.32 to 1.90 | 0.162 |
| Frequency of food purchase | | | 0.489 |
| Weekly | Ref | - | - |
| Biweekly | 0.46 | −0.23 to 1.16 | 0.188 |
| Monthly | −0.07 | −0.90 to 0.75 | 0.864 |
| Weekly food expenditure | | | 0.131 |
| <50 (12.39/11.38) *** | Ref | - | - |
| 50 to 100 (24.77/22.75) | −0.09 | −1.09 to 0.92 | 0.865 |
| >101 (25.02/22.98) | −0.41 | −1.49 to 0.68 | 0.463 |
| I do not know | −0.84 | −2.21 to 0.53 | 0.229 |
| Do you prepare food? | | | 0.041 |
| Never | Ref | - | - |
| Rarely | 1.54 | −0.30 to 3.36 | 0.095 |
| Sometimes | 2.63 | 0.90 to 4.36 | 0.003 |
| Usually | 1.97 | 0.12 to 3.82 | 0.037 |
| Always | 2.50 | 0.56 to 4.38 | 0.012 |
| Household members | | | 0.338 |
| <5 | Ref | - | |
| ≥5 | −0.29 | −0.96 to 0.37 | |

* βaj: Partial regression coefficient is achieved in a multiple linear regression model where the dependent variable is grams of food waste, and the covariates are all those reported in the table. 95% C.I.: confidence interval for βaj, based on transformed values $(g)^{1/3.3}$. ** LMMW: Legal Monthly Minimum Wages. *** COP(USD/EUR): Thousands of Colombian pesos/U.S. dollars/euros. $R^2$ for the model (%): 7.15.

## 4. Discussion

In the present study, food waste behavior was evaluated in university students. It was observed that a significant proportion of the participants reported food waste, with a higher frequency during lunch and breakfast. Most of the food was purchased by parents and prepared at home. The main reasons for waste included food quality considerations and leftovers because excess food was prepared or purchased. The factors associated with food waste included gender, income level, and geographic area of residence, with a higher risk of waste observed in females, students in rural areas, and those with higher incomes. Variables such as household size and socioeconomic status showed no significant association with food waste.

In this study, the amount of food waste per person/day was 22.6 g (S.E. ± 15.5), placing it within the range reported by similar studies [14]. However, the prevalence of waste among university students shows significant variations in the literature, reflecting the influence of different methodologies and socioeconomic contexts on the results obtained. A study conducted in India [15] found that university students waste an average of 31.4% of food. On the other hand, in Valencia, Spain, a study reported a total waste of 14.5%, with higher waste during lunch (18.75%) than at dinner (10.5%), with a statistically significant difference ($p < 0.01$) [16].

These differences in the prevalence of food waste, the different methodologies used to quantify it, and socioeconomic contexts can also be attributed to food supply and quality. Cultural norms and habits related to eating often include a lack of meal planning that can lead to over-purchasing or over-preparation of food at home. At the same time, university dining halls usually serve large portions that are not always fully consumed [14]. Despite these variations, studies agree that food waste in college students is a significant problem that needs to be addressed.

### *4.1. Factors Associated with Food Waste Among University Students*

Studies analyzing the factors associated with food waste in university students have found diverse results. In both the general population [8,17] and the student population [15,16], sociodemographic factors such as age, sex, income, educational level, and household composition have been identified as crucial elements that positively and negatively influence the amount of food wasted.

Our study suggests that sex impacts the amount of food discarded in the household. However, more research is needed to understand these possible differences better since the published results are very diverse. For example, Morata et al. found a higher prevalence of waste in women (15.5%) than in men (11.5%) [16], as have other studies indicating that women waste more food than men [18,19]. This is not in line with Soo-Cheng Chuah et al., who found that, in response to the question "*How much food do you waste in a day?*", females (73.2%) waste less food than males (79.3%) [20]. In addition, that study found that waste disposal behavior differed only between male and female students. At the same time, no differences were observed in the constructs of food waste knowledge and participation in food waste prevention. At the same time, that study revealed no significant association between gender and food planning, food recycling, and waste behaviors among students who were also asked [20]. In 2015, Secondi et al. assessed household food waste behavior in the 27 countries of the European Community (EU-27) using a multilevel analysis that allowed them to jointly consider individual-level and contextual factors as potential variables associated with food waste. While it was not explicitly investigated in the university population, the study did not find an association between food waste and gender [21].

Thus, differences in gender norms and social roles may influence eating behaviors and food waste such that, in specific contexts, women may be more likely to plan their meals and take smaller portions [22]. In Poland, men were found to be more aware of food waste than women, while in Italy, the university curriculum played a significant role in influencing students [13].

Regarding economic income, one study found a correlation between household income and food waste, where households with higher incomes waste more food than those with lower incomes [23]. Although that study did not focus on university students, its results are similar. This relationship could be because students with higher incomes may have more freedom to buy food without regard to budget, which may lead to less purchase planning and more significant purchases of food that will not be consumed, and therefore, more waste is generated. In addition, they may have less time to cook and prepare food, which may lead to greater reliance on processed or prepared foods, which are more likely to be wasted [22].

This study found that students residing in rural areas waste more food than those in urban areas. Although the existing literature on this factor does not focus specifically on the university population, a general population study that examined the influence of sociodemographic, behavioral, and attitudinal factors on the amount of food waste found no clear correlation with area, form, and type of residence [18].

### *4.2. Study Strengths and Limitations*

This study has several notable strengths. First, it is a detailed analysis of food waste practices among university students, a demographic group under-researched compared to the general population. The main findings of this study are relevant to highlight the situation of food waste among university students in Colombia, which serves as a basis for the design and implementation of future educational interventions. The internal validity of the data is supported by the rigorous training process given to the students in terms of filling out the self-reporting format of the amounts of food wasted to minimize information biases.

Including various sociodemographic and economic variables provided a comprehensive understanding of the factors influencing food waste.

However, the study also has some limitations. One of them is the possible lack of external validity given that the sample analyzed represents university students from a public university in Colombia, which may not adequately reflect the diversity of the university population in general. In addition, quantifying wasted food represents a great challenge, mainly considering the self-reported, waste diary-type methodology used, in which students recorded all the food discarded during the day. This technique could be biased by the participants' behavior, who could modify their habits when they know that they are being observed, and therefore, there could be an underestimation of the actual waste or an information bias (memory bias due to self-reporting). In addition, the complexity of measuring food waste also lies in the fact that discarded food can be in different states (raw, cooked, or partially consumed), which complicates its quantification and categorization since waste can include both consumable and non-consumable food (peels and bones).

### 4.3. Implications of This Study and Future Implications

This study has important implications in both the short and long term. In the short term, the findings can inform universities and educational institution authorities about implementing programs and policies to reduce food waste among students, such as education on purchase planning, meal planning, and proper food storage. In the long term, these results can serve as a basis for future research to explore more effective interventions and comprehensive approaches to minimize food waste in the university setting and other similar contexts. These strategies contribute to achieving food sustainability.

## 5. Conclusions

This study identifies several essential factors associated with food waste among university students. First, it was observed that gender significantly influences waste levels, with a higher tendency among women. Additionally, economic income level is related to greater waste generation; students with higher incomes tend to waste more food, possibly due to a lack of planning in their purchases. Moreover, geographic location also proves to be an important factor, as students in rural areas report higher levels of waste compared to those in urban areas. However, no significant associations were found with household size and socioeconomic status.

Regarding the possibilities for reducing food waste in the future, it is crucial to implement educational programs focused on meal planning and responsible purchasing. Raising awareness about food quality and proper storage practices is essential for mitigating waste. Initiatives such as workshops and informative campaigns can promote sustainable practices and highlight the importance of conserving food resources.

The impact of food waste extends beyond economic concerns, significantly affecting the environment and social well-being. This waste contributes to greenhouse gas emissions and the depletion of natural resources, making its reduction a key step toward environmental sustainability. Addressing food waste can help mitigate the effects of climate change and conserve vital resources like water and land.

In higher education institutions, various initiatives and campaigns have begun to be implemented to reduce food waste. These include the establishment of university dining halls that offer appropriate portion sizes and awareness programs about the issue of waste. However, continuous efforts are necessary to foster a culture of waste reduction among students.

Finally, it is recommended that students adopt specific practices to minimize food waste. This includes meal planning, creating shopping lists, and being more conscious of portion sizes. Furthermore, education on food preservation and proper storage can significantly contribute to waste reduction, thus promoting more sustainable consumption behavior.

**Author Contributions:** E.M.G.-D., conceptualization, investigation, methodology, validation, formal analysis; and writing—original draft; O.F.H., methodology, formal analysis, and writing—original draft; D.C.Q.-L., validation, writing—original draft, and review and editing. All authors have read and agreed to the published version of the manuscript.

**Funding:** The Vice Rector's Office for Research and Extension of the Universidad Industrial de Santander, Project code FS202201, funded this study.

**Institutional Review Board Statement:** The study was conducted under the Declaration of Helsinki and approved by the institutional ethics review board Universidad Industrial de Santander, according to Act No. 05 of 2022.

**Informed Consent Statement:** Written informed consent was obtained directly from all participants.

**Data Availability Statement:** The article authors can request the datasets generated and analyzed for this study since they must request that the research ethics committee share the database in an institutional repository.

**Conflicts of Interest:** The authors declare no conflicts of interest.

## References

1. FAO. Organización de las Naciones Unidas Para la Alimentación y la Agricultura. Objetivos del Desarrollo Sostenible. 2015. Available online: https://www.un.org./sustainabledevelopment/es/objetivos-de-desarrollo-sostenible/ (accessed on 1 July 2024).
2. FAO; IFAD; UNICEF; WFP; WHO. *The State of Food Security and Nutrition in the World 2022*; Repurposing Food and Agricultural Policies to Make Healthy Diets More Affordable; FAO: Rome, Italy, 2022. Available online: https://openknowledge.fao.org/items/c0239a36-7f34-4170-87f7-2fcc179ef064 (accessed on 1 July 2024).
3. FAO. La Pérdida y el Desperdicio de Alimentos Debe Reducirse a fin de Aumentar la Seguridad Alimentaria y la Sostenibilidad del Medio Ambiente. 2020. Available online: http://www.fao.org/news/story/es/item/1310444/icode/ (accessed on 1 July 2024).
4. Departamento Nacional de Planeación, DNP. Estudio de Pérdida y Desperdicio de Alimentos en Colombia. Available online: https://colaboracion.dnp.gov.co/CDT/Sinergia/Documentos/Estudio_Perdidas_desperdicios_alimentos_Ficha.pdf (accessed on 1 July 2024).
5. FAO. Organización de las Naciones Unidas para la Alimentación y la Agricultura. Guía práctica para la incorporación de Soluciones a las Pérdidas y los Desperdicios de Alimentos. Bogotá. 2020. Available online: https://online.pubhtml5.com/xlwa/dpak/ (accessed on 1 July 2024).
6. Amorim, A.; Barbosa, A.H.; Sobral, P.J.D.A. Hunger, Obesity, Public Policies, and Food-Based Dietary Guidelines: A Reflection Considering the Socio-Environmental World Context. *Front. Nutr.* **2022**, *8*, 805569. [CrossRef] [PubMed]
7. Conrad, Z.; Niles, M.T.; Neher, D.A.; Roy, E.D.; Tichenor, N.E.; Jahns, L. Relationship between food waste, diet quality, and environmental sustainability. *PLoS ONE* **2018**, *13*, e0195405. [CrossRef] [PubMed]
8. Chia, D.; Yap, C.C.; Wu, S.L.; Berezina, E.; Aroua, M.K.; Gew, L.T. A systematic review of country-specific drivers and barriers to household food waste reduction and prevention. *Waste Manag. Res.* **2024**, *42*, 459–475. [CrossRef] [PubMed]
9. Urugo, M.M.; Teka, T.A.; Gemede, H.F.; Mersha, S.; Tessema, A.; Woldemariam, H.W.; Admassu, H. A comprehensive review of current approaches on food waste reduction strategies. *Compr. Rev. Food Sci. Food Saf.* **2024**, *23*, e70011. [CrossRef] [PubMed]
10. Colombia. Ley 1990 de 2019. *Por Medio de la Cual se Dictan Disposiciones Para la Reducción del Desperdicio de Alimentos en Colombia*; Ministerio de Agricultura y Desarrollo Rural: Bogotá, Colombia, 2019. Available online: https://www.andi.com.co/Uploads/Ley-2019-N0001990_20190802.pdf (accessed on 1 July 2024).
11. FAO. Informe Final. TCP/ARG/3501. Diseño Metodológico Para la Estimación del Desperdicio de Alimentos en la Argentina en las Etapas de Distribución y Comercio Minorista y Consumo en el Hogar. Pérdidas y Desperdicios Cero, Hambre Cero. Buenos Aires. 2017. Available online: https://openknowledge.fao.org/server/api/core/bitstreams/9bcc1594-9fe2-46f5-86e4-0ad025ea32a1/content (accessed on 1 July 2024).
12. Ozanne, L.K.; Ballantine, P.W.; McMaster, A. Understanding Food Waste Produced by University Students: A Social Practice Approach. *Sustainability* **2022**, *14*, 10653. [CrossRef]
13. Miśniakiewicz, M.; Amicarelli, V.; Chrobak, G.; Górka-Chowaniec, A.; Bux, C. Do Living Arrangements and Eating Habits Influence University Students' Food Waste Perception in Italy and Poland? *Sustainability* **2024**, *16*, 2102. [CrossRef]
14. Leal Filho, W.; Ribeiro, P.C.C.; Setti, A.F.F.; Azam, F.M.S.; Abubakar, I.R.; Castillo-Apraiz, J.; Tamayo, U.; Özuyar, P.G.; Frizzo, K.; Borsari, B. Toward food waste reduction at universities. *Environ. Dev. Sustain.* **2023**, *6*, 1–22. [CrossRef] [PubMed]
15. Akhter, S.; Rather, M.I.; Zargar, U.R. Understanding the food waste behavior in university students: An application of the theory of planned behavior. *J. Clean. Prod.* **2024**, *437*, 140632. [CrossRef]
16. Morata, M.P.; González-Santana, R.A.; Blesa, J.; Frígola, A.; Esteve, M.J. Estudio de los hábitos y generación de desperdicios alimentarios de jóvenes estudiantes universitarios. *Nutr. Hosp.* **2020**, *37*, 349–358. [CrossRef]

17. Nikolaus, C.J.; Nickols-Richardson, S.M.; Ellison, B. Wasted Food: A Qualitative Study of U.S. Young Adults' Perceptions, Beliefs and Behaviors. Department of Food Science and Human Nutrition. University of Illinois at Urbana-Champaign. 2018. Available online: https://www.sciencedirect.com/science/article/pii/S0195666317311364 (accessed on 1 July 2024).
18. Koivupuro, H.K.; Hartikainen, H.; Silvennoinen, K.; Katajajuuri, J.M.; Heikintalo, N.; Reinikainen, A.; Jalkanen, L. Influence of socio-demographical, behavioural and attitudinal factors on the amount of avoidable food waste generated in Finnish households. *Int. J. Consum. Stud.* **2012**, *36*, 183–191. [CrossRef]
19. Visschers, V.; Wickli, N.; Siegrist, M. Sorting out food waste behaviour: A survey on the motivators and barriers of self-reported amounts of food waste in households. *J. Environ. Psychol.* **2016**, *45*, 66–78. [CrossRef]
20. Chuah, S.C.; Keshminder, J.S. Food Waste and Disposal Behaviour among University Students. *Adv. Bus. Res. Int. J.* **2020**, *6*, 166–176. Available online: https://ir.uitm.edu.my/id/eprint/47313/1/47313.pdf (accessed on 1 July 2024). [CrossRef]
21. Secondi, L.; Principato, L.; Laureti, T. Household food waste behaviour in EU-27 countries: A multilevel analysis. *Food Policy* **2015**, *56*, 25–40. [CrossRef]
22. Clement, J.; Alenčikienė, G.; Riipi, I.; Starkutė, U.; Čepytė, K.; Buraitytė, A.; Zabulionė, A.; Šalaševičienė, A. Exploring Causes and Potential Solutions for Food Waste among Young Consumers. *Foods* **2023**, *12*, 2570. [CrossRef] [PubMed]
23. Klugman, I. The Correlation Between Household Income and Food Wastage: A Case Study of Food Wastage Quantity and Consumer Attitudes. *Food-Fueled* **2024**, *1*, e00011. [CrossRef]

sustainability

MDPI

*Article*

# From Posts to Action: Leveraging Social Media to Inspire Food Waste Reduction in Hospitality for a Net Zero Future

**Weifeng Chang [1], Mingdi Jiang [2,3,*] and Muhammad Hassan Arshad [4]**

1 School of Culture and Tourism, Kaifeng University, 157 Dongjing Street, Kaifeng 475004, China; 205910316@kfu.edu.cn
2 College of Mobile Communication, Chongqing 400015, China
3 Faculty of Business Management, UiTM Selangor, Puncak Alam Campus, Bandar Puncak Alam 42300, Selangor, Malaysia
4 School of Business Management, Universiti Utara Malaysia, Sintok 06010, Kedah, Malaysia; muhammad_hassan_a@oyagsb.uum.edu.my
* Correspondence: 2021823622@isiswa.uitm.edu.my

**Abstract:** This study explores the contribution of social media to food waste reduction among hospitality customers as part of the global net zero emissions agenda. It does this by using the Stimulus–Organism–Response (SOR) model to examine how social media food waste reduction promotion affects hospitality customer food waste intentions, mediated by social media-driven environmental motivation and sustainability-driven customer happiness. Moreover, gratitude towards food waste initiatives moderates these effects such that motivation and emotional satisfaction are more strongly associated with sustainable behaviors. We find that social media campaigns can effectively elicit cognitive and emotional engagement, which in turn strengthens food waste reduction intentions, using data from 404 hospitality customers in China. Furthermore, the study extends the SOR model to sustainability and consumer behavior literature and provides practical guidance for hospitality businesses to harness social media in engaging customers in sustainability efforts. Organizations can enhance the effectiveness of their environmental campaigns by using emotionally resonant messages and highlighting gratitude. Filling important gaps, this research explores mechanisms driving sustainable behaviors in line with strategies to achieve net zero emissions.

**Keywords:** food waste reduction; social media promotion; sustainability; hospitality industry; net zero emissions; food loss

## 1. Introduction

Climate change is a challenge faced all over the world, affecting the environment, economies, and societies. Net zero emissions are now central to international agendas and as nations seek long-term sustainability [1]. This means concerted action by governments, organizations, and communities for a carbon-neutral future and a healthy planet for future generations [2]. Food waste is one of the biggest and often overlooked contributors to climate change. The FAO (2019) estimates that globally, one-third of food produced for human consumption is wasted, the implications of which are severe from both environmental and economic viewpoints [3]. Approximately 8 to 10 percent of global emissions are water, energy, and the greenhouse gas emissions associated with the waste of these resources [4]. Food decomposition in landfills emits methane, a very powerful greenhouse gas that speeds up global warming [5]. The poor management of food waste is a further problem that exacerbates other problems, such as deforestation, biodiversity loss, and water scarcity [6]. Therefore, minimizing food waste is important for climate mitigation and reaching net zero emissions.

In this context, this study focuses on the usage of social media as a primary tool to change consumer behavior in decreasing food waste. For this study, food wastage primarily refers to food ordered by customers at restaurants and hotels that is not consumed. The reach and user engagement of social media have made it a fantastic place to raise awareness and nudge people to change their behaviors [7]. These platforms are not only used for marketing purposes but also to educate and inspire action to reduce food waste or to take collective responsibility in reaching net zero goals [8]. Social media campaigns can spread awareness, but the hard part is getting awareness to translate into sustained action if convenience or past habits are still in the way. Considering that the hospitality industry is one of the key contributors to food waste [9], this research is conducted to examine how social media food waste reduction promotion may influence customer intention to decrease food waste. Utilized as an enabler for sustainability efforts in hospitality organizations, social media is an important forum to publicize sustainability efforts, engage customers in environmental issues, steer consumer behavior, and contribute to broader net zero emissions goals [10]. This context provides an advantage for brands by allowing them to strengthen their reputation by demonstrating their dedication to sustainability, but if these efforts are viewed as superficial or 'greenwashing', then it can damage brand credibility.

This study examines psychological mechanisms that drive consumer behavior, such as social media-driven environmental motivation and sustainability-driven customer happiness as mediators. In explaining why some consumers respond positively to food waste reduction campaigns, we can explain that consumers motivated by environmental values or happiness from supporting sustainable practices are more likely to act on social media messaging [11]. Reactions to sustainability messages are influenced by personal factors such as gratitude [12]. In this regard, gratitude towards food waste initiatives is expected to moderate the relationship between social media campaigns and customers' intentions to reduce food waste. Customers who experience more gratitude for an organization's efforts are also more likely to respond more positively to communication [13]. Gratitude is a psychological factor that, if better understood, can help design more effective strategies to convince consumers to consume sustainably [14]. This study examines the Chinese hospitality sector, as China plays a key role in the world sustainability agenda due to its environmental impact and the rate of urbanization. China is one of the world's largest emitters of greenhouse gases and also one of the world's largest economies, as well as a major source of environmental degradation, which makes China a key player in the fight against climate change [15]. The country has a large population and growing affluence that has driven up food consumption, leading to a large amount of food waste, estimated at 17–18 million tons per year in urban households and hospitality venues, with a huge carbon footprint and undermining the sustainability effort of the country [16]. With respect to this issue, the Chinese hospitality industry offers a strategic opportunity to promote the reduction in food waste and support the net zero emission agenda. Food consumption patterns and total food waste levels are greatly influenced by hotels and restaurants [17]. The purpose of this research is to investigate how Chinese hospitality organizations can use social media to promote sustainable consumption behaviors and less food waste. Targeting the Chinese market offers lessons for other emerging markets in dealing with a pressing environmental issue of a key global economy and also provides a global contribution to sustainability and carbon neutrality.

Despite increasing focus on sustainability, research gaps persist regarding food waste reduction in the hospitality sector. One key gap is the limited exploration of how digital tools, such as social media, could be used to foster sustainable consumer behaviors. Extensive research has explored how social media can enhance brand awareness [18,19], with studies indicating that platforms such as Facebook and Instagram are effective tools for creating and maintaining consumer engagement [20,21]. However, there remains a limited understanding of how social media can be leveraged to drive sustainability-related behavior change [22,23], particularly in the context of food waste reduction. While efforts have been made to explore general pro-environmental behaviors influenced by digital media, such as recycling or energy conservation [24–26], the specific role of social media in addressing food waste behaviors within the hospitality sector has received scant attention. This study aims to fill this gap by examining how social media food waste reduction promotion impacts hospitality customers' intentions to reduce food waste and how this communication approach fosters sustainable consumption. Additionally, psychological and motivational factors mediating the relationship between sustainability communication and consumer behavior are underexplored in existing research. Studies by Ahmad, et al. [27] and Yuxiang, et al. [28] indicate the importance of emotional and cognitive factors in shaping sustainable actions; yet, the mechanisms through which social media-driven environmental motivation and sustainability-driven customer happiness influence pro-environmental behaviors remain insufficiently studied. Furthermore, personal attributes such as gratitude, which have been shown to significantly amplify pro-social and pro-environmental behaviors [29], are rarely examined in the context of food waste. By introducing gratitude towards food waste initiatives as a moderating variable, this study enriches the literature and demonstrates how personal traits shape consumer receptiveness to sustainability messages, advancing our understanding of how to design impactful communication strategies. In the end, although China has had a huge influence on global sustainability, there are few studies to date on how its hospitality industry can contribute to decreasing food waste and achieving net zero [30]. To fill this gap, this study investigates how hospitality organizations in China can use social media to advocate for food waste reduction, which may be transferrable to other emerging economies and contribute to the global body of sustainability research.

## 2. Literature Review and Theoretical Underpinning

Today, social media advertising is a preferred tool that brands use to manipulate consumer behavior [31]. As digital technologies evolve, so does social media, and the brand–consumer interaction has gone from one-way communication to interactive dialogue and even more engagement [32]. Brands use social media to reach different audiences, advertise their products and services, and change consumer perceptions and behavior in real-time with tailored messages. In particular, it is very useful to promote sustainability initiatives by informing users about environmental activities, educating consumers, and changing their behavior in support of sustainability goals [33]. Creating campaigns that reflect brands' values can shape attitudes and can influence consumers to behave more responsibly. In this study, we examine the influence of social media food waste reduction promotion on hospitality customer food waste intentions. Rini, et al. [34] show that sustainability messages shared through social media can induce eco-friendly behavior, such as cutting back on food waste. Furthermore, hospitality brands can also proactively promote food waste reduction strategies through engaging content that will increase consumer awareness and increase consumer intention to reduce food waste. Although a plethora of existing literature on social media marketing already exists, there is little understanding of how this type of promotion influences the food waste intentions of customers in the hospitality industry. To further underpin this, we employed the Stimulus-Organism-Response (SOR) Model [35] in order to explain this relationship. According to the SOR model, external stimuli (S) (i.e., social media promotions) influence an individual's internal state (O) (i.e., emotions and motivation) that results in a behavioral response (R) (i.e., reduced food waste intentions). Social media food waste reduction promotions act as the stimulus that leads to an internal psychological state, such as environmental motivation and happiness from being able to help support sustainable practices. Thus, the desired response is achieved, i.e., higher intentions to reduce food waste in hospitality settings. This research is best suited for the SOR model because it reflects the way in which social media stimuli affect psychological states and drive behavior [36]. External stimuli in social media food waste reduction promotion create cognitive and emotional reactions in customers, which will then lead to their behavioral intentions. The study applies this model to offer a robust framework to understand how digital stimuli can induce pro-environmental actions for enabling sustainability goals. Therefore, the following hypothesis is proposed:

**H1:** *Social media food waste reduction promotion is positively related to hospitality customer food waste intentions.*

Social media content is valuable because it can provide relevant information and engage with sustainability narratives to generate customers' environmental motivation [37]. Environmental motivation, which is the ability to motivate sustainable consumption through attitudes and values on environmental protection [38], is the main force governing sustainable consumption. The brands that effectively communicate their sustainability initiatives inform their customers about what environmentally responsible actions they are taking. For example, hospitality brands can use social media to post about food waste reduction to build positive attitudes and a stronger sense of environmental responsibility [39]. Social media acts as an external stimulus to induce internal psychological states such as motivation. Consequently, social media food waste reduction promotion is expected to trigger customers' environmental motivation and thus encourage customers to adopt sustainable practices [40]. In this study, we hypothesize that social media food waste reduction promotion will act to increase social media-driven environmental motivation. As an example, hospitality brands can attract customers' concerns about the environment by publishing interesting content such as infographics, videos, and interactive posts that discuss the environmental impacts of food waste and also promote attempts to reduce them [41]. The relationship between social media promotion and environmental motivation explains how social media affects sustainable consumption behaviors. Not only does environmental motivation affect consumers' mindsets, but it also directly impacts their sustainable behavioral intentions. According to research, people who are driven to protect the environment are more inclined to engage in eco-friendly behaviors, such as reducing waste and promoting sustainable products [42]. When dining, customers with higher environmental motivation are more likely to act on their beliefs and reduce food waste in a hospitality context. It indicates that those with stronger intentions to support food waste reduction are more motivated. The expectation in this study is that social media-driven environmental motivation has a direct positive relationship with hospitality customer food waste intentions. Potentially, food waste reduction social media campaigns are also more likely to motivate customers to have more positive intentions of reducing food waste in hospitality settings. In order to judge the effectiveness of sustainability communication to change behavior, this relationship needs to be understood [43]. In addition, environmental motivation serves to mediate the relationship between social media food waste reduction promotion and behavioral intentions. According to the SOR model, social media food waste reduction promotion serves as a stimulus, changing the internal psychological state and environmental motivation, and consequently leads to a behavioral response of food waste reduction intentions. The SOR model supports H2 and H3 by explaining how external stimuli (social media promotions) evoke internal states (environmental motivation), leading to a behavioral response. Research shows that social media can inspire pro-environmental attitudes by fostering awareness and concern for sustainability. Environmental motivation, driven by such stimuli, is a key determinant of eco-friendly behaviors such as reducing food waste. In this way, environmental motivation mediates the relationship between social media promotion and food waste intentions, and also explains the mechanism through which social media affects sustainable consumption behavior. Therefore,

**H2:** *Social media food waste reduction promotion is positively related to social media-driven environmental motivation.*

**H3:** *Social media-driven environmental motivation is positively related to hospitality customer food waste intentions.*

**H4:** *Social media-driven environmental motivation mediates the relationship between social media food waste reduction promotion and hospitality customer food waste intentions.*

How social media content is framed plays a large role in how customers perceive and react to sustainability initiatives [44]. Many organizations use social media to promote their efforts not only to inform customers but also to inspire positive emotional responses that lead to greater engagement and support [45]. Sustainability-driven customer happiness is one key emotional response to the activities supporting environmental sustainability, which is the positive emotional state obtained from supporting activities that contribute to environmental sustainability [46]. If brands communicate their sustainability initiatives correctly, consumers can feel happy for supporting eco-friendly practices [47]. Allen and Allen [48] find that this emotional response can reinforce consumers' intentions to support the brand and its sustainability efforts. Social media food waste reduction promotions are expected to have a positive effect on sustainability-driven customer happiness in this study. Social media can be used by hospitality organizations that promote food waste reduction in order to generate positive emotional responses among customers who resonate with the brand's sustainability values. Customers who perceive these efforts as authentic will most likely experience happiness and satisfaction, which deepens the emotional connection to the brand and its environmental goals [46]. By demonstrating this relationship, we show how sustainability messages driven by social media can generate emotional responses from customers who will support environmental initiatives. There is a significant influence on subsequent behavior from happiness from engaging with a brand's sustainability efforts [49]. According to research, those who are happy tend to engage in behaviors that support the source of their happiness [50]. In the hospitality context, sustainability-driven customer happiness can increase the intention to cut down on food waste because customers are encouraged to behave in alignment with the positive emotions they experience. Therefore, if customers are happy with food waste reduction initiatives, they are more likely to consciously reduce food waste when they visit a particular hospitality services organization. Sustainability-driven customer happiness may positively be related to hospitality customer food waste intentions in the study. Positive emotions from a hospitality brand's food waste reduction efforts can then be transferred into tangible actions, such as actively reducing food waste as customers experience these initiatives [51].

In addition, social media food waste reduction promotions are suggested to be mediated by sustainability-driven customer happiness, influencing hospitality customers' food waste intentions. Social media food waste reduction promotion, viewed through the lens of the SOR model, operates as a stimulus which produces a positive emotional response (happiness). From this emotional response, the resultant behavior is the desired behavior (reducing food waste). The mediating mechanism of sustainability-driven happiness explains the transformation of social media messages into specific consumer actions [52]. Further, H5 and H6 draw from emotional theories in the SOR model, where sustainability-driven happiness serves as an emotional response to social media messaging. Positive emotions reinforce sustainable behaviors, as consumers act in alignment with their happiness derived from supporting eco-friendly practices. Social media creates these emotional connections, which translate into behavioral intentions. It is possible that customers' happiness with food waste reduction campaigns on social media increases the probability of customers supporting these efforts through a reduction in waste. This mediating role further elucidates how hospitality brands can use social media to improve both emotional and behavioral outcomes. Hence,

**H5:** *Social media food waste reduction promotion is positively related to sustainability-driven customer happiness.*

**H6:** *Sustainability-driven customer happiness is positively related to hospitality customer food waste intentions.*

**H7:** *Sustainability-driven customer happiness mediates the relationship between social media food waste reduction promotion and hospitality customer food waste intentions.*

As an emotional state, gratitude motivates people to reward positive actions and efforts with positive responses, sometimes leading to pro-social and pro-environmental behaviors [53]. When people believe that they have been the recipient of someone else's goodwill, they appreciate the experience and want to give back [54]. Gratitude can increase consumers' receptivity to sustainability initiatives by amplifying consumers' likelihood of performing environmentally supportive behaviors [55]. Gratitude can be a powerful tool for hospitality organizations to influence both attitudes and behaviors towards sustainability efforts such as food waste reduction [12]. Customers tend to react positively to promotional messages and endorse the brand's initiatives on the environment when they feel grateful for a brand's real effort in sustainable practices [13]. This study expects the relationship between social media food waste reduction promotion and the internal psychological states (mediators) that determine hospitality customer food waste intentions to be moderated by gratitude. Gratitude is expected to increase the effectiveness of social media messages in increasing customers' environmental motivation [56]. Customers are more likely to be inspired by a hospitality organization's food waste reduction efforts if they feel appreciative of the organization's efforts [53]. This sense of gratitude helps increase the strength of social media messages, making customers more receptive to sustainability content and more likely to translate those feelings into pro-environmental motivations. Therefore, it strengthens the association between social media food waste reduction promotion and environmental motivation, which leads to a higher intention to reduce food waste in hospitality environments. Additionally, gratitude is expected to improve the impact of sustainability-driven customer happiness on social media food waste reduction promotion. If a hospitality brand is grateful for sustainability, customers will be happier engaging with those initiatives. This emotional state reinforces the effect of social media messaging on customer happiness, which leads to higher satisfaction to support the brand's sustainability efforts [57]. People who are happier are more likely to engage in behaviors that are consistent with those positive emotions, such as intentionally reducing food waste, and gratitude makes people happier [58]. Together, this suggests that gratitude is an important driver of the amplification of emotional responses to social media messages, which, in turn, influence customer intentions to support food waste reduction initiatives. Gratitude serves as both a personal disposition that drives sustainable behaviors and a moderator of the psychological mechanisms linking social media promotions to pro-environmental behaviors. Customers' gratitude for an organization's sustainability efforts is intensified when they perceive those efforts as genuine and beneficial, which spurs their motivation and happiness and their intentions to reduce food waste. The moderating effect reveals how emotional factors, such as gratitude, affect the pathway from social media communication to sustainable consumer behaviors in the hospitality sector. Moreover, H8a and H8b integrate gratitude into the SOR framework, where gratitude amplifies motivation and happiness. Studies indicate gratitude enhances receptivity to sustainability efforts, strengthening the impact of emotional and cognitive states on pro-environmental actions. Hence,

**H8a:** *Gratitude towards food waste initiatives moderates the indirect relationship between social media food waste reduction promotion and hospitality customer food waste intentions through social media-driven environmental motivation, such that the indirect relationship is stronger when gratitude is high.*

**H8b:** *Gratitude towards food waste initiatives moderates the indirect relationship between social media food waste reduction promotion and hospitality customer food waste intentions through sustainability-driven customer happiness, such that the indirect relationship is stronger when gratitude is high.*

## 3. Materials and Methods

### 3.1. Participants and Data Collection Procedure

The data for this study were collected from hospitality customers across three major cities in China: Beijing, Shanghai, and Guangzhou. The selection of these cities was based on their status as international business hubs, diverse demographics, and active engagement with sustainability initiatives, and thus are ideal for investigating consumer responses to food waste reduction initiatives in the hospitality sector. The data collection was conducted in person among hotels and restaurants through the use of a structured questionnaire. The participants were given a brief introduction to the study and its purpose and were assured of voluntary participation. The questionnaire was completed by respondents on-site. The study followed ethical guidelines, such as voluntary participation and the right to withdraw at any time without consequence [59–61]. Specifically, the data collection protocols were approved by the ethics committee of Kaifeng University, China (Approval No. SCTK-2024-033, 15 February 2024). Participants' confidentiality and anonymity were assured, and no personally identifiable information was collected [62–64]. However, considering the observational nature of the study and the absence of any medical treatment for the respondents, this study required no formal ethical approval from a local ethics committee. The data were used only for research, which helped to build trust and to receive honest answers. Cohen [65] power analysis was used to determine the sample size to ensure sufficient statistical power. With a medium effect size ($f^2 = 0.15$), a significance level of 0.05, and a statistical power of 0.80, a sample size of approximately 380 was recommended. In order to account for possible low response rates, 600 questionnaires were distributed across the three cities. Of these, 429 were returned, with 404 valid responses after the removal of incomplete or inconsistent responses, exceeding the recommended sample size and providing robust, reliable data for analysis. To ensure that the high frequency of visits to hospitality establishments did not unduly influence responses, the survey design randomized the question order to minimize bias. Respondents were instructed to consider general attitudes toward sustainability rather than specific experiences. Table 1 below presents a summary of the socio-demographic characteristics of the final sample.

**Table 1.** Socio-demographic detail.

| Characteristic | Frequency (*n*) | Percentage (%) |
|---|---|---|
| Gender | | |
| Male | 194 | 48 |
| Female | 210 | 52 |
| Age | | |
| Under 20 | 22 | 5.4 |
| 21–30 | 164 | 40.6 |
| 31–40 | 127 | 31.4 |
| 41–50 | 65 | 16.1 |
| Above 50 | 26 | 6.4 |
| Educational Level | | |
| High School | 49 | 12.1 |
| Diploma | 87 | 21.5 |
| Bachelor's Degree | 193 | 47.8 |
| Master's Degree | 60 | 14.9 |
| Ph.D. or Higher | 15 | 3.7 |
| Monthly Income (RMB) | | |
| Less than 3000 | 65 | 16.1 |
| 3001–5000 | 119 | 29.5 |
| 5001–8000 | 108 | 26.7 |
| 8001–10,000 | 66 | 16.3 |
| Above 10,000 | 46 | 11.4 |
| Frequency of Visits | | |
| 1–2 times per month | 78 | 19.3 |
| 3–5 times per month | 142 | 35.1 |
| 6–8 times per month | 116 | 28.7 |
| More than 8 times | 68 | 16.8 |

### 3.2. Measures

To ensure the reliability and validity of the study variables, established scales from reputable journals were used in operationalizing the variables, and all items were measured on a five-point Likert scale from 1 (Strongly Disagree) to 5 (Strongly Agree). The social media food waste reduction promotion was assessed using a four-item scale adapted from Herman, et al. [66]. An example item is as follows: "This hospitality services organization uses clear and compelling messages in social media posts to promote minimizing food waste". Social media-driven environmental motivation was measured using a five-item scale adopted from Ali, et al. [67]. A sample item is as follows: "I enjoy engaging with new sustainable food practices promoted by hospitality services on social media". Sustainability-driven customer happiness was measured using a five-item scale adopted from Ruo-Fei, et al. [46]. An example item is as follows: "I am happy to support this hospitality organization's sustainable food practices". Gratitude toward food waste initiatives was measured using a three-item scale adopted from Kim and Park [68]. An example item is as follows: "I am grateful that this hospitality organization is trying to educate people on how to practice responsible food and reduce food waste". Finally, a five-item scale adopted from Bell and Ulhas [69] was used to assess hospitality customer food waste intentions. A sample item is as follows: "I strive to minimize the amount of food waste I produce during my visits to hospitality services".

To mitigate method bias, procedural remedies were applied during questionnaire design and data collection. Items were rephrased for clarity, and their order was randomized to reduce patterned responses [70]. Social desirability bias was addressed by ensuring anonymity and emphasizing that there were no right or wrong answers to encourage honesty. Harman's Single-Factor Test was conducted to check for common method bias [71–73]. Results showed that the largest factor accounted for less than 40% of the variance, suggesting that bias was not a significant concern and the data were reliable for analysis.

## 4. Data Analysis

The measurement model was reliable according to the analysis conducted in SMART-PLS, as all indicator loadings for the variables were above the recommended threshold of 0.70. The internal consistency was also above the 0.70 threshold [74,75] for each construct, and the composite reliability was also high. Good convergent validity was confirmed through the average variance extracted (AVE), exceeding the 0.50 threshold for each construct [76,77]. Table 2 summarizes these findings.

The Fornell-Larcker criterion and Heterotrait-Monotrait (HTMT) ratio were used to test discriminant validity (Table 3). The Fornell-Larcker criterion compares the square root of the average variance extracted from each construct to its correlations. To confirm the discriminant validity, the square root of the AVE is higher than the correlation between the constructs; the diagonal elements (square root of AVEs) are greater than the off-diagonal correlations in Table 3 [78–81]. Furthermore, all HTMT values in the table are below 0.90, suggesting discriminant validity, and it is also observed that all values are below 0.90. For example, consider the square root of the AVE for GFWI (0.915) exceeds the correlations with other constructs (e.g., 0.528 with HCFWI). Furthermore, we confirm the constructs are distinct, with HTMT for GFWI and SDCH being 0.661. Both tests also render adequate discriminant validity.

The result of hypothesis testing showed significant relationships among the variables that indicate SMFWP, SMDEM, and SDCH shape HCFWI. Social media promotions had a significant effect on the path from SMFWP to HCFWI (Beta = 0.320) and thus significantly influenced customers' intention to reduce food waste. Furthermore, the SMDEM to SMFWP path was also significant (Beta = 0.421), showing that social media leads to sustainable behavior. Results provided support for the relationship between SMDEM and HCFWI (Beta = 0.175), indicating that environmentally motivated customers have stronger food waste reduction intentions. Confirming that gratitude increased environmental motivation to reduce food waste, SMDEM (Beta = 0.071) was also mediated through GFWI. Results for SMFWP showed a positive effect on SDCH (Beta = 0.415), and SDCH was found to be significantly related to HCFWI (Beta = 0.428). This indicated a positive relationship between customer happiness due to supporting sustainability and food waste reduction intentions. Additionally, GFWI moderated the effect of SMDEM and SDCH on HCFWI (Beta = 0.017; Beta = 0.057), such that environmental motivation and happiness increased their influence on food waste intentions. The model explanatory power is shown by the R-square values. The model explained 40.9% of the variance for HCFWI. The model explained 43.3% of the variance in SDCH and 38.2% of the variance in SMDEM. Table 4 includes more detail and Figure 1 represents the structural model of this study.

**Table 2.** Quality indicators of the measurement model.

| Construct | Abbreviation | Item | Loading | Composite Reliability | Average Variance Extracted (AVE) |
|---|---|---|---|---|---|
| Gratitude Towards Food Waste Initiatives | GFWI | GFWI1 | 0.905 | 0.94 | 0.838 |
| | | GFWI2 | 0.923 | | |
| | | GFWI3 | 0.918 | | |
| Hospitality Customer Food Waste Intentions | HCFWI | HCFWI1 | 0.932 | 0.975 | 0.888 |
| | | HCFWI2 | 0.943 | | |
| | | HCFWI3 | 0.953 | | |
| | | HCFWI4 | 0.942 | | |
| | | HCFWI5 | 0.941 | | |
| Sustainability-Driven Customer Happiness | SDCH | SDCH1 | 0.925 | 0.948 | 0.785 |
| | | SDCH2 | 0.738 | | |
| | | SDCH3 | 0.918 | | |
| | | SDCH4 | 0.937 | | |
| | | SDCH5 | 0.896 | | |
| Social Media-Driven Environmental Motivation | SMDEM | SMDEM1 | 0.822 | 0.925 | 0.711 |
| | | SMDEM2 | 0.846 | | |
| | | SMDEM3 | 0.851 | | |
| | | SMDEM4 | 0.844 | | |
| | | SMDEM5 | 0.852 | | |
| Social Media Food Waste Reduction Promotion | SMFWP | SMFWP1 | 0.763 | 0.865 | 0.616 |
| | | SMFWP2 | 0.763 | | |
| | | SMFWP3 | 0.82 | | |
| | | SMFWP4 | 0.792 | | |

**Table 3.** Discriminant validity.

| Construct | GFWI | HCFWI | SDCH | SMDEM | SMFWP |
|---|---|---|---|---|---|
| Correlations and Square Root of AVEs | | | | | |
| GFWI | 0.915 | | | | |
| HCFWI | 0.528 | 0.942 | | | |
| SDCH | 0.604 | 0.77 | 0.886 | | |
| SMDEM | 0.618 | 0.708 | 0.758 | 0.843 | |
| SMFWP | 0.476 | 0.715 | 0.656 | 0.655 | 0.785 |
| HTMT Ratios | | | | | |
| GFWI | | | | | |
| HCFWI | 0.565 | | | | |
| SDCH | 0.661 | 0.811 | | | |
| SMDEM | 0.69 | 0.747 | 0.829 | | |
| SMFWP | 0.446 | 0.586 | 0.581 | 0.614 | |

**Table 4.** The results of structural model.

| Hypothesis | Beta Coefficient | S.D. | t-Statistics | *p*-Value | Lower CI | Upper CI | R-Square |
|---|---|---|---|---|---|---|---|
| SDCH → HCFWI (H7) | 0.428 | 0.071 | 6.001 | 0.000 | 0.294 | 0.570 | HCFWI ($R^2 = 0.409$) |
| SMDEM → HCFWI (H3) | 0.175 | 0.064 | 2.740 | 0.006 | 0.042 | 0.297 | |
| SMFWP → HCFWI (H1) | 0.320 | 0.044 | 7.230 | 0.000 | 0.249 | 0.420 | |
| SMFWP → SDCH (H5) | 0.415 | 0.035 | 11.772 | 0.000 | 0.337 | 0.485 | SDCH ($R^2 = 0.433$) |
| SMFWP → SMDEM (H2) | 0.421 | 0.036 | 11.715 | 0.000 | 0.344 | 0.484 | SMDEM ($R^2 = 0.382$) |
| Moderating Effect 1 (H8b) | 0.057 | 0.016 | 3.653 | 0.000 | 0.030 | 0.090 | - |
| SMFWP → SDCH → HCFWI (H7) | 0.178 | 0.038 | 4.688 | 0.000 | 0.113 | 0.267 | - |
| GFWI → SMDEM → HCFWI (H4) | 0.071 | 0.027 | 2.648 | 0.008 | 0.017 | 0.122 | - |
| Moderating Effect 2 (H8a) | 0.017 | 0.008 | 2.102 | 0.036 | 0.005 | 0.035 | - |

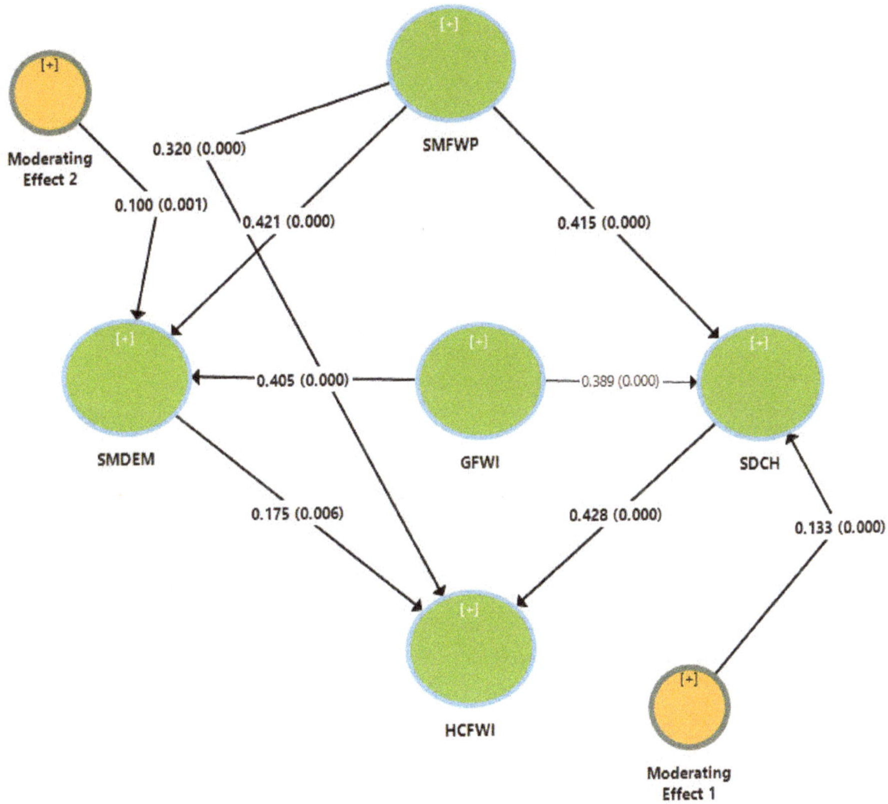


**Figure 1.** The structural model, showing the hypothesized relationships.

## 5. Discussion

This study identified several critical direct effects of food waste reduction in the hospitality industry. Social media campaigns are positively related to hospitality customer food waste intentions and, hence, encourage customers to reduce food waste. This aligns with recent work that social media can foster pro-environmental behaviors [33]. Further, the above discussion also aligns with Lau, et al. [82], who acknowledged the role of social media in shaping customer behavior in the context of food waste. Social media food waste reduction promotion also directly influences sustainability-driven customer happiness because customers feel emotional happiness in supporting social media-promoted sustainability efforts. This is in line with previous work from Ruo-Fei, et al. [46], who showed that emotional responses may be a driver of pro-environmental actions. Furthermore, a positive relationship between social media-driven environmental motivation and hospitality customer food waste intentions implies that those customers with higher environmental motivation are likely to engage in sustainable behavior [42]. This study also helps to understand how mediating effects affect food waste reduction behaviors. Social media food waste reduction promotion and hospitality customer food waste intentions were significantly mediated by social media-driven environmental motivation, meaning that the more food waste reduction social media campaigns drive environmental motivation, the more food waste reduction will occur. Motivation is identified as a critical bridge between awareness and action. Similarly, sustainability-driven customer happiness mediated the relationship between social media food waste reduction promotion and food waste intentions. This implies that, in addition to informing, campaigns also promote happiness that leads to sustainable action, thus emphasizing the dual impact of cognitive (motivation) and emotional (happiness) factors in encouraging the sustainable behavior of customers. The findings of this study align with recent research emphasizing the role of social media in driving pro-environmental behaviors. For example, Teoh, et al. [41] highlighted how interactive social media campaigns can significantly influence food waste reduction behaviors by engaging users with visually compelling content. Similarly, Jenkins, et al. [83] demonstrated that social media's ability to disseminate educational materials enhances consumers' motivation for sustainable food practices. Our results corroborate these findings by showing that social media-driven environmental motivation and sustainability-driven happiness are significant predictors of food waste reduction intentions.

Gratitude towards food waste initiatives also had a significant moderating role, enhancing the relationship between mediators and hospitality customer food waste intentions. The effects of social media-driven environmental motivation and sustainability-driven customer happiness were amplified by gratitude, increasing their effect on food waste reduction intentions. This is consistent with Tam [53] research on gratitude as a motivator of prosocial behavior and extends it by demonstrating that gratitude increases the effectiveness of sustainability campaigns. Those customers who feel grateful for a hospitality brand's sustainability efforts are more likely to be motivated and emotionally engaged and, therefore, have stronger intentions to reduce food waste. Gratitude promotes cognitive and emotional processes in the hospitality sector to drive sustainable behavior. Furthermore, this study advances the literature by introducing gratitude as a moderator, extending the work of Tam [53], who explored gratitude's role in amplifying pro-social behaviors. These contributions emphasize the importance of integrating cognitive and emotional factors into sustainability-focused social media campaigns to achieve meaningful behavioral change in the hospitality sector. These findings validate the SOR model. Social media food waste reduction promotion acts as the external stimulus, triggering internal states, social media-driven environmental motivation, and sustainability-driven customer happiness that lead to food waste reduction behaviors. Gratitude further amplifies these internal states, strengthening their impact on behavior. By integrating cognitive (motivation) and emotional (happiness, gratitude) factors, this study offers a more complete understanding of how social media campaigns influence sustainable consumption in hospitality. The SOR model effectively captures the interaction between external stimuli, psychological states, and behavior, offering key insights into what drives pro-environmental actions.

### 5.1. Theoretical Implications

This research is important theoretically and can help push the frontier in terms of how social media can spur responsible behaviors, in particular in the case of food waste reduction in the hospitality sector. This study extends the application of the SOR model to sustainability research to show that external stimuli, such as social media food waste reduction promotion, can trigger both cognitive and emotional internal states, social media-driven environmental motivation and sustainability-driven customer happiness, that lead to concrete behavioral outcomes such as reduced food waste. The integration of cognitive and emotional factors into the SOR framework in this sophisticated way constitutes a major step forward over previous models that fail to recognize the emotional involvement required for continuous pro-social behavior. In a social media context, this study makes important theoretical contributions by demonstrating that digital platforms are not just communication tools but strong behavioral change drivers. The study demonstrates that social media campaigns aimed at promoting sustainability, specifically food waste reduction, can have far-reaching psychological effects on consumers, which in turn affects their motivation and their emotional satisfaction. This research fills a critical gap in the literature by focusing on the role of social media to generate not only awareness but also emotional engagement. This demonstrates that social media can serve as a powerful stimulus that provides the impetus for complex emotional and cognitive responses (i.e., motivation and happiness), which lead to meaningful behavioral outcomes (i.e., less food waste). This insight highlights the changing role of social media in sustainability marketing and its ability to influence other pro-environmental behaviors. In addition, gratitude towards food waste initiatives is added as a key moderating factor that theoretically disrupts the existing theoretical paradigms, emphasizing the role of a positive emotional state in amplifying the effect of social media campaigns. Trusting the SOR model to focus on external stimuli and internal processing, this research reveals that gratitude is an amplifier of the former two, which enhances motivation and happiness and, in turn, motivates consumers to act on their environmental intentions. This is a major contribution to the sustainability and consumer behavior literature that emphasizes the importance of emotional resonance in stimulating sustainable consumer actions, particularly in the hospitality industry, which has been overlooked by environmental psychology and marketing research. Additionally, this study fills a key gap in the literature by specifically studying the hospitality sector and food waste reduction and by focusing more deeply on how targeted social media campaigns can change consumer behavior in this context. The models suggest that responsible behaviors are not purely rational decisions and are influenced by emotional factors, such as happiness and gratitude, which have been ignored in previous models. Therefore, this research offers a robust theoretical base for future work on the role of emotions in sustainability marketing and a pathway to future work incorporating the integration of emotional and cognitive responses into broader behavioral theories.

### 5.2. Practical Implications

These research implications have valuable practical implications for the hospitality industry, especially in terms of sustainability and reducing food waste. We found that social media campaigns can indeed be a powerful instrument to influence customer behavior to more sustainable practices such as saving food waste. Social media platforms provide a great opportunity for hospitality organizations to design specific campaigns that not only raise awareness of sustainability efforts but also evoke customer emotions (motivation and happiness) to encourage customers to participate in sustainability efforts. These campaigns should be consistent, informative, and emotional and should target tangible benefits in reducing food waste for the environment and for society as a whole. For example, the hospitality sector is urged to incorporate gratitude in its messaging, such as saying 'thank you' to customers for helping to keep the industry afloat. Findings from this study suggest that gratitude enhances the influence of social media-driven motivation and happiness on plans to reduce food waste. For instance, hospitality organizations can show appreciation for customers performing sustainability actions through personal thank you messages or incentives for sustainable behavior, such as discounts or rewards for customers who decide to reduce waste through portion control or participation in recycling programs. It can contribute to increasing the emotional relationship between the customer and the brand and to greater loyalty and more pro-environmental behavior. In addition, hospitality organizations require their sustainability communications to engage both cognitively and emotionally. To support the organization's efforts in this case, we can use engaging content such as videos, infographics, and interactive posts, which not only inform but also make them happy and fulfilled to support the organization. In fact, hospitality services can enhance their probability of long-term behavioral change by customers' emotional satisfaction with working for sustainability initiatives. In addition to this, they can also organize educational workshops or events that talk about how important it is to reduce food waste, both online and in person, to a bigger audience. Lastly, hospitality businesses should measure and monitor the impact of their social media campaigns on customers' behavior in order to continuously adjust their strategy. Data analytics tools can be used by companies to find out which types of messages and content are most effective in inducing the desired food waste reduction behaviors and to adjust their communications to have the greatest impact. By doing this, hospitality organizations can lead the global sustainability movement and help the organization's own environmental goals and wider efforts to achieve net zero emissions.

### 5.3. Limitations and Future Research Directions

This study offers valuable insights, but some limitations must be acknowledged. First, the data were collected from hospitality customers in three major Chinese cities, which may limit the generalizability to other regions and cultures where attitudes towards sustainability and food waste may differ. Future research should explore these dynamics in other countries to account for potential cultural variations. Second, the focus on the hospitality sector and food waste limits the broader application of these findings. Further research should examine how social media influences pro-environmental behaviors in other industries, such as retail or tourism, to see if the same mechanisms of motivation, happiness, and gratitude apply. Additionally, the use of cross-sectional data prevents establishing causal relationships between variables. Longitudinal studies could confirm how sustained exposure to social media campaigns impacts behaviors over time. The reliance on self-reported data also introduces the risk of social desirability bias, as respondents may overstate pro-environmental intentions. Future studies could use objective behavioral measures, such as actual food waste reduction data. Finally, exploring other psychological factors, such as environmental guilt or pride, and examining the effects of different social media content types (user-generated vs. brand-generated), could provide a fuller understanding of how digital platforms drive sustainability behaviors.

## 6. Conclusions

This research provides valuable insights into the role of social media in promoting pro-environmental behaviors, specifically in the context of food waste reduction in the hospitality sector. By applying the SOR model, this study demonstrates that social media food waste reduction promotion can effectively influence hospitality customer food waste intentions through the mediating effects of social media-driven environmental motivation and sustainability-driven customer happiness. The findings underscore the importance of emotional and cognitive factors, showing that both motivation and emotional satisfaction play crucial roles in driving sustainable behaviors. Furthermore, the moderating role of gratitude towards food waste initiatives highlights how positive emotional responses can amplify the impact of social media campaigns on customer behavior. This study contributes to both the theoretical understanding of consumer behavior in sustainability contexts and offers practical recommendations for hospitality organizations to leverage social media as a tool to engage customers in reducing food waste. By incorporating emotionally resonant messages and expressing gratitude for customer participation in sustainability initiatives, hospitality businesses can enhance the effectiveness of their environmental campaigns. While the study is not without limitations, it opens several avenues for future research, including examining other sectors, expanding geographic scope, and employing longitudinal designs to better understand the long-term effects of social media-driven pro-environmental behavior. Overall, this research emphasizes the transformative potential of social media in promoting sustainability and the critical role of emotional engagement in achieving meaningful behavioral change.

**Author Contributions:** Conceptualization, W.C. and M.J.; methodology, M.J. and M.H.A.; software, M.J. and M.H.A.; formal analysis, W.C.; writing—original draft preparation, W.C.; writing—review and editing, M.J. All authors have read and agreed to the published version of the manuscript.

**Funding:** This research was funded by Science and Technology Research Program of Chongqing Municipal Education Commission (Project No.: KJQN202202405).

**Institutional Review Board Statement:** This study was approved by the ethics committee of Kaifeng University, China (Approval No. SCTK-2024-033, 15 February 2024).

**Informed Consent Statement:** Informed consent was obtained from all subjects involved in the study.

**Data Availability Statement:** The original contributions presented in this study are included in the article. Further inquiries can be directed to the corresponding author.

**Conflicts of Interest:** The authors declare no conflict of interest.

## References

1. Allen, M.R.; Friedlingstein, P.; Girardin, C.A.; Jenkins, S.; Malhi, Y.; Mitchell-Larson, E.; Peters, G.P.; Rajamani, L. Net zero: Science, origins, and implications. *Annu. Rev. Environ. Resour.* **2022**, *47*, 849–887. [CrossRef]
2. Rajamani, L. The 2015 Paris Agreement: Interplay between hard, soft and non-obligations. *J. Environ. Law* **2016**, *28*, 337–358. [CrossRef]
3. FAO. Tackling Food Loss and Waste: A Triple Win Opportunity. Available online: https://www.fao.org/newsroom/detail/FAO-UNEP-agriculture-environment-food-loss-waste-day-2022/en (accessed on 20 May 2024).
4. EPA. Global Greenhouse Gas Overview. Available online: https://www.epa.gov/ghgemissions/global-greenhouse-gas-overview (accessed on 20 May 2024).
5. Liu, C.; Ahmad, N.; Jiang, M.; Arshad, M.Z. Steering the path to safer food: The role of transformational leadership in food services to combat against foodborne illness. *J. Retail. Consum. Serv.* **2024**, *81*, 103958. [CrossRef]
6. Crippa, M.; Solazzo, E.; Guizzardi, D.; Monforti-Ferrario, F.; Tubiello, F.N.; Leip, A. Food systems are responsible for a third of global anthropogenic GHG emissions. *Nat. Food* **2021**, *2*, 198–209. [CrossRef]
7. Hertwig, R.; Grüne-Yanoff, T. Nudging and boosting: Steering or empowering good decisions. *Perspect. Psychol. Sci.* **2017**, *12*, 973–986. [CrossRef]
8. Zaman, U. Seizing momentum on climate action: Nexus between net-zero commitment concern, destination competitiveness, influencer marketing, and regenerative tourism intention. *Sustainability* **2023**, *15*, 5213. [CrossRef]
9. Filimonau, V.; Delysia, A. Food waste management in hospitality operations: A critical review. *Tour. Manag.* **2019**, *71*, 234–245. [CrossRef]
10. Everitt, H.; van der Werf, P.; Seabrook, J.A.; Gilliland, J.A. Household food wasting in a net-zero energy neighbourhood: Analyzing relationships between household food waste and pro-environmentalism. *Can. Geogr./Géographe Can.* **2024**, *68*, 489–502. [CrossRef]
11. Ahmed, S.; Stewart, A.; Smith, E.; Warne, T.; Byker Shanks, C. Consumer perceptions, behaviors, and knowledge of food waste in a rural American state. *Front. Sustain. Food Syst.* **2021**, *5*, 734785. [CrossRef]
12. Septianto, F.; Kemper, J.A.; Northey, G. Thanks, but no thanks: The influence of gratitude on consumer awareness of food waste. *J. Clean. Prod.* **2020**, *258*, 120591. [CrossRef]
13. Septianto, F.; Kemper, J.; Northey, G.; Andonopoulos, V.; Barbera, M. Let Us Give Thanks: How Gratitude and Message Framing Can Reduce Food Waste. *Adv. Consum. Res.* **2019**, *47*, 684–685.
14. Sansone, R.A.; Sansone, L.A. Gratitude and well being: The benefits of appreciation. *Psychiatry* **2010**, *7*, 18. [PubMed]
15. Lin, X.; Yang, R.; Zhang, W.; Zeng, N.; Zhao, Y.; Wang, G.; Li, T.; Cai, Q. An integrated view of correlated emissions of greenhouse gases and air pollutants in China. *Carbon Balance Manag.* **2023**, *18*, 9. [CrossRef] [PubMed]
16. Chloe, W. Food Waste in China and How the Government is Combating It. Available online: https://earth.org/food-waste-in-china (accessed on 22 May 2024).
17. Pirani, S.I.; Arafat, H.A. Reduction of food waste generation in the hospitality industry. *J. Clean. Prod.* **2016**, *132*, 129–145. [CrossRef]
18. Heskiano, H.; Syah, T.Y.R.; Hilmy, M.R. Social media marketing relations, brand awareness to brand loyalty through the brand image. *J. Multidiscip. Acad.* **2020**, *4*, 208–214.
19. Aljumah, A.; Nuseir, M.; Refae, G. Examining the effect of social media interaction, E-WOM, and public relations: Assessing the mediating role of brand awareness. *Int. J. Data Netw. Sci.* **2023**, *7*, 467–476. [CrossRef]
20. Ahmad, N.; Ahmad, A.; Lewandowska, A.; Han, H. From screen to service: How corporate social responsibility messages on social media shape hotel consumer advocacy. *J. Hosp. Mark. Manag.* **2024**, *33*, 384–413. [CrossRef]
21. Zhou, X.; Ahmad, N.; Lho, L.H.; Han, H. Social ripple: Unraveling the impact of customer relationship management via social media on consumer emotions and behavior. *Soc. Behav. Personal. Int. J.* **2023**, *51*, 1–12. [CrossRef]

22. Li, Y.; Ahmad, N.; Lewandowska, A. Eco-Chic stays: The influence of green communication on guest booking intentions. *J. Vacat. Mark.* **2024**, 13567667241293792. [CrossRef]
23. Bonnici, T.; Briguglio, M.; Spiteri, G.W. Humor helps: An experimental analysis of pro-environmental social media communication. *Sustainability* **2023**, *15*, 5157. [CrossRef]
24. Ahmad, N.; Samad, S.; Mahmood, S. Sustainable pathways: The intersection of CSR, hospitality and the United Nations' sustainable development goals. *Curr. Issues Tour.* **2024**, *27*, 3924–3943. [CrossRef]
25. Gupta, S.; Nawaz, N.; Alfalah, A.A.; Naveed, R.T.; Muneer, S.; Ahmad, N. The relationship of CSR communication on social media with consumer purchase intention and brand admiration. *J. Theor. Appl. Electron. Commer. Res.* **2021**, *16*, 1217–1230. [CrossRef]
26. Ahmad, N.; Naveed, R.T.; Scholz, M.; Irfan, M.; Usman, M.; Ahmad, I. CSR communication through social media: A litmus test for banking consumers' loyalty. *Sustainability* **2021**, *13*, 2319. [CrossRef]
27. Ahmad, N.; Ahmad, A.; Siddique, I. Responsible Tourism and Hospitality: The Intersection of Altruistic Values, Human Emotions, and Corporate Social Responsibility. *Adm. Sci.* **2023**, *13*, 105. [CrossRef]
28. Yuxiang, L.; Ahmad, N.; Linda, H.; Heesup, H. From boardroom to breakroom: Corporate social responsibility, happiness, green self-efficacy, and altruistic values shape sustainable behavior. *Soc. Behav. Personal. Int. J.* **2024**, *52*, 1–14.
29. Ahmad, N.; Samad, S.; Han, H. Travel and Tourism Marketing in the age of the conscious tourists: A study on CSR and tourist brand advocacy. *J. Travel Tour. Mark.* **2023**, *40*, 551–567. [CrossRef]
30. Kurniawan, T.A.; Liang, X.; Goh, H.H.; Othman, M.H.D.; Anouzla, A.; Al-Hazmi, H.E.; Chew, K.W.; Aziz, F.; Ali, I. Leveraging food waste for electricity: A low-carbon approach in energy sector for mitigating climate change and achieving net zero emission in Hong Kong (China). *J. Environ. Manag.* **2024**, *351*, 119879. [CrossRef]
31. De Fano, D.; Schena, R.; Russo, A. Empowering plastic recycling: Empirical investigation on the influence of social media on consumer behavior. *Resour. Conserv. Recycl.* **2022**, *182*, 106269. [CrossRef]
32. Rachmad, Y.E. Social Media Marketing Mediated Changes In Consumer Behavior From E-Commerce To Social Commerce. *Int. J. Econ. Manag. Res.* **2022**, *1*, 227–242. [CrossRef]
33. Kong, H.M.; Witmaier, A.; Ko, E. Sustainability and social media communication: How consumers respond to marketing efforts of luxury and non-luxury fashion brands. *J. Bus. Res.* **2021**, *131*, 640–651. [CrossRef]
34. Rini, L.; Schouteten, J.J.; Faber, I.; Frøst, M.B.; Perez-Cueto, F.J.; De Steur, H. Social media and food consumer behavior: A systematic review. *Trends Food Sci. Technol.* **2023**, *143*, 104290. [CrossRef]
35. Mehrabian, A.; Russell, J.A. *An Approach to Environmental Psychology*; The MIT Press: Cambridge, MA, USA, 1974.
36. Talwar, S.; Kaur, P.; Kumar, S.; Salo, J.; Dhir, A. The balancing act: How do moral norms and anticipated pride drive food waste/reduction behaviour? *J. Retail. Consum. Serv.* **2022**, *66*, 102901. [CrossRef]
37. Kumar, A.; Pandey, M. Social media and impact of altruistic motivation, egoistic motivation, subjective norms, and ewom toward green consumption behavior: An empirical investigation. *Sustainability* **2023**, *15*, 4222. [CrossRef]
38. Guo, L.; Hu, X.; Wei, X.; Cai, X. The influence of personal motivation and environmental stimuli on customer participation and engagement behavior: The mediating role of experience evaluation. *J. Hosp. Tour. Technol.* **2020**, *11*, 643–666. [CrossRef]
39. Sutinen, U.-M.; Närvänen, E. Constructing the food waste issue on social media: A discursive social marketing approach. *J. Mark. Manag.* **2022**, *38*, 219–247. [CrossRef]
40. Matharu, M.; Gupta, N.; Swarnakar, V. Efforts are made but food wastage is still going on: A study of motivation factors for food waste reduction among household consumers. *Asia Pac. J. Bus. Adm.* **2022**, *14*, 244–264. [CrossRef]
41. Teoh, C.W.; Koay, K.Y.; Chai, P.S. The role of social media in food waste prevention behaviour. *Br. Food J.* **2022**, *124*, 1680–1696. [CrossRef]
42. Nguyen, N.P.; Mogaji, E. A theoretical framework for the influence of green marketing communication on consumer behaviour in emerging economies. In *Green Marketing in Emerging Economies: A Communications Perspective*; Palgrave Macmillan: Cham, Switzerland, 2022; pp. 253–274.
43. Chung, C.D.; Gao, Y.L.; Leung, D. Corporate social responsibility communications on social media and consumers' brand engagement: A case study of hotels in Hong Kong. In *New Technology and Mediated Chinese Tourists*; Routledge: Abingdon, UK, 2023; pp. 61–79.
44. Wibowo, A.; Chen, S.-C.; Wiangin, U.; Ma, Y.; Ruangkanjanases, A. Customer behavior as an outcome of social media marketing: The role of social media marketing activity and customer experience. *Sustainability* **2020**, *13*, 189. [CrossRef]
45. Brodie, R.J.; Hollebeek, L.D.; Jurić, B.; Ilić, A. Customer engagement: Conceptual domain, fundamental propositions, and implications for research. *J. Serv. Res.* **2011**, *14*, 252–271. [CrossRef]
46. Ruo-Fei, S.; Zeng, J.-Y.; Jin, C.-H. The role of consumer'social capital on ethical consumption and consumer happiness. *Sage Open* **2022**, *12*, 21582440221095026. [CrossRef]
47. Acampora, A.; Preziosi, M.; Lucchetti, M.C.; Merli, R. The role of hotel environmental communication and guests' environmental concern in determining guests' behavioral intentions. *Sustainability* **2022**, *14*, 11638. [CrossRef]
48. Allen, M.; Allen, M. Understanding pro-environmental behavior: Models and messages. In *Strategic Communication for Sustainable Organizations: Theory and Practice*; Springer: Cham, Switzerland, 2016; pp. 105–137.

49. Ravina Ripoll, R.; Romero-Rodríguez, L.M.; Ahumada-Tello, E. Guest editorial: Happiness management: Key factors for sustainability and organizational communication in the age of Industry 4.0. *Corp. Gov. Int. J. Bus. Soc.* **2022**, *22*, 449–457. [CrossRef]
50. Cachero-Martínez, S.; García-Rodríguez, N.; Salido-Andrés, N. Because I'm happy: Exploring the happiness of shopping in social enterprises and its effect on customer satisfaction and loyalty. *Manag. Decis.* **2024**, *62*, 492–512. [CrossRef]
51. Attiq, S.; Chau, K.Y.; Bashir, S.; Habib, M.D.; Azam, R.I.; Wong, W.-K. Sustainability of household food waste reduction: A fresh insight on youth's emotional and cognitive behaviors. *Int. J. Environ. Res. Public Health* **2021**, *18*, 7013. [CrossRef]
52. Sirieix, L.; Lála, J.; Kocmanová, K. Understanding the antecedents of consumers' attitudes towards doggy bags in restaurants: Concern about food waste, culture, norms and emotions. *J. Retail. Consum. Serv.* **2017**, *34*, 153–158. [CrossRef]
53. Tam, K.-P. Gratitude to nature: Presenting a theory of its conceptualization, measurement, and effects on pro-environmental behavior. *J. Environ. Psychol.* **2022**, *79*, 101754. [CrossRef]
54. Emmons, R.A.; Mishra, A. Why gratitude enhances well-being: What we know, what we need to know. *Des. Posit. Psychol. Tak. Stock Mov. Forw.* **2011**, *248*, 262.
55. Sun, J.; Ma, B.; Wei, S. Same gratitude, different pro-environmental behaviors? Effect of the dual-path influence mechanism of gratitude on pro-environmental behavior. *J. Clean. Prod.* **2023**, *415*, 137779. [CrossRef]
56. Syropoulos, S.; Watkins, H.M.; Shariff, A.F.; Hodges, S.D.; Markowitz, E.M. The role of gratitude in motivating intergenerational environmental stewardship. *J. Environ. Psychol.* **2020**, *72*, 101517. [CrossRef]
57. Septianto, F.; Garg, N. The impact of gratitude (vs. pride) on the effectiveness of cause-related marketing. *Eur. J. Mark.* **2021**, *55*, 1594–1623. [CrossRef]
58. Witvliet, C.v.; Richie, F.J.; Root Luna, L.M.; Van Tongeren, D.R. Gratitude predicts hope and happiness: A two-study assessment of traits and states. *J. Posit. Psychol.* **2019**, *14*, 271–282. [CrossRef]
59. Ahmad, N.; Ullah, Z.; Ryu, H.B.; Ariza-Montes, A.; Han, H. From corporate social responsibility to employee well-being: Navigating the pathway to sustainable healthcare. *Psychol. Res. Behav. Manag.* **2023**, *16*, 1079–1095. [CrossRef] [PubMed]
60. Liu, Y.; Cherian, J.; Ahmad, N.; Han, H.; de Vicente-Lama, M.; Ariza-Montes, A. Internal corporate social responsibility and employee burnout: An employee management perspective from the healthcare sector. *Psychol. Res. Behav. Manag.* **2023**, *16*, 283–302. [CrossRef] [PubMed]
61. Fu, B.; Ahmad, N.; Lho, L.H.; Han, H. Triple-E effect: Corporate ethical responsibility, ethical values, and employee emotions in the healthcare sector. *Soc. Behav. Personal. Int. J.* **2023**, *51*, 1–14. [CrossRef]
62. Ahmad, N.; Han, H.; Kim, M. Elevated emotions, elevated ideas: The CSR-employee creativity nexus in hospitality. *J. Serv. Theory Pract.* **2024**, *34*, 891–914. [CrossRef]
63. Ahmad, N.; Samad, S.; Han, H. Charting new terrains: How CSR initiatives shape employee creativity and contribute to UN-SDGs in a knowledge-driven world. *J. Innov. Knowl.* **2024**, *9*, 100557. [CrossRef]
64. Ahmad, N.; Ahmad, A.; Siddique, I. Beyond self-interest: How altruistic values and human emotions drive brand advocacy in hospitality consumers through corporate social responsibility. *Corp. Soc. Responsib. Environ. Manag.* **2024**, *31*, 2439–2453. [CrossRef]
65. Cohen, J. Statistical power analysis. *Curr. Dir. Psychol. Sci.* **1992**, *1*, 98–101. [CrossRef]
66. Herman, L.E.; Udayana, I.B.N.; Farida, N. Young generation and environmental friendly awareness: Does it the impact of green advertising? *Bus. Theory Pract.* **2021**, *22*, 159–166. [CrossRef]
67. Ali, F.; Ashfaq, M.; Begum, S.; Ali, A. How "Green" thinking and altruism translate into purchasing intentions for electronics products: The intrinsic-extrinsic motivation mechanism. *Sustain. Prod. Consum.* **2020**, *24*, 281–291. [CrossRef]
68. Kim, J.; Park, T. How corporate social responsibility (CSR) saves a company: The role of gratitude in buffering vindictive consumer behavior from product failures. *J. Bus. Res.* **2020**, *117*, 461–472. [CrossRef]
69. Bell, A.E.; Ulhas, K.R. Working to reduce food waste: Investigating determinants of food waste amongst Taiwanese workers in factory cafeteria settings. *Sustainability* **2020**, *12*, 9669. [CrossRef]
70. Ahmad, N.; Ullah, Z.; Arshad, M.Z.; waqas Kamran, H.; Scholz, M.; Han, H. Relationship between corporate social responsibility at the micro-level and environmental performance: The mediating role of employee pro-environmental behavior and the moderating role of gender. *Sustain. Prod. Consum.* **2021**, *27*, 1138–1148. [CrossRef]
71. Peng, J.; Samad, S.; Comite, U.; Ahmad, N.; Han, H.; Ariza-Montes, A.; Vega-Muñoz, A. Environmentally specific servant leadership and employees' Energy-specific pro-environmental behavior: Evidence from healthcare sector of a developing economy. *Int. J. Environ. Res. Public Health* **2022**, *19*, 7641. [CrossRef]
72. Xu, L.; Mohammad, S.J.; Nawaz, N.; Samad, S.; Ahmad, N.; Comite, U. The role of CSR for de-carbonization of hospitality sector through employees: A leadership perspective. *Sustainability* **2022**, *14*, 5365. [CrossRef]
73. Ahmad, N.; Ullah, Z.; AlDhaen, E.; Han, H.; Ariza-Montes, A.; Vega-Muñoz, A. Fostering advocacy behavior of employees: A corporate social responsibility perspective from the hospitality sector. *Front. Psychol.* **2022**, *13*, 865021. [CrossRef]
74. Ahmad, N.; Ullah, Z.; AlDhaen, E.; Siddique, I. Promoting the advocacy behavior of customers through corporate social responsibility: The role of brand admiration. *Bus. Soc. Rev.* **2023**, *128*, 367–386. [CrossRef]
75. Han, H.; Al-Ansi, A.; Chua, B.-L.; Ahmad, N.; Kim, J.J.; Radic, A.; Bobby Ryu, H. Reconciling civilizations: Eliciting residents' attitude and behaviours for international Muslim tourism and development. *Curr. Issues Tour.* **2022**, *9*, 1463–1481. [CrossRef]

76. Chen, J.; Ghardallou, W.; Comite, U.; Ahmad, N.; Ryu, H.B.; Ariza-Montes, A.; Han, H. Managing hospital employees' burnout through transformational leadership: The role of resilience, role clarity, and intrinsic motivation. *Int. J. Environ. Res. Public Health* **2022**, *19*, 10941. [CrossRef]
77. Ahmad, N.; Ullah, Z.; AlDhaen, E.; Han, H.; Scholz, M. A CSR perspective to foster employee creativity in the banking sector: The role of work engagement and psychological safety. *J. Retail. Consum. Serv.* **2022**, *67*, 102968. [CrossRef]
78. Ahmad, N.; Ullah, Z.; AlDhaen, E.; Han, H.; Araya-Castillo, L.; Ariza-Montes, A. Fostering hotel-employee creativity through micro-level corporate social responsibility: A social identity theory perspective. *Front. Psychol.* **2022**, *13*, 853125. [CrossRef] [PubMed]
79. Fu, Q.; Cherian, J.; Ahmad, N.; Scholz, M.; Samad, S.; Comite, U. An inclusive leadership framework to foster employee creativity in the healthcare sector: The role of psychological safety and polychronicity. *Int. J. Environ. Res. Public Health* **2022**, *19*, 4519. [CrossRef] [PubMed]
80. Deng, Y.; Cherian, J.; Ahmad, N.; Scholz, M.; Samad, S. Conceptualizing the role of target-specific environmental transformational leadership between corporate social responsibility and pro-environmental behaviors of hospital employees. *Int. J. Environ. Res. Public Health* **2022**, *19*, 3565. [CrossRef]
81. Guan, X.; Ahmad, N.; Sial, M.S.; Cherian, J.; Han, H. CSR and organizational performance: The role of pro-environmental behavior and personal values. *Corp. Soc. Responsib. Environ. Manag.* **2023**, *30*, 677–694. [CrossRef]
82. Lau, E.A.; Rukmana, A.Y.; Uhai, S.; Mokodenseho, S.; Tapaningsih, W.I.D.A. Mapping Research on the Influence of Social Media on Consumer Food Behavior a Bibliometric Approach. *Eastasouth J. Soc. Sci. Humanit.* **2024**, *1*, 84–94. [CrossRef]
83. Jenkins, E.L.; Brennan, L.; Molenaar, A.; McCaffrey, T.A. Exploring the application of social media in food waste campaigns and interventions: A systematic scoping review of the academic and grey literature. *J. Clean. Prod.* **2022**, *360*, 132068. [CrossRef]

*sustainability*

MDPI

*Article*

# The Environmental and Economic Dynamics of Food Waste and Greenhouse Gas Emissions: A Causal Time Series Analysis from 2000 to 2022

**Salim Yılmaz [1,*], Ahmet Murat Günal [2], Gizem Köse [3] and Murat Baş [3]**

1 Department of Health Management, Faculty of Health Sciences, Acibadem Mehmet Ali Aydinlar University, Istanbul 34752, Turkey
2 Department of Nutrition and Dietetics, Faculty of Health Sciences, Halic University, Istanbul 34060, Turkey; ahmetmuratgunal@halic.edu.tr
3 Department of Nutrition and Dietetics, Faculty of Health Sciences, Acibadem Mehmet Ali Aydinlar University, Istanbul 34752, Turkey; gizem.kose@acibadem.edu.tr (G.K.); murat.bas@acibadem.edu.tr (M.B.)
* Correspondence: salim.yilmaz@acibadem.edu.tr

**Abstract:** Food loss and waste pose significant social, economic, and environmental challenges worldwide, threatening food security and hindering sustainable development. While developing countries primarily face losses during production and storage, developed nations struggle with waste driven by consumer habits, spoilage, and overstocking, particularly in markets, restaurants, and homes. This study was aimed to analyze the complex relationships between food loss, waste, and various economic and environmental variables. The study examined the effects of variables such as education expenditures, food security, food prices, greenhouse gas emissions, and carbon emissions per capita on food losses and waste. These analyses shed light on the development of sustainable food policies at both national and global levels. Interventions to reduce food loss and waste will not only optimize food production and consumption processes but will also support a sustainable management of resources. As a result, this study aimed to understand the long-term effects of food loss and waste on economic growth, environmental sustainability, and social welfare. The findings of the study are of great importance in terms of directing future policies and aim to be an important guide in the transition to sustainable food systems.

**Keywords:** sustainable food systems; greenhouse gas emissions; food loss and waste

## 1. Introduction

Food loss and waste stands out as one of the biggest social, economic, and environmental problems considering the increasing population and limited natural resources worldwide. According to the Food and Agriculture Organization of the United Nations (FAO) [1], approximately one-third of the food produced globally is wasted or lost every year. This rate shows a serious problem not only for developing countries but also for developed countries [2]. The high levels of food loss and waste jeopardize food security worldwide and also make it difficult to achieve sustainable development goals [3]. From food production to consumption, there may be many losses. While food losses that occur during the production, harvest, transportation, and storage processes are a fundamental problem in developing countries, consumer habits and waste culture constitute the main causes of food losses in developed countries [4–6]. Waste that occurs during the consumption stage is caused by reasons such as spoilage, damaged packaging, misinterpretation of expiration dates, overproduction, and overstocking especially in markets, restaurants, and homes. These problems reduce food supply efficiency and negatively affect price stability.

The impact of food losses and waste is not limited to the economic issues [7]; it also leads to serious environmental problems [4]. The waste of natural resources such as energy, water, and soil used in food production creates great pressure on ecosystems. In addition, greenhouse gas emissions (GHG) resulting from the sending of discarded food to landfills contribute to climate change and make it difficult to achieve environmental sustainability goals [8]. In this context, reducing food losses and waste provides economic gain and contributes to reduce the carbon footprint and combating climate change [9].

The scope and methodology of the present study have been well-designed to increase the reliability of the results. In this context, long-term and short-term relationships between variables were examined using advanced econometric methods [10–13]. The analyses conducted aimed to reveal guiding information to policy makers on the subject by comprehensively addressing the relationships of food loss and waste with different variables. Additionally, this study offers a unique contribution to the literature by employing a time series approach to explore the long-term and lagged relationships between food loss and waste (FLW) and key economic and environmental factors, including education expenditure, greenhouse gas emissions, food insecurity, GDP per capita, and food prices. Unlike many existing studies that rely on cross-sectional or panel data, our analysis captures temporal dynamics and causal pathways, providing deeper insights into the evolving nature of these interactions. Additionally, the integration of structural break analysis allows us to identify critical shifts in FLW trends, shedding light on the impact of external shocks, such as economic downturns, on sustainability outcomes. These methodological and analytical advancements position the study as a valuable resource for policymakers and researchers working towards sustainable food systems.

## 2. Materials and Methods

The aim of the study was to analyze the dynamic relationships between food loss and waste and greenhouse gas emission emissions and their effects on the economic, environmental, and social consequences of these factors by analyzing the dynamic relationships between variables that are critical for sustainability and to examine the relationships of food loss and waste with various indicators such as economic growth through gross domestic product per capita, the share of education expenditures in total government expenditures, greenhouse gas emissions, food prices, and food insecurity. Thus, it was aimed to support the road maps for societies to reduce food loss and waste and greenhouse gas emissions and to achieve sustainable development goals by revealing causal relationships. The study elaborates on the dynamic effects of social problems such as the impact of food price fluctuations on GHG emissions and the impact of food insecurity on food loss and waste. Through cointegration analyses including structural breaks, the long-term co-movement tendencies of food loss and waste and greenhouse gas emissions were examined. It was envisaged that changes in consumer behavior due to high food prices may contribute to sustainable food production. In this framework, the relationships between variables such as education expenditures, food insecurity, greenhouse gas emissions, population growth, and economic growth were evaluated with food loss and wastage from both short- and long-term perspectives, causality on each other were investigated, and ways to reduce food loss and wastage were discussed from the outputs obtained. In addition, the study also investigates whether the share of education expenditures has a time-dependent decreasing effect on food loss and food waste and evaluates its long-term effects.

The research was a longitudinal and causality explanatory time series study based on secondary data analysis. Table 1 provides information on the variables used and the sources.

**Table 1.** The definitions of the variables.

| Code | Variable Name | Source | Link | Year Range |
|---|---|---|---|---|
| FLW | Food Loss and Waste | FAO | https://www.fao.org/platform-food-loss-waste/flw-data/en/, accessed on 30 October 2024 | 1965–2022 |
| FP | Annual FAO Food Price Indices | FAO | https://www.fao.org/worldfoodsituation/foodpricesindex/en/, accessed on 30 October 2024 | 1990–2024 |
| GDP | GDP Per Capita | Our World in Data from World Bank (2023) | https://ourworldindata.org/grapher/gdp-per-capita-worldbank?time=latest, accessed on 30 October 2024 | 1990–2022 |
| POP | Population, total | World Bank (2022) | https://data.worldbank.org/indicator/SP.POP.TOTL, accessed on 30 October 2024 | 1960–2023 |
| ES | Education spending as a share of total government expenditure | Our World in Data from World Bank World Bank (2024) | https://ourworldindata.org/grapher/share-of-education-in-government-expenditure, accessed on 30 October 2024 | 2000–2022 |
| GHG | Greenhouse gas emissions (tones) | Jones et al. (2024)—with major processing by Our World in Data. | https://ourworldindata.org/greenhouse-gas-emissions, accessed on 30 October 2024 | 1850–2022 |
| FI | Prevalence of Severe Food Insecurity (%) | World Bank (2024) | https://microdata.worldbank.org/index.php/catalog/6103/study-description, accessed 30 October 2024 | 1999–2022 |

Among the variables obtained in Table 1, data from 2000 to 2022, which was the common year range, were included in the analysis. Since the share of education expenditures in total government expenditures was until 2021, the year 2022 was estimated, and the autoregressive integrated moving average (ARIMA) model was constructed for this. It was stated that the Food Loss and Waste variable includes transportation, storage, shelling, threshing, drying, harvesting, and winnowing activities. The sources from which the other variables were obtained, the coding and explanations of the variables, are presented in Table 1.

Figure 1 shows the level values and changes of the variables used between 2000 and 2022.

Food Insecurity Index, Food Loss and Waste Index, Food Price Per Capita, total GHG, GHG Per Capita (divided by population in the same year), population, and GDP Per Capita variables are given for the period between 2000 and 2022. Educational Spending's total government expenditures are given in Figure 1.

For the variable Educational Spending's share in total government expenditures, data from 2000 to 2021 were obtained from the source. However, since the 2022 data needed to be available as in other variables in order to establish a causal relationship with other variables, the 2022 data were estimated from its own historical data with the ARIMA model. For this, the Kwiatkowski–Phillips–Schmidt–Shin (KPSS) test was used from the stationarity assumption test, and it was determined that the data were not stationary in the first stage (Lagrange Multiplier [LM] test statistics$_1$ = 0.48616; Truncation lag$_1$ = 2; and $p_1$ = 0.04478). Thereupon, logarithmic transformation and first-degree difference were taken, and the stationarity problem was eliminated (LM test statistics$_2$ = 0.072683; Truncation lag$_2$ = 2; and $p_2$ = 0.10000). When the existence of seasonality was examined, it was thought that there may be a repetitive pattern in the function graph, and frequency spectrum analysis was performed (frequency bandwidth = 0.012), and when examined together with the autocorrelation function graph, a weak and insignificant autocorrelation (~0.2 < 0.4[sig.cutoff]) was determined; a pattern that changes in intensity and repeats itself in a variable manner in 2 to 3 years and an insignificant seasonality was observed. Accordingly, instead of a seasonal autoregressive integrated moving average (SARIMA), the autoregressive integrated moving average (ARIMA) model was established on the data from which the logarithmic transformation difference was taken, and the 2022 education expenditures were estimated by applying the inverse logarithmic transformation to the obtained estimation result:

$$y_t = Educational\ Spendingt\ and\ z_t = \log(y_t)\ for\ (2000 \leq t \leq 2021)$$

$$\Delta z_t = z_t - z_{t-1} = \log(y_t) - \log(y_{t-1})\ for\ (2001 \leq t \leq 2021)$$

$$\Delta z_t = c + \varnothing_1 \Delta z_{t-1} + \varnothing_2 \Delta z_{t-2} + \ldots + \epsilon_t\ for\ (2001 \leq t \leq 2021),\ \text{so}$$

$$\Delta \hat{z}_{2022} = c + \varnothing_1 \Delta z_t + \varnothing_2 \Delta z_{t-2} + \ldots + \epsilon_{t+1}\ and$$

$$\hat{y}_{2022} = \exp(z_{2021} + \Delta \hat{z}_{2022})$$

$$\hat{y}_{2022} = \exp\left(\log(y_{2021} + \Delta \hat{z}_{2022}\right) = y_{2021} * \exp(\Delta \hat{z}_{2022})$$

$$\hat{y}_{2022} = \exp(\log(y_{2021}) + 1.4213)$$

$$\hat{y}_{2022} = 4.1424$$

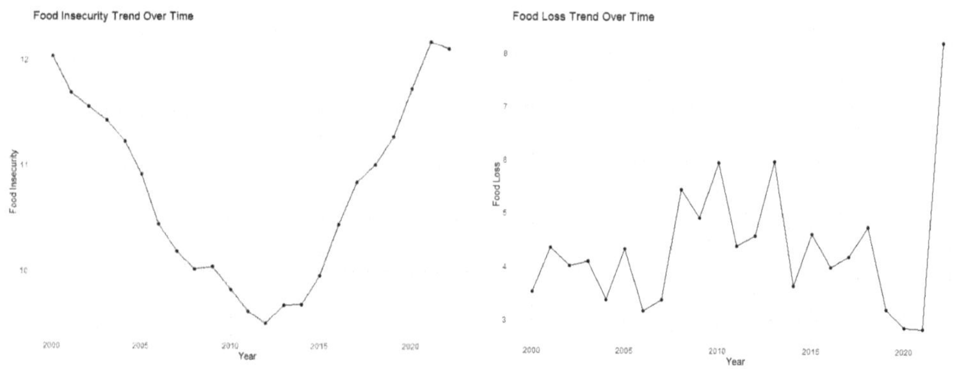


**Figure 1.** *Cont.*

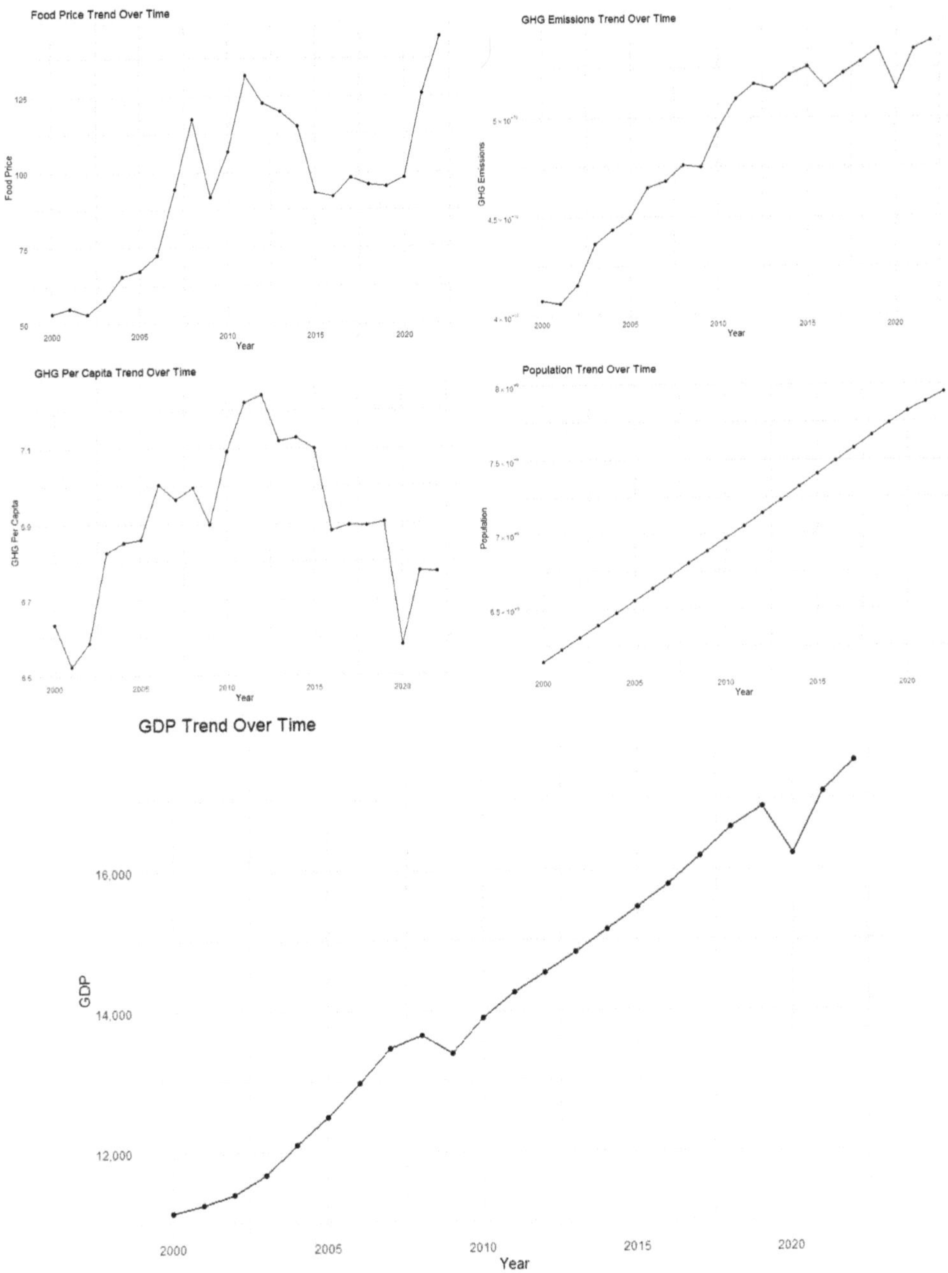


**Figure 1.** Change of variables between 2000 and 2022.

An ARIMA model was created using the data of Educational Spending's share in total government expenditures for the years 2000–2021, and a forecast was made for the year 2022. In order to evaluate the accuracy of the model, the estimated value was compared with the actual value, and the mean absolute error (MAE) and root mean square error (RMSE) were calculated. The results obtained based on the comparisons of the test and training datasets including the median year, the last two years, and the last year data showed that the MAE and RMSE values varied between 0.0345 and 0.0093 and remained at low levels. Thus, it was determined that the model's predictions had a high accuracy, and the error rates were minimal, and therefore the new data were reliable.

After the prediction, the education expenditure ratio in GDP between 2000 and 2022 is given in Figure 2.

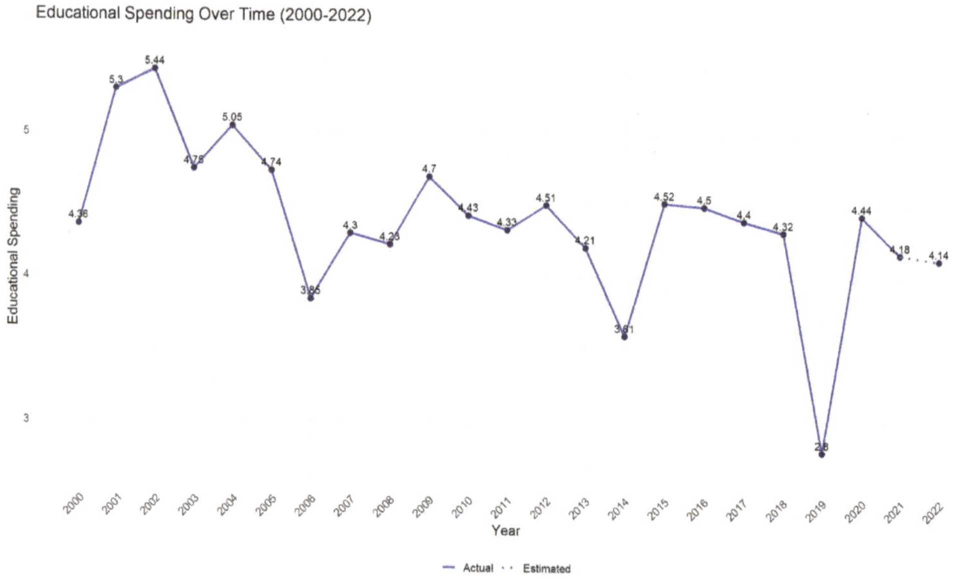


**Figure 2.** Trend of Educational Spending as a share of total government expenditure (2000–2022) with 2022 estimate.

Figure 2 showed the Educational Spending share of total government expenditures between 2000 and 2022, and 2022 was shown as estimated data with a dashed line. While rapid increases were seen in 2000–2002 and 2004, a sharp decline in 2019, thought to be due to the coronavirus-19 outbreak, was notable.

We used the R 4.4.1 program in the analysis of the data. In the assumption tests, the White test and Breusch–Pagan test were used for homoscedasticity control. Normality control was performed with the Jarque–Bera test. The Breusch–Godfrey test was used for autocorrelation control. The Kwiatkowski–Phillips–Schmidt–Shin test was used in unit root analyses. The Bai–Perron test was used to control multiple structural breaks, and the Chow test was used to control point breaks. The existence of cointegration was checked with the Bound test from the Wald test family. In determining the lag length, the Akaike Information Criterion (AIC) suggestion and the size of the explained variance were evaluated. In addition, the GHG total and Population Wald-Bound test were scaled to correspond to two-digit numbers. Toda–Yamamoto and Granger tests were used in causality tests. The Granger causality test is a widely used test to examine the causality relationship between two variables over time and evaluates the causality relationship by analyzing whether the past values of one series can explain the current values of another series [13]. On the other hand, the Granger test is sensitive to restrictions such as cointegration, different levels of stationarity, and the presence or absence of structural break; therefore the Toda–Yamamoto method was used as a complement when there is cointegration, structural break, or different levels of stationarity [11,13]. The test method developed by Toda and Yamamoto (1995) was an improved version of the traditional Granger causality test and allows causality analyses to be performed regardless of whether there is cointegration between the series [11]. The Toda–Yamamoto approach increases the lag length by considering the maximum degree of integration (dmax) of the model and then tests the causality relationships using the Wald test [10]. In the study, the Toda–Yamamoto method was preferred in some models in order to accurately detect short-term causality relationships between the series by considering cointegration and structural breaks between the variables. Since there was a total of 23 years of observation series, instead of Vector Error Correction Models in short-term series, Autoregressive Distributed Lag (ARDL), Unrestricted Error Correction Model (UECM), and Restricted Error Correction Model (RECM) models, which are effective in the short-term, were used [14]. Error Correction Term (ECT) was used in the RECM model. The ARDL model is a flexible model used to examine short-term and long-term relationships together and was preferred especially in cases where there may be a cointegration relationship between variables [15]. While ARDL allows analysis of series at different stationarity levels (only I(0) and I(1)), it was preferred to examine dynamic relationships in the short- and long-term. Then, the UECM model, which is a derivative of the ARDL model, was used to examine the balance dynamics and especially to observe how the long-term balance is corrected by short-term shocks [16]. The RECM model, as a restricted version of the UECM, was used to examine the dynamics of short-term changes in the long run [12].

The hypotheses determined for the research are as follows:

**H1a.** *The share of education expenditure in total government expenditure has a significant effect on Food Loss and Waste.*

**H1b.** *The share of education expenditure in total government expenditure is a causal determinant of Food Loss and Waste.*

**H2a.** *Food Insecurity has a significant effect on Food Loss and Waste.*

**H2b.** *Food Insecurity is a causal determinant of Food Loss and Waste.*

**H3a.** *Food Loss and Waste has a significant effect on Food Insecurity.*

**H3b.** *Food Loss and Waste is a causal determinant of Food Insecurity.*

**H4a.** *Food Loss and Waste has a significant effect on Food Price.*

**H4b.** *Food Loss and Waste is a causal determinant of Food Price.*

**H5a.** *Food Loss and Waste has a significant effect on greenhouse gas emissions (GHG).*

**H5b.** *Food Loss and Waste is a causal factor of greenhouse gas emissions (GHG).*

**H6a.** *Food Loss and Waste has a significant effect on greenhouse gas emissions per capita (GHG Per Capita).*

**H6b.** *Food Loss and Waste is a causal factor of greenhouse gas emissions per capita (GHG Per Capita).*

**H7a.** *Food Insecurity has a significant effect on greenhouse gas emissions (GHG).*

**H7b.** *Food Insecurity is a causal factor of greenhouse gas emissions (GHG).*

**H8a.** *Food Insecurity has a significant effect on greenhouse gas emissions per capita (GHG Per Capita).*

**H8b.** *Food Insecurity is a causal factor of greenhouse gas emissions per capita (GHG Per Capita).*

**H9a.** *Food Price has a significant effect on greenhouse gas emissions (GHG).*

**H9b.** *Food Price is a causal factor of greenhouse gas emissions (GHG).*

**H10a.** *Food Price has a significant effect on greenhouse gas emissions per capita (GHG Per Capita).*

**H10b.** *Food Price is a causal factor of greenhouse gas emissions per capita (GHG Per Capita).*

**H11a.** *Population has a significant effect on greenhouse gas emissions (GHG).*

**H11b.** *Population is a causal factor of greenhouse gas emissions (GHG).*

**H12a.** *GDP Per Capita has a significant effect on Food Loss and Waste.*

**H12b.** *GDP Per Capita is a causal factor of Food Loss and Waste.*

The equations of the research models are listed according to the models as follows:

Model 1:

$$FLW_t = \alpha_0 + \sum_{i=1}^{2} \beta_i FLW_{t-i} + \sum_{j=1}^{3} \gamma_j ES_{t-j} + \epsilon_t$$

$$FLW_t = \beta_0 + \sum_{i=1}^{2} \alpha_{1i} FLW_{t-i} + \sum_{j=3}^{dmax} \beta_{2j} FLW_{t-j} + \sum_{i=1}^{3} \theta_{1i} ES_{t-i} + \sum_{j=3}^{dmax} \gamma_{2j} ES_{t-j} + \epsilon_t$$

Model 2:

$$FLW_t = \alpha_0 + \sum_{i=1}^{4} \beta_i FLW_{t-i} + \sum_{j=1}^{3} \gamma_j ES_{t-j} + \delta_1 Dummy_{before2012} + \delta_2 Dummy_{after2012} + \epsilon_t$$

$$FLW_t = \beta_0 + \sum_{i=1}^{4} \alpha_{1i} FLW_{t-i} + \sum_{j=4}^{dmax} \beta_{2j} FLW_{t-j} + \sum_{i=1}^{3} \theta_{1i} ES_{t-i} + \sum_{j=3}^{dmax} \gamma_{2j} ES_{t-j} + \delta_1 Dummy_{before2012} + \delta_2 Dummy_{after2012} + \epsilon_t$$

Model 3:

$$FI_t = \alpha_0 + \sum_{i=1}^{3} \beta_i FI_{t-i} + \sum_{j=1}^{3} \gamma_j FLW_{t-j} + \delta_1 Dummy_{before2012} + \delta_1 Dummy_{after2012} + \epsilon_t$$

$$FI_t = \beta_0 + \sum_{i=1}^{3} \alpha_{1i} FI_{t-i} + \sum_{j=3}^{dmax} \beta_{2j} FI_{t-j} + \sum_{i=1}^{3} \theta_{1i} FLW_{t-i} + \sum_{j=3}^{dmax} \gamma_{2j} FLW_{t-j} + \delta_1 Dummy_{before2012} + \delta_2 Dummy_{after2012} + \epsilon_t$$

Model 4:

$$FP_t = \alpha_0 + \sum_{i=1}^{3} \beta_i FP_{t-i} + \sum_{j=1}^{3} \gamma_j FLW_{t-j} + \epsilon_t$$

$$FP_t = \beta_0 + \sum_{i=1}^{3} \alpha_{1i} FP_{t-i} + \sum_{j=3}^{dmax} \beta_{2j} FP_{t-j} + \sum_{i=1}^{3} \theta_{1i} FLW_{t-i} + \sum_{j=3}^{dmax} \gamma_{2j} FLW_{t-j} + \epsilon_t$$

Model 5:

$$GHG_t = \alpha_0 + \sum_{i=1}^{2} \beta_i GHG_{t-i} + \gamma FLW_t + \epsilon_t$$

$$\Delta GHG_t = \alpha_0 + \sum_{i=1}^{1} \beta_i \Delta GHG_{t-i} + \gamma \Delta FLW_t + \lambda (GHG_{t-1} - \phi FLW_{t-1}) + \epsilon_t$$

$$\Delta GHG_t = \alpha_0 + \sum_{i=1}^{1} \beta_i \Delta GHG_{t-i} + \gamma \Delta FLW_t + \lambda ECT_{t-1} + \epsilon_t$$

$$GHG_t = \beta_0 + \sum_{i=1}^{2} \alpha_i GHG_{t-i} + \sum_{j=0}^{0} \theta_{1j} FLW_{t-j} + \epsilon_t$$

Model 6:

$$GHGPerCapita_t = \alpha_0 + \sum_{i=1}^{3} \beta_i GHGPerCapita_{t-i} + \sum_{j=1}^{3} \gamma_j FLW_{t-j} + \delta_1 Dummy_{before2012} + \delta_2 Dummy_{after2012} + \epsilon_t$$

$$\Delta GHGPerCapita_t = \alpha_0 + \sum_{i=1}^{2} \beta_i \Delta GHGPerCapita_{t-i} + \sum_{j=1}^{3} \gamma_j \Delta FLW_{t-j} + \lambda (GHGPerCapita_{t-1} - \phi FLW_{t-1}) + \delta_1 Dummy_{before2012} + \delta_2 Dummy_{after2012} + \epsilon_t$$

$$\Delta GHGPerCapita_t = \alpha_0 + \sum_{i=1}^{2} \beta_i \Delta GHGPerCapita_{t-i} + \sum_{j=1}^{3} \gamma_j \Delta FLW_{t-j} + \lambda ECT_{t-1} + \delta_1 Dummy_{before2012} + \delta_2 Dummy_{after2012} + \epsilon_t$$

$$GHGPerCapita_t = \beta_0 + \sum_{i=1}^{3} \alpha_{1i} GHGPerCapita_{t-i} + \sum_{j=3}^{dmax} \beta_{2j} GHGPerCapita_{t-j} + \sum_{i=1}^{3} \theta_{1i} FLW_{t-i} + \sum_{j=3}^{dmax} \gamma_{2j} FLW_{t-j} + \delta_1 Dummy_{before2012} + \delta_2 Dummy_{after2012} + \epsilon_t$$

Model 7:

$$GHG_t = \alpha_0 + \sum_{i=1}^{2} \beta_i GHG_{t-i} + \sum_{j=1}^{2} \gamma_j FI_{t-j} + \delta_1 Dummy_{before2012} + \delta_2 Dummy_{after2012} + \epsilon_t$$

$$\Delta GHG_t = \alpha_0 + \sum_{i=1}^{1} \beta_i \Delta GHG_{t-i} + \sum_{j=1}^{2} \gamma_j \Delta FI_{t-j} + \lambda (GHG_{t-1} - \phi FI_{t-1}) + \delta_1 Dummy_{before2012} + \delta_2 Dummy_{after2012} + \epsilon_t$$

$$\Delta GHG_t = \alpha_0 + \sum_{i=1}^{1} \beta_i \Delta GHG_{t-i} + \sum_{j=1}^{2} \gamma_j \Delta FI_{t-j} + \lambda ECT_{t-1} + \delta_1 Dummy_{before2012} + \delta_2 Dummy_{after2012} + \epsilon_t$$

$$GHG_t = \beta_0 + \sum_{i=1}^{2} \alpha_{1i} GHG_{t-i} + \sum_{j=2}^{dmax} \beta_{2j} GHG_{t-j} + \sum_{i=1}^{2} \theta_{1i} FI_{t-i} + \sum_{j=2}^{dmax} \gamma_{2j} FI_{t-j} + \delta_1 Dummy_{before2012} + \delta_2 Dummy_{after2012} + \epsilon_t$$

Model 8:

$$GHGPerCapita_t = \alpha_0 + \sum_{i=1}^{2} \beta_i GHGPerCapita_{t-i} + \sum_{j=1}^{3} \gamma_j FP_{t-j} + \delta_1 Dummy_{before2012} + \delta_2 Dummy_{after2012} + \epsilon_t$$

$$\Delta GHGPerCapita_t = \alpha_0 + \sum_{i=1}^{1} \beta_i \Delta GHGPerCapita_{t-i} + \sum_{j=1}^{3} \gamma_j \Delta FP_{t-j} + \lambda(GHGPerCapita_{t-1} - \phi FP_{t-1}) + \delta_1 Dummy_{before2012} + \delta_2 Dummy_{after2012} + \epsilon_t$$

$$\Delta GHGPerCapita_t = \alpha_0 + \sum_{i=1}^{1} \beta_i \Delta GHGPerCapita_{t-i} + \sum_{j=1}^{3} \gamma_j \Delta FP_{t-j} + \lambda ECT_{t-1} + \delta_1 Dummy_{before2012} + \delta_2 Dummy_{after2012} + \epsilon_t$$

$$GHGPerCapita_t = \beta_0 + \sum_{i=1}^{2} \alpha_{1i} GHGPerCapita_{t-i} + \sum_{j=2}^{dmax} \beta_{2j} GHGPerCapita_{t-j} + \sum_{i=1}^{3} \theta_{1i} FP_{t-i} + \sum_{j=3}^{dmax} \gamma_{2j} FP_{t-j} + \delta_1 Dummy_{before2012} + \delta_2 Dummy_{after2012} + \epsilon_t$$

Model 9:

$$GHG_t = \alpha_0 + \sum_{i=1}^{2} \beta_i GHG_{t-i} + \gamma FLW_t + \epsilon_t$$

$$\Delta GHG_t = \alpha_0 + \sum_{i=1}^{1} \beta_i \Delta GHG_{t-i} + \gamma \Delta FLW_t + \lambda(GHG_{t-1} - \phi FLW_{t-1}) + \epsilon_t$$

$$\Delta GHG_t = \alpha_0 + \sum_{i=1}^{1} \beta_i \Delta GHG_{t-i} + \gamma \Delta FLW_t + \lambda ECT_{t-1} + \epsilon_t$$

$$GHG_t = \beta_0 + \sum_{i=1}^{2} \alpha_{1i} GHG_{t-i} + \sum_{i=0}^{0} \theta_{1i} FLW_{t-j} + \epsilon_t$$

Model 10:

$$GHGPerCapita_t = \alpha_0 + \sum_{i=1}^{2} \beta_i GHGPerCapita_{t-i} + \sum_{j=1}^{3} \gamma_j FP_{t-j} + \delta_1 Dummy_{before2012} + \delta_2 Dummy_{after2012} + \epsilon_t$$

$$\Delta GHGPerCapita_t = \alpha_0 + \sum_{i=1}^{1} \beta_i \Delta GHGPerCapita_{t-i} + \sum_{j=1}^{3} \gamma_j \Delta FP_{t-j} + \lambda(GHGPerCapita_{t-1} - \phi FP_{t-1}) + \delta_1 Dummy_{before2012} + \delta_2 Dummy_{after2012} + \epsilon_t$$

$$\Delta GHGPerCapita_t = \alpha_0 + \sum_{i=1}^{1} \beta_i \Delta GHGPerCapita_{t-i} + \sum_{j=1}^{3} \gamma_j \Delta FP_{t-j} + \lambda ECT_{t-1} + \delta_1 Dummy_{before2012} + \delta_2 Dummy_{after2012} + \epsilon_t$$

$$GHGPerCapita_t = \beta_0 + \sum_{i=1}^{2} \alpha_{1i} GHGPerCapita_{t-i} + \sum_{j=2}^{dmax} \beta_{2j} GHGPerCapita_{t-j} + \sum_{i=1}^{3} \theta_{1i} FP_{t-i} + \sum_{j=3}^{dmax} \gamma_{2j} FP_{t-j} + \delta_1 Dummy_{before2012} + \delta_2 Dummy_{after2012} + \epsilon_t$$

Model 11:

$$GHG_t = \alpha_0 + \sum_{i=1}^{2} \beta_i GHG_{t-i} + \gamma POP_t + \epsilon_t$$

$$\Delta GHG_t = \alpha_0 + \sum_{i=1}^{1} \beta_i \Delta GHG_{t-i} + \gamma \Delta POP_t + \lambda (GHG_{t-1} - \phi POP_{t-1}) + \epsilon_t$$

$$\Delta GHG_t = \alpha_0 + \sum_{i=1}^{1} \beta_i \Delta GHG_{t-i} + \gamma \Delta POP_t + \lambda ECT_{t-1} + \epsilon_t$$

$$GHG_t = \beta_0 + \sum_{i=1}^{2} \alpha_i GHG_{t-i} + \sum_{j=0}^{0} \theta_{1j} POP_{t-j} + \epsilon_t$$

Model 12:

$$FLW_t = \alpha_0 + \sum_{i=1}^{2} \beta_i FLW_{t-i} + \sum_{j=1}^{1} \gamma_j GDPPerCapita_{t-j} + \epsilon_t$$

$$FLW_t = \beta_0 + \sum_{i=1}^{2} \alpha_{1i} FLW_{t-i} + \sum_{j=2}^{dmax} \beta_{2j} FLW_{t-j} + \sum_{i=1}^{1} \theta_{1i} GDPPerCapita_{t-i} + \sum_{j=1}^{dmax} \gamma_{2j} GDPPerCapita_{t-j} + \epsilon_t$$

In the present study, it was aimed to observe the trend effects that change depending on the break as well as the long- and short-term relationships between the variables through the relevant models, and the breaks were in 2012 and were the same for both GHG Per Capita and Food Insecurity. We also interpreted the effects of the break points analyzed with trend dummy variables. While a 10% margin of error was accepted in the cointegration and causality tests and long-term balance relationships and causal trends were evaluated, we used a 5% margin of error in order to provide a stronger level of significance in the models in general. While examining the short-term and long-term dynamic relationships with these approaches, we determined the existence or absence of direct causal relationships between the variables. In some models, the lag length could not be included in the independent variable according to the AIC and explained variance, and therefore the short-term could be examined. In addition, in some models, long-term dynamic relationships could not be examined, and cointegration could not be interpreted because H0 hypothesis could not be rejected as a result of the Bound test. The equations given above were created in this direction, and the analysis was performed accordingly.

## 3. Results

For the selection of causality analysis and suitability for the ARDL model, the stationarity of the variables was tested with unit root analyses (Table 1).

In unit root analyses, the Kwiatkowski–Phillips–Schmidt–Shin test was selected according to trend or fixed existence, and Educational Spending and Greenhouse Gas Emission were made stationary by taking the first difference, and the others were determined to be stationary as a result of the test. Accordingly, the use of variables in the models and the selection of causality analyses were carried out (Table 2).

Cointegration analyses were examined according to models in Table 2, while p is the lag of the dependent variable, q represents the lag of the independent variable, and these were selected by evaluating them together according to the AIC and R2 of the models. In addition, cointegration was examined with the Wald-Bound test, and Bai–Perron's structural breaks were checked with the Chow test, and if there was no break, and if there was cointegration, the Granger causality was selected; if there was no structural break or cointegration, the Toda–Yamamoto causality analysis was selected. If there was cointegration, the ARDL/UECM-RECM models were selected; otherwise, only the ARDL model was used. According to this information, no cointegration was found between the share of Educational Spending in total government expenditures and Food Loss, and accordingly, the ARDL and Toda–Yamamoto methods were selected. A break was detected between Food Insecurity and Food Loss, but no cointegration was found; ARDL and Toda–Yamamoto methods were proposed. There was a break between Food Loss and Food Insecurity, but there was no cointegration; the ARDL and Toda–Yamamoto methods were proposed. No cointegration was found between Food Loss and Food Price; the ARDL and Toda–Yamamoto methods were used. There was cointegration between Food Loss and GHG; the ARDL, UECM, and RECM models were selected, and the Granger causality method was preferred. Break and cointegration were detected between Food Loss and GHG Per Capita, and the ARDL or UECM-RECM and Toda–Yamamoto methods were selected. It was found that there was a break and significant cointegration between Food Insecurity and GHG, and the ARDL, UECM, and RECM methods and the Toda–Yamamoto method were selected. It was seen that there was a break and strong cointegration between Food Insecurity and GHG Per Capita, and accordingly the ARDL or UECM-RECM models and the Toda–Yamamoto causality method were applied. Significant cointegration was detected between Food Price and GHG, and since it was understood that there was no structural break, the ARDL and UECM-RECM models were used, and Granger was selected as the causality analysis method. Since there was a break and cointegration between Food Price and GHG Per Capita, the ARDL and UECM-RECM models were preferred, and the Toda–Yamamoto causality method was preferred. Cointegration was detected between Population and GHG, and since it was determined that there was no structural break, the ARDL and UECM-RECM models were used, and the Granger causality method was preferred. It was seen that there was no cointegration between GDP Per Capita and Food Loss; therefore the ARDL model and the Toda–Yamamoto method were used (Table 3).

**Table 2.** Unit root tests.

| Variable Codes | Result | KPSS Test Result ($p$) for $H_A$ on Non-Stationary | Stationary Hypothesis Result |
|---|---|---|---|
| FLW | Stationary | Level: 0.1379, $p \geq 0.1$ | $H_0$ cannot be rejected. |
| FP | Stationary | Level: 0.1400, $p = 0.061$ | $H_0$ cannot be rejected. |
| GDP Per Capita | Stationary | Level: 0.0636, $p \geq 0.1$ | $H_0$ cannot be rejected. |
| POP | Stationary | Level: 0.1043, $p \geq 0.1$ | $H_0$ cannot be rejected. |
| ES | First-Degree Differencing than Stationary | Level: 0.0706, $p \geq 0.1$ | $H_0$ cannot be rejected after diff. |
| GHG | First-Degree Differencing than Stationary | Level: 0.0740, $p \geq 0.1$ | $H_0$ cannot be rejected after diff. |
| GHG Per Capita | Stationary | Level: 0.2379, $p \geq 0.1$ | $H_0$ cannot be rejected. |
| FI | Stationary | Level: 0.2230, $p \geq 0.1$ | $H_0$ cannot be rejected. |

Kwiatkowski–Phillips–Schmidt–Shin test.

**Table 3.** Cointegration and breakpoint analysis results.

| No. | Models | p | q | Wald-Bound Test | | Bai–Perron and Chow Test | Result | Suggested Model | Causality Method |
|---|---|---|---|---|---|---|---|---|---|
| | | | | F Value | $p$-Value | | | | |
| 1 | FoodLoss <- EducationalSpending | 2 | 3 | 2.8281 | 0.2000 | No breakpoint | No cointegration | ARDL | Toda–Yamamoto |
| 2 | FoodLoss <- FoodInsecurity | 4 | 3 | 1.3500 | 0.7564 | Breakpoint | No cointegration | ARDL | Toda–Yamamoto |
| 3 | FoodInsecurity <- FoodLoss | 3 | 3 | 1.5663 | 0.6508 | Breakpoint | No cointegration | ARDL | Toda–Yamamoto |
| 4 | FoodPrice <- FoodLoss | 3 | 3 | 1.3928 | 0.6674 | No breakpoint | No cointegration | ARDL | Toda–Yamamoto |
| 5 | GHG <- FoodLoss | 2 | 0 | 5.9296 | 0.0062 * | No breakpoint | Cointegration | ARDL/UECM-RECM | Granger |

**Table 3.** *Cont.*

| No. | Models | p | q | Wald-Bound Test | | Bai–Perron and Chow Test | Result | Suggested Model | Causality Method |
|---|---|---|---|---|---|---|---|---|---|
| | | | | F Value | *p*-Value | | | | |
| 6 | GHG Per Capita <- FoodLoss | 3 | 3 | 3.8785 | 0.0877 * | Breakpoint | Cointegration | ARDL/UECM-RECM | Toda–Yamamoto |
| 7 | GHG <- FoodInsecurity | 2 | 2 | 8.0017 | 0.0008 * | Breakpoint | Cointegration | ARDL/UECM-RECM | Toda–Yamamoto |
| 8 | GHG Per Capita <- FoodInsecurity | 1 | 3 | 8.7465 | 0.0001 * | Breakpoint | Cointegration | ARDL/UECM-RECM | Toda–Yamamoto |
| 9 | GHG <- FoodPrice | 2 | 0 | 9.2837 | 0.0004 * | No breakpoint | Cointegration | ARDL/UECM-RECM | Granger |
| 10 | GHG Per Capita <- FoodPrice | 2 | 3 | 4.4151 | 0.0142 | Breakpoint | Cointegration | ARDL/UECM-RECM | Toda–Yamamoto |
| 11 | GHG <- Population | 2 | 0 | 5.5857 | 0.0097 | No breakpoint | Cointegration | ARDL/UECM-RECM | Granger |
| 12 | FoodLoss <- GDPPerCapita | 2 | 1 | 2.3421 | 0.3158 | No breakpoint | No cointegration | ARDL | Toda–Yamamoto |

p: dependent variable and optimal lag for AIC or $R^2$ decision; q: independent variable and optimal lag for AIC or $R^2$ decision; *: significant at 10% margin of error.

Table 4 shows the modeling of temporal dynamics between the share of Educational Spending in total government expenditures and Food Loss and Waste. The 3-year lagged effect of Educational Spending (β = −1.4737) is significant (*p* = 0.0151) and decreasing on Food Loss and Waste. However, the model can explain 12.22% of the variance and gives insignificant results (*p* = 0.2723).

**Table 4.** Model 1: the effect of Educational Spending as a share of total government expenditure on Food Loss and Waste.

| | Models | Independent Variable | β | SE | t | *p* |
|---|---|---|---|---|---|---|
| Model 1: Food Loss and Waste <- Educational Spending | ARDL | Intercept (Food Loss and Waste) | 6.5724 | 3.9256 | 1.6740 | 0.1180 |
| | | L (Food $Loss_{(1)}$) | −0.1944 | 0.3498 | −0.5560 | 0.5879 |
| | | L (Food $Loss_{(2)}$) | 0.1377 | 0.3749 | 0.3670 | 0.7192 |
| | | Educational Spending | 0.4043 | 0.6420 | 0.6300 | 0.5398 |
| | | L (Educational $Spending_{(1)}$) | 0.5243 | 0.6061 | 0.8650 | 0.4027 |
| | | L (Educational $Spending_{(2)}$) | 0.1204 | 0.5412 | 0.2220 | 0.8275 |
| | | **L (Educational $Spending_{(3)}$)** | **−1.4737** | **0.5271** | **−2.7960** | **0.0151 *** |

* significant at 5% margin of error; $R^2$: 0.3994; Adj $R^2$: 0.1222; F-statistic = 1.441; *p*-value = 0.2723.

In Table 5, the modelings of the temporal dynamics between Food Insecurity and Food Loss and Waste were shown. Food Insecurity, which showed a decreasing trend until 2012, resulted in a rapidly increasing trend since 2012, so dummy variables were included in the model due to the structural break and trend cycle of 2012. The instantaneous effect of Food Insecurity on Food Loss and Waste is statistically limited in significance and decreasing (β = −5.4931 and *p* = 0.0654). The model, which examined long-term relationships and included the trend, was insufficient to explain the variance.

In Table 6, the modeling of the temporal dynamics was established between Food Insecurity and Food Loss and Waste. Food Insecurity, which showed a decreasing trend until 2012, has shown a rapidly increasing trend since 2012, so the structural break and trend cycle of 2012 were included in the model. The 1-year lagged effect of Food Insecurity on itself was found to be increasing (β = 1.3836) and significant (*p* = 0.0014). The instantaneous effect of Food Loss and Waste tends to decrease Food Insecurity but has limited significance (*p* = 0.0850).

**Table 5.** Model 2: the impact of Food Insecurity on Food Loss and Waste.

| | Models | Independent Variable | β | SE | t | $p$ |
|---|---|---|---|---|---|---|
| Model 2: Food Loss and Waste <- Food Insecurity | ARDL | Intercept (Food Loss and Waste) | 3.4024 | 1.6497 | 2.0620 | 0.0731 |
| | | L (Food $Loss_{(1)}$) | −0.7763 | 0.6029 | −1.2880 | 0.2339 |
| | | L (Food $Loss_{(2)}$) | −0.4976 | 0.5697 | −0.8730 | 0.4079 |
| | | L (Food $Loss_{(3)}$) | −0.0812 | 0.6288 | −0.1290 | 0.9005 |
| | | L (Food $Loss_{(4)}$) | −0.1461 | 0.7886 | −0.1850 | 0.8576 |
| | | Food Insecurity | −5.4931 | 2.5747 | −2.1340 | 0.0654 |
| | | L (Food $Insecurity_{(1)}$) | 6.9392 | 4.3233 | 1.6050 | 0.1471 |
| | | L (Food $Insecurity_{(2)}$) | −2.7969 | 4.9105 | −0.5700 | 0.5846 |
| | | L (Food $Insecurity_{(3)}$) | −0.9204 | 3.2458 | −0.2840 | 0.7839 |
| | | Trend $Effect_{(before\ 2012)}$ | −0.0348 | 0.2235 | −0.1560 | 0.8800 |
| | | Trend $Effect_{(after\ 2012)}$ | −0.2097 | 0.3531 | −0.5940 | 0.5690 |

$R^2$: 0.4816; Adj $R^2$: −0.1665; F-statistic = 0.7431; $p$-value = 0.6763.

**Table 6.** Model 3: the impact of Food Loss and Waste on Food Insecurity.

| | Models | Independent Variable | β | SE | t | $p$ |
|---|---|---|---|---|---|---|
| Model 3: Food Insecurity <- Food Loss and Waste | ARDL | Intercept (Food Insecurity) | 2.0336 | 1.4054 | 1.4470 | 0.1785 |
| | | **L (Food $Insecurity_{(1)}$)** | **1.3836** | **0.3167** | **4.3690** | **0.0014 *** |
| | | L (Food $Insecurity_{(2)}$) | −0.8464 | 0.4704 | −1.8000 | 0.1021 |
| | | L (Food $Insecurity_{(3)}$) | 0.2855 | 0.3225 | 0.8850 | 0.3969 |
| | | Food Loss | −0.0591 | 0.0309 | −1.9120 | 0.0850 |
| | | L (Food $Loss_{(1)}$) | −0.0476 | 0.0617 | −0.7720 | 0.4582 |
| | | L (Food $Loss_{(2)}$) | 0.0066 | 0.0600 | 0.1100 | 0.9148 |
| | | L (Food $Loss_{(3)}$) | 0.0349 | 0.0568 | 0.6140 | 0.5528 |
| | | Trend $Effect_{(before\ 2012)}$ | −0.0033 | 0.0149 | −0.2200 | 0.8304 |
| | | Trend $Effect_{(after\ 2012)}$ | 0.0518 | 0.0351 | 1.4790 | 0.1700 |

* significant at 5% margin of error; for ARDL Model: $R^2$: 0.9815; Adj $R^2$: 0.9648; F-statistic = 58.81; $p$-value = $1.774 \times 10^{-7}$.

The temporal dynamics established between Food Price and Food Loss and Waste were shown in Table 7. The 1-year delayed effect of Food Price on itself was found to be increasing (β = 0.8022) and significant ($p$ = 0.0143). The instantaneous and delayed effects of Food Loss and Waste were not found to be significant ($p$ > 0.05).

In Table 8, the temporal dynamics established between GHG and Food Loss and Waste were evaluated. Although Food Loss and Waste increased GHG instantly, it had no significant effect ($p$ = 0.3381). GHG was affected by its previous period and had a positive increasing cumulative effect (β = 0.5591 and $p$ = 0.0241). In the long-term, it has been determined that GHG may resist fluctuations in the previous period and has a corrective significant effect on its trend in the current period (β = −0.6261 and $p$ = 0.0271). Apart from all these, the cointegration coefficient of Food Loss and Waste and GHG was positive (β = 38.6367) but does not have sufficient statistical significance ($p$ = 0.3713).

**Table 7.** Model 4: effect of Food Loss and Waste on Food Price.

| | Models | Independent Variable | β | SE | t | *p* |
|---|---|---|---|---|---|---|
| **Model 4: Food Price <- Food Loss and Waste** | ARDL | Intercept (Food Price) | 3.1739 | 2.97044 | 1.068 | 0.3063 |
| | | **L (Food $Price_{(1)}$)** | **0.8022** | **0.2803** | **2.862** | **0.0143 *** |
| | | L (Food $Price_{(2)}$) | −0.3854 | 0.3662 | −1.053 | 0.3133 |
| | | L (Food $Price_{(3)}$) | 0.3878 | 0.2833 | 1.369 | 0.1961 |
| | | Food Loss | 3.1516 | 3.3118 | 0.952 | 0.3601 |
| | | L (Food $Loss_{(1)}$) | −3.5135 | 4.0124 | −0.876 | 0.3984 |
| | | L (Food $Loss_{(2)}$) | −3.4758 | 4.3728 | −0.795 | 0.4421 |
| | | L (Food $Loss_{(3)}$) | 1.8831 | 4.7804 | 0.394 | 0.7006 |

* significant at 5% margin of error; for ARDL Model: $R^2$: 0.7445; Adj $R^2$: 0.5955; F-statistic = 4.996; *p*-value = 0.007495.

**Table 8.** Model 5: the impact of Food Loss and Waste on GHG.

| | Models | Independent Variable | β | SE | t | *p* |
|---|---|---|---|---|---|---|
| **Model 5: GHG <- Food Loss and Waste** | ARDL | **Intercept (GHG)** | **0.6275** | **0.2388** | **2.6280** | **0.0176 *** |
| | | **L ($GHG_{(1)}$)** | **0.5591** | **0.2258** | **2.4760** | **0.0241 *** |
| | | L ($GHG_{(2)}$) | 0.3148 | 0.2116 | 1.4880 | 0.1551 |
| | | Food Loss | 0.0162 | 0.0164 | 0.9860 | 0.3381 |
| | UECM | **Intercept (GHG)** | **0.6275** | **0.2388** | **2.6280** | **0.0176 *** |
| | | **L (d_$GHG_{(1)}$)** | **−0.1261** | **0.0495** | **−2.5470** | **0.0208 *** |
| | | L ($FoodLoss_{(2)}$) | 0.0162 | 0.0164 | 0.9860 | 0.3381 |
| | | d (L(d_$GHG_{(1)}$) | −0.3148 | 0.2116 | −1.4880 | 0.1551 |
| | RECM | **Intercept (GHG)** | **0.0599** | **0.0202** | **2.9680** | **0.0086 *** |
| | | **L(ECT)** | **−0.6261** | **0.2588** | **−2.4190** | **0.0271 *** |
| | | d (Food Loss) | 0.0052 | 0.0130 | 0.3990 | 0.6951 |
| | Cointegration Coefficient | | 38.6367 | 38.6367 | 42.3303 | 0.9127 |

* significant at 5% margin of error; for ARDL Model: $R^2$: 0.9535; Adj $R^2$: 0.9453; F-statistic = 116.3; *p*-value = $1.575 \times 10^{-11}$.

The temporal dynamics established between GHG Per Capita and Food Loss and Waste are shown in Table 9. In addition, since GHG Per Capita followed an increasing trend before 2012 and a decreasing trend after 2012, the trend effect was also included in the model. GHG Per Capita's positive significant effect on itself with a 1-year delay was determined in the model (β = 0.8975). In addition, the inverse relationship with the trend before 2012 tended to turn into a positive relationship after 2012. GHG Per Capita tends to correct its instantly changing dynamics in the long-term (β = −1.2616), and this situation is statistically significant ($p = 0.0006$). It was also determined that changes in Food Loss and Waste tend to positively affect GHG Per Capita (β = 0.0281), but this was of limited statistical significance ($p = 0.0688$). It was determined that GHG Per Capita and Food Loss and Waste act together in the long-term (β = 42.9738), but the statistical significance was not sufficient ($p = 0.2420$).

**Table 9.** Model 6: the impact of Food Loss and Waste on GHG Per Capita.

| | Models | Independent Variable | β | SE | t | *p* |
|---|---|---|---|---|---|---|
| Model 6: GHG Per Capita <- Food Loss and Waste | ARDL | Intercept (GHG Per Capita) | −4.1631 | 4.1548 | −1.0020 | 0.3457 |
| | | **L (GHG Per Capita$_{(1)}$)** | **0.8975** | **0.2740** | **3.2760** | **0.0113 *** |
| | | L (GHG Per Capita$_{(2)}$) | 0.1064 | 0.3611 | 0.2950 | 0.7758 |
| | | L (GHG Per Capita$_{(3)}$) | 0.2549 | 0.3637 | 0.7010 | 0.5033 |
| | | L (GHG Per Capita$_{(4)}$) | 0.3189 | 0.3561 | 0.8950 | 0.3967 |
| | | Food Loss | −0.0272 | 0.0458 | −0.5930 | 0.5696 |
| | | L (Food Loss$_{(1)}$) | −0.0024 | 0.0455 | −0.0530 | 0.9587 |
| | | L (Food Loss$_{(2)}$) | 0.0718 | 0.0399 | 1.7980 | 0.1098 |
| | | L (Food Loss$_{(3)}$) | 0.0077 | 0.0390 | 0.1980 | 0.8480 |
| | | Trend Effect$_{(before\ 2012)}$ | 0.0321 | 0.0157 | 2.0440 | 0.0752 |
| | | Trend Effect$_{(after\ 2012)}$ | −0.0291 | 0.0280 | −1.0400 | 0.3288 |
| | UECM | **Intercept (GHG Per Capita)** | **4.6672** | **1.4366** | **3.2490** | **0.0087 *** |
| | | **L (GHGPerCapita$_{(1)}$)** | **−0.7544** | **0.2379** | **−3.1720** | **0.0100 *** |
| | | L (FoodLoss$_{(1)}$) | 0.1469 | 0.0786 | 1.8680 | 0.0913 |
| | | Trend Effect$_{(before\ 2012)}$ | −0.0025 | 0.0100 | −0.2550 | 0.8039 |
| | | Trend Effect$_{(after\ 2012)}$ | 0.0205 | 0.0143 | 1.4330 | 0.1823 |
| | | d (L(GHG Per Capita$_{(1)}$)) | −0.2222 | 0.2303 | −0.9650 | 0.3574 |
| | | d (L(GHG Per Capita$_{(2)}$)) | −0.1413 | 0.2995 | −0.4720 | 0.6471 |
| | | d (Food Loss) | 0.0294 | 0.0235 | 1.2500 | 0.2399 |
| | | d (L(Food Loss$_{(1)}$)) | −0.0865 | 0.0590 | −1.4670 | 0.1731 |
| | | d (L(Food Loss$_{(2)}$)) | −0.0379 | 0.0403 | −0.9420 | 0.3683 |
| | RECM | Intercept (GHG Per Capita) | −0.0065 | 0.0461 | −0.1410 | 0.8898 |
| | | **L (ECT)** | **−1.2616** | **0.2855** | **−4.4180** | **0.0006 *** |
| | | d (Food Loss) | 0.0281 | 0.0142 | 1.9710 | 0.0689 |
| | | Trend Effect$_{(before\ 2012)}$ | −0.0060 | 0.0069 | −0.8620 | 0.4030 |
| | | Trend Effect$_{(after\ 2012)}$ | 0.0086 | 0.0085 | 1.0100 | 0.3297 |
| | Cointegration Coefficient | | 42.97382 | 42.9738 | 35.7407 | 1.2024 |

* significant at 5% margin of error; for ARDL Model: $R^2$: 0.762; Adj $R^2$: 0.5477; F-statistic = 3.556; *p*-value = 0.03037.

The temporal dynamics established between GHG and Food Insecurity are shown in Table 10. The instantaneous effect of Food Insecurity on GHG was significant and decreasing (β = −0.3815 and $p$ = 0.0244). Short-term changes in Food Insecurity were also negatively related to GHG. In the long-term model, the trend after the 2012 break created a positive relationship between GHG and Food Insecurity, but this effect was determined to be of limited significance ($p$ = 0.0815). In addition, GHG and Food Insecurity tend to move together in the long-term (β = 1.0237), but this relationship was not statistically significant enough.

In Table 11, the temporal dynamics were established between GHG Per Capita and Food Insecurity. The instantaneous decreasing effect of Food Insecurity on GHG Per Capita was determined (β = −0.4918 and $p$ = 0.0009). The structural break in 2012 revealed a negative significant relationship between them (β = −0.0151 and $p$ = 0.0332). In addition, dynamic changes in Food Insecurity tend to decrease GHG Per Capita (β = −0.3815 and $p$ = 0.0244). This situation was also valid in the long-term, and Food Insecurity creates a decreasing effect on GHG Per Capita (β = −0.3450 and $p$ = 0.0338). In addition, GHG Per Capita and Food Insecurity have a tendency to move negatively in the long run (β = −2.0822), but its statistical significance was not sufficient ($p$ = 0.1320).

**Table 10.** Model 7: Food Insecurity's Effect on GHG.

| | Models | Independent Variable | β | SE | t | $p$ |
|---|---|---|---|---|---|---|
| Model 7: GHG <- Food Insecurity | ARDL | Intercept (GHG) | 3.8839 | 3.7308 | 1.0410 | 0.3168 |
| | | L ($GHG_{(1)}$) | 0.3202 | 0.2937 | 1.0900 | 0.2954 |
| | | L ($GHG_{(2)}$) | 0.2388 | 0.3146 | 0.7590 | 0.4614 |
| | | **Food Insecurity** | **−0.3815** | **0.1499** | **−2.5450** | **0.0244 *** |
| | | L (Food $Insecurity_{(1)}$) | 0.3803 | 0.2079 | 1.8290 | 0.0904 |
| | | L (Food $Insecurity_{(2)}$) | −0.1663 | 0.1403 | −1.1850 | 0.2572 |
| | | Trend $Effect_{(before\ 2012)}$ | −0.0009 | 0.0068 | −0.1280 | 0.9004 |
| | | Trend $Effect_{(after\ 2012)}$ | 0.0529 | 0.0579 | 0.9140 | 0.3771 |
| | UECM | Intercept (GHG) | 3.8839 | 3.7308 | 1.0410 | 0.3168 |
| | | L ($d\_GHG_{(1)}$) | −0.4410 | 0.4226 | −1.0430 | 0.3158 |
| | | L (Food $Insecurity_{(1)}$) | −0.1675 | 0.1725 | −0.9710 | 0.3492 |
| | | Trend $Effect_{(before\ 2012)}$ | −0.0009 | 0.0068 | −0.1280 | 0.9004 |
| | | Trend $Effect_{(after\ 2012)}$ | 0.0529 | 0.0579 | 0.9140 | 0.3771 |
| | | d (L($d\_GHG_{(1)}$) | −0.2388 | 0.3146 | −0.7590 | 0.4614 |
| | | **d (Food Insecurity)** | **−0.3815** | **0.1499** | **−2.5450** | **0.0244 *** |
| | | d (L(Food $Insecurity_{(1)}$)) | 0.1663 | 0.1403 | 1.1850 | 0.2572 |
| | RECM | **Intercept (GHG)** | **4.8631** | **0.1222** | **39.7820** | **$<2 \times 10^{-16}$ *** |
| | | L (ECT) | 0.3905 | 0.9692 | 0.4030 | 0.6927 |
| | | d (Food Insecurity) | 0.4008 | 0.3115 | 1.2870 | 0.2177 |
| | | Trend $Effect_{(before\ 2012)}$ | −0.0010 | 0.0222 | −0.0460 | 0.9637 |
| | | Trend $Effect_{(after\ 2012)}$ | 0.0436 | 0.0233 | 1.8680 | 0.0815 |
| | | Cointegration Coefficient | 1.023663 | 1.0237 | 2.5722 | 0.3980 |

* significant at 5% margin of error; for ARDL Model: $R^2$: 0.9711; Adj $R^2$: 0.9555; F-statistic = 62.33; $p$-value = $5.464 \times 10^{-9}$.

In Table 12, the temporal dynamics were established between GHG and Food Price. Food Price had an increasing effect on GHG in the short-term (β = 0.4681 and $p$ = 0.0304). It also had an increasing effect momentarily (β = 0.0025 and $p$ = 0.0237). In this model, GHG has been found to return to its own trend in the long-term and to be less affected and corrected by changing dynamics in the long-term (β = −0.7300 and $p$ = 0.0092). In the long-term, Food Price and GHG tend to move together (β = 294.8786), and this is of limited significance ($p$ = 0.0711).

In Table 13, the temporal dynamics were established between Food Price and GHG Per Capita. Structurally, the 2012 break has created a significant effect on the relationship between Food Price and GHG Per Capita, turning it negative (β = −0.0498 and $p$ = 0.0033). While short-term changes in Food Price have created a significant effect that makes it difficult for GHG Per Capita to return to its trend (β = 0.0060 and $p$ = 0.0157), GHG Per Capita has a corrective effect that quickly returns to its trend after short-term changing dynamics in the system (β = −1.0554 and $p$ = 0.0025). In the long-term, the changing dynamics of Food Price have created a significant effect that reduces GHG Per Capita (β = −1.1380 and $p$ = 0.0008). On the other hand, in the long run, GHG Per Capita and Food

Price move together (β = 378.9154), but this does not have a sufficient statistical significance ($p$ = 0.1557).

**Table 11.** Model 8: Food Insecurity's impact on GHG Per Capita.

| | Models | Independent Variable | β | SE | t | $p$ |
|---|---|---|---|---|---|---|
| Model 8: GHG Per Capita <- Food Insecurity | ARDL | **Intercept (GHG)** | **1.4228** | **2.4004** | **5.9280** | **0.0001 *** |
| | | L ($GHG_{(1)}$) | −0.5426 | 0.2509 | −2.1620 | 0.0515 |
| | | **Food Insecurity** | **−0.4918** | **0.1122** | **−4.3820** | **0.0009 *** |
| | | L (Food $Insecurity_{(1)}$) | 0.0168 | 0.2019 | 0.0830 | 0.9351 |
| | | L (Food $Insecurity_{(2)}$) | 0.3737 | 0.2174 | 1.7190 | 0.1113 |
| | | L (Food $Insecurity_{(3)}$) | −0.2278 | 0.1187 | −1.9200 | 0.0789 |
| | | **Trend $Effect_{(before\ 2012)}$** | **−0.0151** | **0.0063** | **−2.4060** | **0.0332 *** |
| | | Trend $Effect_{(after\ 2012)}$ | 0.0127 | 0.0149 | 0.8540 | 0.4098 |
| | UECM | Intercept (GHG) | 3.8839 | 3.7308 | 1.0410 | 0.3168 |
| | | L ($GHG_{(1)}$) | −0.4410 | 0.4226 | −1.0430 | 0.3158 |
| | | L (Food $Insecurity_{(1)}$) | −0.1675 | 0.1725 | −0.9710 | 0.3492 |
| | | Trend $Effect_{(before\ 2012)}$ | −0.0009 | 0.0068 | −0.1280 | 0.9004 |
| | | Trend $Effect_{(after\ 2012)}$ | 0.0529 | 0.0579 | 0.9140 | 0.3771 |
| | | d (L($GHG_{(1)}$)) | −0.2388 | 0.3146 | −0.7590 | 0.4614 |
| | | **d (L(Food Insecurity))** | **−0.3815** | **0.1499** | **−2.5450** | **0.0244 *** |
| | | d (L(Food $Insecurity_{(1)}$)) | 0.1663 | 0.1403 | 1.1850 | 0.2572 |
| | RECM | Intercept (GHG) | −0.0049 | 0.0633 | −0.0770 | 0.9399 |
| | | L (ECT) | −1.0240 | 0.6114 | −1.6750 | 0.1162 |
| | | **d (Food Insecurity)** | **−0.3450** | **0.1467** | **−2.3520** | **0.0338 *** |
| | | Trend $Effect_{(before\ 2012)}$ | −0.0037 | 0.0113 | −0.3290 | 0.7471 |
| | | Trend $Effect_{(after\ 2012)}$ | 0.0086 | 0.0112 | 0.7660 | 0.4564 |
| | Cointegration Coefficient | | −2.082163 | −2.0822 | 1.3310 | −1.5644 |

* significant at 5% margin of error; for ARDL Model: $R^2$: 0.9026; Adj $R^2$: 0.8458; F-statistic = 15.89, $p$-value = $3.417 \times 10^{-5}$.

**Table 12.** Model 9: Food Price Impact on GHG.

| | Models | Independent Variable | β | SE | t | $p$ |
|---|---|---|---|---|---|---|
| Model 9: GHG<- Food Price | ARDL | **Intercept (GHG)** | **0.9464** | **0.2432** | **3.8910** | **0.0012 *** |
| | | **L ($GHG_{(1)}$)** | **0.4681** | **0.1982** | **2.3620** | **0.0304 *** |
| | | L ($GHG_{(2)}$) | 0.3053 | 0.1778 | 1.7180 | 0.1040 |
| | | **Food Price** | **0.0025** | **0.0010** | **2.4840** | **0.0237 *** |
| | UECM | **Intercept (GHG)** | **0.9464** | **0.2432** | **3.8910** | **0.0012 *** |
| | | **L ($GHG_{(1)}$)** | **−0.2266** | **0.0618** | **−3.6680** | **0.0019 *** |
| | | **Food Price** | **0.0025** | **0.0010** | **2.4840** | **0.0237 *** |
| | | d (L($GHG_{(3)}$)) | −0.3053 | 0.1778 | −1.7180 | 0.1040 |
| | RECM | **Intercept (GHG)** | **0.0501** | **0.0183** | **2.7360** | **0.0141 *** |
| | | **L (ECT)** | −0.7300 | 0.2486 | −2.9360 | 0.0092 * |
| | | d (FoodPrice) | 0.0024 | 0.0012 | 1.9660 | 0.0659 |
| | Cointegration Coefficient | | 294.8786 | 294.8786 | 155.5032 | 1.8963 |

* significant at 5% margin of error; for ARDL Model: $R^2$: 0.964; Adj $R^2$: 0.9576; F-statistic = 151.6, $p$-value = $1.826 \times 10^{-12}$.

**Table 13.** Model 10: Food Price's impact on GHG Per Capita.

| | Models | Independent Variable | β | SE | t | *p* |
|---|---|---|---|---|---|---|
| **Model 10: GHG Per Capita <- Food Price** | **ARDL** | **Intercept (GHG Per Capita)** | **6.9200** | **1.7464** | **3.9620** | **0.0022 *** |
| | | L (GHG Per Capita$_{(1)}$) | −0.0109 | 0.2535 | −0.0430 | 0.9665 |
| | | L (GHG Per Capita $_{(2)}$) | −0.0445 | 0.2089 | −0.2130 | 0.8352 |
| | | Food Price | 0.0030 | 0.0019 | 1.5650 | 0.1458 |
| | | L (Food Price$_{(1)}$) | 0.0004 | 0.0026 | 0.1420 | 0.8896 |
| | | L (Food Price$_{(2)}$) | 0.0003 | 0.0025 | 0.1000 | 0.9218 |
| | | L (Food Price$_{(3)}$) | 0.0024 | 0.0019 | 1.2500 | 0.2371 |
| | | Trend Effect$_{(before\ 2012)}$ | 0.0083 | 0.0087 | 0.9490 | 0.3630 |
| | | **Trend Effect$_{(after\ 2012)}$** | **−0.0498** | **0.0133** | **−3.7350** | **0.0033 *** |
| | **UECM** | **Intercept (GHG Per Capita)** | **6.9200** | **1.7464** | **3.9620** | **0.0022 *** |
| | | **L (GHG Per Capita$_{(1)}$)** | **−1.0554** | **0.2705** | **−3.9010** | **0.0025 *** |
| | | **L (Food Price$_{(1)}$)** | **0.0060** | **0.0021** | **2.8530** | **0.0157 *** |
| | | Trend Effect$_{(before\ 2012)}$ | 0.0083 | 0.0087 | 0.9490 | 0.3630 |
| | | **Trend Effect$_{(after\ 2012)}$** | **−0.0498** | **0.0133** | **−3.7350** | **0.0033 *** |
| | | d (L(GHG Per Capita$_{(1)}$) | 0.0445 | 0.2089 | 0.2130 | 0.8352 |
| | | d (Food Price) | 0.0030 | 0.0019 | 1.5650 | 0.1458 |
| | | d (L(Food Price$_{(1)}$)) | −0.0027 | 0.0022 | −1.2290 | 0.2448 |
| | | d (L(Food Price$_{(2)}$)) | −0.0024 | 0.0019 | −1.2500 | 0.2371 |
| | **RECM** | **Intercept (GHG Per Capita)** | **−0.6904** | **0.1965** | **−3.5140** | **0.0031 *** |
| | | L (ECT) | −0.0065 | 0.0450 | −0.1450 | 0.8865 |
| | | **d (Food Price)** | **−1.1380** | **0.2659** | **−4.2790** | **0.0008 *** |
| | | **Trend Effect$_{(before\ 2012)}$** | **0.0035** | **0.0015** | **2.2890** | **0.0381 *** |
| | | Trend Effect$_{(after\ 2012)}$ | −0.0025 | 0.0074 | −0.3390 | 0.7399 |
| | Cointegration Coefficient | | 378.9154 | 378.9154 | 257.7324 | 1.4702 |

* significant at 5% margin of error; for ARDL Model: $R^2$: 0.7306; Adj $R^2$: 0.5347; F-statistic = 3.73, *p*-value = 0.02354.

The temporal dynamics were established between Population and GHG in Table 14. It was found that GHG has a significant cumulative increasing effect on itself (β = 0.6059 and $p = 0.0135$). In addition, GHG tends to correct itself by quickly returning to its trend in short-term changes with limited statistical significance (β = −0.5375 and $p = 0.0559$). In the long-term, it was found that there is a tendency for Population and GHG to move together (β = 6.3145), but this situation was not statistically significant ($p = 0.5351$).

In Table 15, the temporal dynamics established between GDP Per Capita and Food Loss and Waste. No statistically significant relationship was found in the two-year lagged model ($p > 0.05$). The instantaneous effect of GDP on Food Loss and Waste was found to be increasing (β = 0.3114) but was not found to have a sufficient statistical significance ($p = 0.770$). In addition, the model is far from being significant in explaining the variance ($p = 0.9624$).

**Table 14.** Model 11: Population impact on GHG.

| | Models | Independent Variable | β | SE | t | *p* |
|---|---|---|---|---|---|---|
| Model 11: GHG <- Population | ARDL | Intercept (GHG) | 0.5268 | 0.2974 | 1.7710 | 0.0945 |
| | | **L ($GHG_{(1)}$)** | **0.6059** | **0.2198** | **2.7570** | **0.0135 *** |
| | | L ($GHG_{(2)}$) | 0.1770 | 0.2289 | 0.7730 | 0.4500 |
| | | Population | 0.0851 | 0.1285 | 0.6620 | 0.5167 |
| | UECM | Intercept (GHG) | 0.5268 | 0.2974 | 1.7710 | 0.0945 |
| | | L (d_$GHG_{(1)}$) | −0.2171 | 0.1614 | −1.3450 | 0.1963 |
| | | Population | 0.0851 | 0.1285 | 0.6620 | 0.5167 |
| | | d (L($GHG_{(1)}$) | −0.1770 | 0.2289 | −0.7730 | 0.4500 |
| | RECM | Intercept (GHG) | 0.1026 | 0.2789 | 0.3680 | 0.7175 |
| | | L (ECT) | −0.5375 | 0.2619 | −2.0520 | 0.0559 |
| | | d (Population) | −0.5106 | 3.3843 | −0.1510 | 0.8818 |
| | **Cointegration Coefficient** | | 6.314508 | 6.3145 | 10.0202 | 0.6302 |

* significant at 5% margin of error; for ARDL Model: $R^2$: 0.9521; Adj $R^2$: 0.9437; F-statistic = 112.7; *p*-value = $2.032 \times 10^{-11}$.

**Table 15.** Model 12: GDP Per Capita's impact on Food Loss and Waste.

| | Models | Independent Variable | β | SE | t | *p* |
|---|---|---|---|---|---|---|
| Model 12: Food Loss and Waste <- GDP Per Capita | ARDL | Intercept (FoodLoss) | 3.2608 | 3.2967 | 0.9890 | 0.3370 |
| | | L($FoodLoss_{(1)}$) | −0.0014 | 0.3612 | −0.0040 | 0.9970 |
| | | L($FoodLoss_{(2)}$) | −0.1181 | 0.3711 | −0.3180 | 0.7540 |
| | | GDP | 0.3114 | 1.0483 | 0.2970 | 0.7700 |
| | | L($GDP_{(1)}$) | −0.2042 | 1.0442 | −0.1960 | 0.8470 |

$R^2$: 0.03508; Adj $R^2$: −0.2062; F-statistic = 0.1454; *p*-value = 0.9624.

When the causality analysis results were examined, it was found that Educational Spending, the share of total government expenditure, has no significant causality on Food Loss and Waste ($p = 0.1598$). Food Insecurity was a significant cause of Food Loss and Waste ($p < 0.001$). Food Loss and Waste had no causal effect on Food Insecurity ($p = 0.5134$). Again, Food Loss and Waste had no significant causality on Food Price ($p = 0.5652$). Also, Food Loss and Waste did not have a direct causal effect on GHG ($p = 0.8496$). On the other hand, Food Loss and Waste had a significant causality on GHG Per Capita ($p = 0.0219$). Food Insecurity had no causal effect on GHG ($p = 0.5805$). But Food Insecurity was a significant cause of GHG Per Capita ($p < 0.001$). Food Price did not have a direct causal effect on GHG ($p = 0.3193$). Also, Food Price had no causal effect on GHG Per Capita ($p = 0.9115$). Population was a significant cause of GHG ($p = 0.0171$). GDP Per Capita was a significant cause of Food Loss and Waste ($p = 0.0939$) (Table 16).

**Table 16.** Causality analysis results.

| No. | Model | $\chi^2$/F | df | *p* | Result |
|---|---|---|---|---|---|
| 1 | Food Loss and Waste <- Educational Spending | $7.9363_{\chi 2}$ | 5 | 0.1598 | $H_0$ could not be rejected. |
| 2 | Food Loss and Waste <- Food Insecurity | $76.3294_{\chi 2}$ | 7 | <0.001 *** | $H_0$ was rejected. |
| 3 | Food Insecurity <- Food Loss and Waste | $5.2403_{\chi 2}$ | 7 | 0.5134 | $H_0$ could not be rejected. |
| 4 | Food Price <- Food Loss and Waste | $4.8349_{\chi 2}$ | 6 | 0.5652 | $H_0$ could not be rejected. |

**Table 16.** *Cont.*

| No. | Model | $\chi^2$/F | df | *p* | Result |
|---|---|---|---|---|---|
| 5 | GHG <- Food Loss and Waste | $0.1639_F$ | $df_1$:2; $df_2$:30 | 0.8496 | $H_0$ could not be rejected. |
| 6 | GHG Per Capita <- Food Loss and Waste | $14.7953_{\chi 2}$ | 6 | 0.0219 ** | $H_0$ was rejected. |
| 7 | GHG <- Food Insecurity | $2.8660_{\chi 2}$ | 4 | 0.5805 | $H_0$ could not be rejected. |
| 8 | GHG Per Capita <- Food Insecurity | $41.2964_{\chi 2}$ | 7 | <0.001 *** | $H_0$ was rejected. |
| 9 | GHG <- Food Price | $1.1863_F$ | $df_1$:2; $df_2$: 30 | 0.3193 | $H_0$ could not be rejected. |
| 10 | GHG Per Capita <- Food Price | $2.0872_{\chi 2}$ | 6 | 0.9116 | $H_0$ could not be rejected. |
| 11 | GHG <- Population | $4.6773_F$ | $df_1$:2; $df_2$:30 | 0.0171 ** | $H_0$ was rejected. |
| 12 | Food Loss and Waste <- GDP Per Capita | $6.3950_{\chi 2}$ | 3 | 0.0939 * | $H_0$ was rejected. |

$\chi^2$:Toda–Yamamoto technique; F: Granger causality; *: $p < 0.1$; **: $p < 0.05$; ***: $p < 0.01$.

## 4. Discussion

This study explores the complex relationships between food loss and waste (FLW), greenhouse gas (GHG) emissions, and economic indicators such as education spending, food insecurity, GDP per capita, and food prices across short- and long-term frameworks. The analysis identifies several critical insights into how these variables interact, underscoring both direct and indirect impacts on sustainability efforts.

This study identifies a significant three-year lagged effect of education spending on reducing FLW, although evidence for a direct causal relationship remains insufficient. This finding suggests that sustainability issues, particularly food waste reduction, may not yet be a key focus within the educational sector. However, the delayed impact underscores the potential for long-term benefits from strengthened education-based sustainability programs. Greater emphasis on sustainability principles in education, particularly regarding food resource management, could lead to more immediate and potentially causal reductions in FLW. Previous research supports the role of education in food loss reduction [17,18] and facilitating food waste reduction [19]. Future studies using longer datasets could clarify whether the observed lagged relationship evolves into a definitive causal link, reinforcing the importance of education as a strategic driver in promoting sustainable food systems.

The relationship between food insecurity and FLW reveals complex dynamics. While no significant direct impact of food insecurity on FLW was found, evidence suggests a minor immediate reduction in FLW during heightened food insecurity, likely due to precautionary behaviors to conserve resources. Diana et al. [20] also observed reduced household food waste with increased food insecurity, though income significantly influenced this relationship.

A seven-year lagged model suggests that FLW may contribute to worsening food insecurity over time by reducing resource availability. Studies in Iran and China support this, showing that FLW reduction enhances food security by increasing resource access [21,22]. Extended datasets are needed to fully capture these long-term interactions.

Food insecurity exhibits a self-reinforcing trend, where higher levels persist and worsen over time. Wang et al. [23] highlighted this path dependency, noting transitions to high insecurity are more likely than reversals. These findings underscore the vicious cycle of food insecurity.

A structural break in 2012, marking a shift from declining to rising food insecurity, coincided with a stronger alignment between food insecurity and FLW trends. While FLW does not directly cause food insecurity, the causative link from food insecurity to FLW and the post-2012 trends emphasize the need for interventions to reduce FLW. Targeted efforts could mitigate food insecurity's impact and improve resource stability [4].

In line with prior studies [5,6], our findings suggest that FLW has a minimal impact on food prices with no clear causal relationship, though there is a slight, positive lagged effect after one year, which might reflect cumulative impacts. This observation aligns with Aschemann-Witzel et al. [24], who noted that food waste reduction does not directly lower food prices due to the indirect nature of these economic effects. However, the lagged effect suggests a potential cumulative impact, aligning with arguments by Buzby and Hyman [25] that FLW influences food supply dynamics, which might impact prices over the long-term. These cumulative effects likely emerge from broader economic fluctuations and input costs (e.g., fuel and labor), supporting the complex relationship between FLW and market stability [1].

Regarding GDP per capita, while no direct dynamic relationship with FLW was observed, a limited level of causality was detected based on prior FLW levels and a two-year lagged effect from GDP per capita. This indicates that GDP per capita may, to a limited extent, act as a contributing factor to FLW. Moreover, although not statistically significant, higher GDP per capita values show an immediate reactive association with increased FLW, suggesting that economic growth could indirectly lead to higher waste through increased production and consumption patterns [26].

Our findings indicate a positive cointegration between FLW and GHG emissions without a statistically significant causal relationship in the temporal analysis. However, a positive cointegration coefficient suggests a long-term tendency for them to move together, warranting further research with longer datasets. This may relate to food production's resource-intensive nature, as Scherhaufer et al. [7] noted. FLW shows a short-term tendency to increase GHG emissions, though this is not statistically significant, indicating stronger influences from other factors [27]. On the other hand, a significant causal relationship was found between FLW and GHG emissions per capita in a six-year lagged model. The results also showed that FLW and GHG per capita tend to move together in the long-term. Additionally, while the long-term effects were limited in significance, FLW was found to have an increasing effect on GHG per capita as it grows [2].

An important observation from the findings is that the GHG system demonstrates resilience to fluctuations in prior periods, indicating a tendency to maintain long-term stability. Moreover, GHG emissions per capita demonstrated a lagged self-reinforcing effect, with a corrective role in maintaining its overall trend. This underscores the necessity of swift and decisive measures to slow the upward trend in GHG emissions [28].

The study's identification of a structural break in 2012 corresponds with a decoupling between GHG emissions and population growth aligning with the literature suggesting that advancements in low-emission agricultural technologies and energy efficiency may help mitigate per capita emissions [29]. However, the resilience of GHG levels and their tendency to stabilize despite fluctuations imply that, without stringent policy efforts, natural reductions in emissions may be insufficient to achieve environmental targets [8]. Thus, the findings underscore the need for systematic reduction efforts in FLW and GHG emissions to ensure environmental sustainability and resource adequacy amidst growing populations. Our findings suggest a complex and dynamic relationship between food insecurity and total GHG emissions. Short-term reductions associated with rising food insecurity, contrast with a weak positive correlation in the long-term without statistical significance. This aligns with research by Vermeulen et al. [30], who found that food scarcity often coincides with reduced food production and emissions due to lower economic activity and consumption. Conversely, as food insecurity becomes more persistent, increasing demand for affordable food production might drive higher GHG emissions, reflecting similar findings by

Niles et al. [31]. The co-movement post-2012 suggests that socioeconomic and environmental stressors may exacerbate one another [32], highlighting the need for integrated responses to food insecurity and emissions.

In contrast, food insecurity emerges as a causal factor for per capita GHG emissions, displaying a significant negative association in the short- and long-term. The 2012 structural break amplifies this dynamic, with food insecurity continuing to rise while per capita GHG emissions decline. This inverse relationship likely reflects the compounded effects of economic hardship caused by food insecurity. Reduced access to food may lower per capita consumption and waste, thereby decreasing emissions. Additionally, rapid population growth relative to production capacity may exacerbate food scarcity, further suppressing per capita consumption and emissions [33].

The analysis reveals a long-term co-movement between GHG emissions and food prices, though food prices are not a direct cause of GHG emissions. In the short-term, rising food prices are positively associated with GHG emissions, which aligns with theories in environmental economics, particularly concerning the impact of fossil fuel prices on food production costs and related emissions [9]. Over time, higher food prices may incentivize energy-intensive production expansions and indirectly raise GHG emissions due to the carbon intensity of production processes [34].

A positive long-term association between food prices and per capita GHG emissions is also observed, without food prices being the direct cause. Interestingly, the study finds that long-term fluctuations in food prices tend to decrease per capita GHG emissions. This inverse effect may be attributed to reduced consumer access to food as prices rise, leading to decreased per capita consumption. Additionally, elevated food prices may encourage a shift toward lower-cost and possibly lower-emission production and consumption practices, like less animal-based and more plant-based foods, indirectly contributing to reduced emissions on a per capita basis [35].

The analysis reveals a long-term co-movement between GHG emissions and population growth, with the latter emerging as a significant driver in a two-year lagged model. Population growth amplifies GHG emissions cumulatively, reinforced by the self-perpetuating nature of emissions. These findings underscore the environmental impact of demographic pressure and highlight the need for strategies to reduce emissions in high-growth regions [36]. Population increases generally lead to greater resource consumption and waste production, amplifying GHG levels [37].

The complex dynamics between food loss and waste (FLW) and environmental and economic factors require a detailed understanding. While this study addresses many critical variables, there are potential effects of hidden or unobserved factors, such as economic growth and economic downturns during and prior to the pandemic. Economic growth, as a general driver of increased production capacity, consumption, and waste generation, may indirectly influence FLW. Furthermore, economic contraction periods, such as the global economic slowdown experienced during the COVID-19 pandemic, could have impacted food production, distribution, and consumer behavior, thereby altering FLW dynamics. Disruptions in supply chains and reduced purchasing power during economic recessions may have led to changes in food consumption patterns and waste generation.

Although the present study does not directly model economic growth or downturns, the inclusion of GDP per capita partially accounts for broader economic trends. However, we acknowledge that GDP alone may not fully capture the multifaceted impacts of economic fluctuations on FLW. Future research could integrate additional economic indicators, such as unemployment rates, inflation, or trade disruptions, to provide a more comprehensive perspective on these dynamics.

## 5. Limitations

The study faced several limitations primarily due to the restricted period of the dataset and the relatively small sample size of 23 observations. These constraints hindered the use of more robust models that could provide stronger results over the longer time series. The limited dataset also restricted the comprehensive modeling of lag lengths and impeded a deeper exploration of the dynamic relationships within the time series. Additionally, the use of estimated data for Educational Spending in 2022 introduced a minor degree of uncertainty that could slightly affect the accuracy of the findings. Another limitation was the focus on examining the temporal effects of a single independent variable at a time. This approach restricted the inclusion of external factors and other determinants in the models, limiting the ability to evaluate relationships between variables more holistically.

Nevertheless, the study hypotheses and analytical methods were appropriately designed to align with the dataset's structure, ensuring the findings were as reliable as possible despite the limited observations. Expanding the time series and increasing the number of observations in future research would enhance the robustness of the modeling and improve the generalizability of the results.

## 6. Conclusions

This study highlights the complex relationships between food loss and waste, greenhouse gas emissions, and economic indicators such as education expenditure, food insecurity, GDP per capita, and food prices. By emphasizing their direct and indirect impacts on sustainability, the findings underscore the necessity of targeted efforts to reduce FLW and GHG emissions. Long-term datasets and interdisciplinary approaches prove invaluable for understanding these dynamics, supporting resource security, and ensuring environmental sustainability.

The analysis reveals significant delayed effects of education expenditure and GDP per capita on FLW, highlighting the importance of long-term policy planning. The identification of a six-year lagged causal relationship between FLW and per capita GHG emissions provides new insights into the environmental impacts of food waste. Additionally, the incorporation of structural break analysis offers critical insights into temporal shifts, such as the post-2012 alignment of food insecurity and FLW trends, reflecting broader economic and social disruptions.

In light of these findings, the study presents actionable policy recommendations. Sustainability must be integrated into education programs without exception, as education-based interventions have the potential to yield long-term reductions in FLW. Furthermore, addressing per capita food waste requires not only fostering economic growth but also balancing social inequalities. Policies such as increasing incentives for sustainable agricultural technologies and implementing subsidy programs can enhance food security. Finally, the delayed effects of food waste on per capita GHG emissions emphasize the need for robust waste management strategies. Practices such as recycling food waste, utilizing it for energy production (e.g., biogas), and converting it into organic fertilizers are effective approaches to reduce these emissions.

While the study adopts a global perspective using time series analysis, future research could explore country-specific dynamics to provide more localized insights. Country-specific datasets and methods would better capture the unique cultural, economic, and political factors influencing FLW. However, modeling multiple countries simultaneously would require panel data analysis, a method distinct from the time series approach used here. Such complementary frameworks could further enrich the policy implications of these findings and support the development of sustainable food systems at both national and global levels.

**Author Contributions:** Conceptualization, S.Y. and A.M.G.; methodology, S.Y.; software, S.Y.; validation, A.M.G., G.K. and M.B.; formal analysis, S.Y.; investigation, S.Y., A.M.G. and G.K.; resources, S.Y. and A.M.G.; data curation, S.Y. and A.M.G.; writing—original draft preparation, A.M.G. and G.K.; writing—review and editing, A.M.G. and G.K.; visualization, M.B.; supervision, A.M.G. and G.K.; project administration, S.Y., A.M.G. and M.B.; funding acquisition, S.Y. All authors have read and agreed to the published version of the manuscript.

**Funding:** The APC was funded by Acibadem Mehmet Ali Aydinlar University.

**Institutional Review Board Statement:** Not applicable.

**Informed Consent Statement:** Not applicable.

**Data Availability Statement:** The data utilized in this study are openly accessible at the following link: https://www.dosya.tc/server2/9ww6b0/dataset.xlsx.html, accessed on 4 January 2025.

**Acknowledgments:** We would like to express our gratitude to Our World in Data for making a comprehensive dataset accessible to researchers and the public, which has significantly facilitated the scope of this research.

**Conflicts of Interest:** The authors declare no conflicts of interest.

## References

1. Food and Agriculture Organization of the United Nations. *The State of Food and Agriculture 2019: Moving Forward on Food Loss and Waste Reduction*; FAO: Rome, Italy, 2019.
2. Zhu, J.; Luo, Z.; Sun, T.; Li, W.; Zhou, W.; Wang, X.; Fei, X.; Tong, H.; Yin, K. Cradle-to-grave emissions from food loss and waste represent half of total greenhouse gas emissions from food systems. *Nat. Food* **2023**, *4*, 247–256. [CrossRef] [PubMed]
3. Tchonkouang, R.D.; Onyeaka, H.; Miri, T. From waste to plate: Exploring the impact of food waste valorisation on achieving zero hunger. *Sustainability* **2023**, *15*, 10571. [CrossRef]
4. Wani, N.R.; Rather, R.A.; Farooq, A.; Padder, S.A.; Baba, T.R.; Sharma, S.; Mubarak, N.M.; Khan, A.H.; Singh, P.; Ara, S. New insights in food security and environmental sustainability through waste food management. *Environ. Sci. Pollut. Res.* **2024**, *31*, 17835–17857. [CrossRef] [PubMed]
5. Bellemare, M.F.; Çakir, M.; Peterson, H.H.; Novak, L.; Rudi, J. On the measurement of food waste. *Am. J. Agric. Econ.* **2017**, *99*, 1148–1158. [CrossRef]
6. Sheahan, M.; Barrett, C.B. Food loss and waste in Sub-Saharan Africa. *Food Policy* **2017**, *70*, 1–12. [CrossRef]
7. Scherhaufer, S.; Moates, G.; Hartikainen, H.; Waldron, K.; Obersteiner, G. Environmental impacts of food waste in Europe. *Waste Manag.* **2018**, *77*, 98–113. [CrossRef]
8. Höhne, N.; Den Elzen, M.; Escalante, D. Regional GHG reduction targets based on effort sharing: A comparison of studies. *Clim. Policy* **2014**, *14*, 122–147. [CrossRef]
9. Tariq, G.; Sun, H.; Ali, I.; Ali, S.; Shah, Q. Influence of access to clean fuels and technology, food production index, consumer price index, and income on greenhouse gas emissions from food system: Evidence from developed countries. *Environ. Sci. Pollut. Res.* **2023**, *30*, 59528–59539. [CrossRef]
10. Toda, H.Y.; Yamamoto, T. Statistical inference in vector autoregressions with possibly integrated processes. *J. Econom.* **1995**, *66*, 225–250. [CrossRef]
11. Zapata, H.O.; Rambaldi, A.N. Monte Carlo evidence on cointegration and causation. *Oxf. Bull. Econ. Stat.* **1997**, *59*, 285–298. [CrossRef]
12. Engle, R.F.; Granger, C.W. Co-integration and error correction: Representation, estimation, and testing. *Econom. J. Econom. Soc.* **1987**, *55*, 251–276. [CrossRef]
13. Granger, C.W. Investigating causal relations by econometric models and cross-spectral methods. *Econom. J. Econom. Soc.* **1969**, *37*, 424–438. [CrossRef]
14. Narayan, P.K. The saving and investment nexus for China: Evidence from cointegration tests. *Appl. Econ.* **2005**, *37*, 1979–1990. [CrossRef]
15. Pesaran, M.H.; Shin, Y.; Smith, R.J. Bounds testing approaches to the analysis of level relationships. *J. Appl. Econom.* **2001**, *16*, 289–326. [CrossRef]
16. Banerjee, A.; Dolado, J.; Mestre, R. Error-correction mechanism tests for cointegration in a single-equation framework. *J. Time Ser. Anal.* **1998**, *19*, 267–283. [CrossRef]
17. Chrisendo, D.; Piipponen, J.; Heino, M.; Kummu, M. Socioeconomic factors of global food loss. *Agric. Food Secur.* **2023**, *12*, 23. [CrossRef]

18. Delgado, L.; Schuster, M.; Torero, M. On the origins of food loss. *Appl. Econ. Perspect. Policy* **2021**, *43*, 750–780. [CrossRef]
19. Mariam, N.; Nina, L.; Fabian, B.; Florence, Z.; Simone, A.; Ulf, S.; Daniel, F. The Food Waste Lab: Improving food waste reduction behavior through education. *J. Clean. Prod.* **2022**, *370*, 133447. [CrossRef]
20. Diana, R.; Martianto, D.; Baliwati, Y.F.; Sukandar, D.; Hendriadi, A. Prevalence of household food insecurity and its association with food waste. *J. Egypt. Public Health Assoc.* **2024**, *99*, 21. [CrossRef] [PubMed]
21. Fami, H.S.; Aramyan, L.H.; Sijtsema, S.J.; Alambaigi, A. The relationship between household food waste and food security in Tehran city: The role of urban women in household management. *Ind. Mark. Manag.* **2021**, *97*, 71–83. [CrossRef]
22. Durán-Sandoval, D.; Durán-Romero, G.; López, A.M. Achieving the food security strategy by quantifying food loss and waste. A case study of the chinese economy. *Sustainability* **2021**, *13*, 12259. [CrossRef]
23. Wang, D.; Andrée, B.P.J.; Chamorro, A.F.; Spencer, P.G. Transitions into and out of food insecurity: A probabilistic approach with panel data evidence from 15 countries. *World Dev.* **2022**, *159*, 106035. [CrossRef]
24. Aschemann-Witzel, J.; De Hooge, I.; Amani, P.; Bech-Larsen, T.; Oostindjer, M. Consumer-related food waste: Causes and potential for action. *Sustainability* **2015**, *7*, 6457–6477. [CrossRef]
25. Buzby, J.C.; Hyman, J. Total and per capita value of food loss in the United States. *Food Policy* **2012**, *37*, 561–570. [CrossRef]
26. Blagoeva, N.; Georgieva, V.; Dimova, D. Relationship between GDP and Municipal Waste: Regional Disparities and Implication for Waste Management Policies. *Sustainability* **2023**, *15*, 15193. [CrossRef]
27. Poore, J.; Nemecek, T. Reducing food's environmental impacts through producers and consumers. *Science* **2018**, *360*, 987–992. [CrossRef]
28. Van Boven, L.; Burgess, M.G. Introduction to topical collection: Social science and sustainability technology. *Clim. Change* **2024**, *177*, 69. [CrossRef]
29. Clark, M.A.; Domingo, N.G.; Colgan, K.; Thakrar, S.K.; Tilman, D.; Lynch, J.; Azevedo, I.L.; Hill, J.D. Global food system emissions could preclude achieving the 1.5 and 2 C climate change targets. *Science* **2020**, *370*, 705–708. [CrossRef] [PubMed]
30. Vermeulen, S.J.; Campbell, B.M.; Ingram, J.S. Climate change and food systems. *Annu. Rev. Environ. Resour.* **2012**, *37*, 195–222. [CrossRef]
31. Niles, M.T.; Ahuja, R.; Barker, T.; Esquivel, J.; Gutterman, S.; Heller, M.C.; Mango, N.; Portner, D.; Raimond, R.; Tirado, C. Climate change mitigation beyond agriculture: A review of food system opportunities and implications. *Renew. Agric. Food Syst.* **2018**, *33*, 297–308. [CrossRef]
32. Cottrell, R.S.; Nash, K.L.; Halpern, B.S.; Remenyi, T.A.; Corney, S.P.; Fleming, A.; Fulton, E.A.; Hornborg, S.; Johne, A.; Watson, R.A. Food production shocks across land and sea. *Nat. Sustain.* **2019**, *2*, 130–137. [CrossRef]
33. Springmann, M.; Clark, M.; Mason-D'Croz, D.; Wiebe, K.; Bodirsky, B.L.; Lassaletta, L.; De Vries, W.; Vermeulen, S.J.; Herrero, M.; Carlson, K.M. Options for keeping the food system within environmental limits. *Nature* **2018**, *562*, 519–525. [CrossRef] [PubMed]
34. Khan, A.A.; Cui, F.-J.; Ullah, S.; Ali, K. Food production and food-related $CO_2$ emissions in China: An examination of the role of economic governance; insights from innovative fourier approach. *Environ. Dev. Sustain.* **2024**, 1–24. [CrossRef]
35. Chen, J.; Zhong, F. The Role of Income and Food Prices in Diet-Related Greenhouse Gas Emissions in China: A Path towards a Sustainable Diet. *Foods* **2022**, *11*, 4051. [CrossRef]
36. Purohit, S. Demographic Transition Model and Population Growth of India-Implications and Assessments. *J. Environ. Sci. Public Health* **2023**, *7*, 176–184. [CrossRef]
37. Jie, H.; Khan, I.; Alharthi, M.; Zafar, M.W.; Saeed, A. Sustainable energy policy, socio-economic development, and ecological footprint: The economic significance of natural resources, population growth, and industrial development. *Util. Policy* **2023**, *81*, 101490. [CrossRef]

*sustainability*

MDPI

*Article*

# Using Dimensionality Reduction Methods to Explore the Social, Cultural and Geographical Reasons Behind Food Waste in the European Union

**Anikó Zseni [1,*], András Horváth [2] and Gergely Zoltán Macher [1]**

[1] Department of Applied Sustainability, Albert Kázmér Mosonmagyaróvár Faculty of Agricultural and Food Sciences, Széchenyi István University, Egyetem tér 1, 9026 Győr, Hungary
[2] Department of Mathematics and Physics, Faculty of Informatics and Electrical Engineering, Széchenyi István University, Egyetem tér 1, 9026 Győr, Hungary; horvatha@sze.hu
* Correspondence: zseniani@sze.hu

**Abstract:** The paper investigates disparities in food waste generation across European Union countries between 2020 and 2022, focusing on spatial and sustainability dimensions. It utilizes data for six key food waste parameters and a broad range of environmental, social and economic indicators. A combination of statistical methods, including correlation analysis, cluster analysis and Principal Component Analysis, uncovers multivariate patterns and identifies groups of countries with similar food waste characteristics and related factors. The paper highlights the temporal and spatial dynamics of food waste over the three-year period, particularly in light of the COVID-19 pandemic. While the total volume of food waste remained relatively stable across the EU, notable shifts occurred in waste sources. Household food waste peaked in 2021, likely due to increased time spent at home during pandemic-related lockdowns. Conversely, waste from retail, restaurants and food service sectors showed a consistent increase. The paper identifies non-trivial correlations between food waste and socio-economic variables, suggesting that differences in food waste generation across EU countries are influenced by a complex interplay of factors, including policy effectiveness, cultural practices, consumer behaviour and economic conditions. This comprehensive analysis of food waste patterns across EU countries and over time offers valuable insights for policymakers aiming to reduce waste and promote sustainability.

**Keywords:** food waste; household waste; socio-economic factors; principal component analysis; spatial analysis

## 1. Introduction

The onset of the COVID-19 pandemic in early 2020 triggered unforeseen disruptions in global food systems [1,2]. Initial lockdowns led to rapid shifts in food availability, characterized by empty shelves in certain regions and significant surpluses of unsold produce in others [3,4]. This public health crisis quickly tested the resilience of the entire food supply chain [5]. Our recent food waste studies that utilized correlation analysis and cluster analysis [6], as well as Principal Component Analysis and one- and multidimensional linear regression [7], reveal a contrasting dynamic, in which the pandemic not only exposed systemic vulnerabilities but also prompted rapid adaptation and innovative responses, while reflecting and reinforcing existing patterns of socio-economic disparity.

The relationship between the COVID-19 pandemic and food waste proved to be multifaceted [8]. Specifically, supply chains experienced interruptions, consumer behaviours changed rapidly, and established patterns of production, distribution and consumption were disrupted [9,10]. Initial global reports indicated widespread disturbances, with farms facing challenges in delivering products to markets, processors encountering obstacles and retailers struggling to adapt to fluctuating demand [3,11,12]. Consequently, food that decays quickly, such as dairy, fruits and vegetables, was especially vulnerable. In fact, prior studies have indicated that some producers were compelled to dispose of substantial amounts of produce that could not be stored or redirected [13–15].

In contrast to the consistency observed in food waste patterns at earlier stages of the supply chain, household behaviours exhibited greater variability [16,17]. Initially, responses to the crisis involved widespread stockpiling, motivated by uncertainty rather than scarcity, leading to increased expiration of unused food [18,19]. However, as the pandemic progressed, a shift towards more careful meal planning and home cooking was observed, coupled with a decreased reliance on dining out [20,21]. Simultaneously, the closure of hospitality and institutional food service sectors resulted in substantial stock losses. To mitigate these losses [22], food donation networks and community-based programs expanded their operations, redirecting surplus food to vulnerable populations [23]. Food waste generation is influenced not only by proximal disruptions but also by deeper social, cultural and geographical determinants. For example, differences in diet, family size and composition, and urban versus rural residential contexts have been shown to influence the quantity and quality of waste.

Our ongoing work takes these observations further, exploring how crises like COVID-19 shape not only waste patterns but also the resilience and adaptability of food systems. Early results suggest that pandemic-era waste patterns reflected pre-existing socio-economic differences as much as they did the virus's direct impact. This paper seeks to address existing research gaps by analysing how varied policy frameworks and cultural practices across European Union member states interact with food supply chains, thereby influencing food waste outcomes. Furthermore, it employs advanced dimensionality reduction techniques to uncover latent data structures, overcoming the limitations of previous studies that often lacked multivariate statistical approaches for disentangling complex interactions. The paper uses a comprehensive approach to explore the relationships between structural vulnerabilities, community-level coping mechanisms and socio-economic inequalities, which collectively shape both the intensity of disruptions and the rate of recovery. Initial results indicate that waste trends identified during the pandemic were not exclusively a direct consequence of the virus; instead, they mirrored and, in certain instances, intensified pre-existing disparities grounded in the social, cultural and economic frameworks of individual countries.

Against this backdrop, the paper analyses food waste trends in the European Union; by using six key indicators and applying dimensionality reduction methods, it explores how national trajectories cluster or diverge, and how social, cultural and geographical factors may have shaped these patterns. To better understand these patterns, this study focuses exclusively on EU member states, as coherent and comparable datasets for other regions are not available; this limitation is acknowledged as an important constraint of the research. Two guiding hypotheses inform our approach:

- **H-1:** Variations in food waste generation across EU member states can be linked to underlying social, cultural and geographical characteristics, which influence both the scale and type of waste produced.
- **H-2:** Changes in consumer behaviour, supply chain dynamics and institutional practices during the study period reflect deeper structural drivers that predate the pandemic but were brought into sharper focus by it.

By identifying these relationships, this research aims to contribute actionable insights for designing targeted waste reduction strategies, strengthening resilience and fostering sustainable food systems that are better equipped to withstand future crises. The paper underscores the innovative approach of distinguishing system-level factors from individual behaviours using Principal Component Analysis, thereby enabling policymakers to devise more effective interventions and tackle the underlying causes of food waste rather than merely its manifestations.

## 2. Literature Review

### *2.1. Food Waste: Definition, Origin and Effect*

The EU Waste Framework Directive [24] provides a definition of food waste as all food, as defined in the General Food Law [25], which has become waste. Food is considered as a whole and applies throughout the entire food supply chain, from production through to consumption. Food includes inedible parts which are not separated from edible parts during production. According to this definition, it is not technically possible to completely eliminate food waste for several reasons: consumers cannot typically consume or reuse certain inedible parts (e.g., bones), some food must be discarded for safety reasons, and preventing food waste should not compromise food safety, or animal or human health. The EU's definition of food waste excludes items that were not classified as "food" at the time of being discarded or removed from the supply chain (e.g., losses before harvest or slaughter) and does not encompass non-waste material such as surplus donated foods or by-products used for other purposes (e.g., animal feed). Food wastage occurs at every stage of the supply chain, including primary production; processing and manufacturing; retail and distribution; restaurants and food services; and households, both within homes and in out-of-home settings. Therefore, stages relating to residential properties/restaurants/services together constitute what is termed the "consumption" stage [26].

Food waste can be attributed to a number of causes, including losses in food production and processing, outdated quality standards, over-packaging, consumer perceptions and the influence of market trends [27]. In addition, food waste is often a consequence of consumer habits, convenience and economic considerations [28]. Overproduction and inefficient distribution systems, consumer attitudes and preferences [29], and lack of awareness and education about food waste can also play a significant role [29–32]. Furthermore, improper storage and handling, confusion over expiry dates (marked as "use by" or "best before" dates) and food labelling, as well as advertising and marketing strategies that encourage overconsumption, can also contribute to an increase in food waste. Lastly, cultural and social norms and aesthetic preferences for foods further compound this issue.

The effectiveness of food waste reduction initiatives is influenced by socio-cultural acceptance and public awareness, emphasizing the need for targeted communication strategies and policy instruments that consider local contexts.

### *2.2. Overview of EU Action for Preventing Food Waste*

Preventing food waste became a specific political priority at the EU level in 2015, reflecting EU commitments made in the context of the 2030 Sustainable Development Agenda to reduce food waste by 50% and decrease food loss by the year 2030. The EU has implemented a specific strategy to decrease the amount of food that is lost or wasted that involves both regulatory and non-regulatory actions, first as part of the 2015 Circular Economy Action Plan and, more recently, under the 2020 Farm to Fork Strategy (a strategy that connects food production on farms directly to consumption by minimizing the distances food travels and the processes it undergoes between its origin in the fields and the final consumer). Since 2015, the Commission has been working on initiatives to clarify and harmonize relevant legislation. They have also supported the development and sharing of best practices and solutions to reduce food waste across the EU. Article 9 (Prevention of Waste) of 2018/851, amending Directive 2008/98/EC on waste, requires member states to implement measures aimed at decreasing food waste at all stages of the food supply chain [24]. It also promotes food donation and other forms of redistribution for human consumption, giving priority to human use over animal feed and the conversion of surplus food into non-food products [24]. Several guidelines, recommendations and informational publications have been published recently. While the European Union provides a standardized policy framework, its efficacy and practical application vary considerably among member nations. These disparities result in differing outcomes and highlight the significant impact of contextual factors on food waste dynamics.

The European Commission Directorate-General for Environment published guidelines on the preparation of food waste prevention programmes. This document is intended to serve as a valuable tool for waste management organizations, businesses, institutions, local authorities and environmental protection agencies. It covers the European policy framework concerning bio-waste, with a specific focus on food waste, and applies the waste hierarchy to food waste prevention and includes discussions on methods for measuring food waste, setting targets and implementing prevention strategies [33].

The EU Platform on Food Losses and Food Waste was established in 2016 to support all stakeholders in defining measures to prevent food waste, sharing best practices and evaluating progress made over time [34]. In 2019, the Commission adopted a common methodology for measuring food waste as a basis for monitoring food waste in the EU countries [35].

As part of the Circular Economy Action Plan, the European Commission has implemented EU 2017/C 361/01 Guidelines on food donations [36] to promote the efficient collection and distribution of safe and edible food to individuals facing food insecurity. This guidance clarifies provisions in EU legislation and helps remove barriers to food redistribution within the current EU regulatory framework. The main aspects involve following food safety and hygiene laws, keeping food at proper storage and transport temperatures, and permitting the donation of imperfect or expired foods.

The 2018/C 133/02 Guidelines for the feed use of food no longer intended for human consumption [37] address issues regarding feed utilization of food which is no longer intended for human consumption for commercial reasons or due to problems of manufacturing or certain defects, but which can be used in animal nutrition without compromising animal or public health. It helps prevent the nutrients in food from being composted, converted into biogas, or disposed of through incineration or landfill.

The recommendations for action in food waste prevention were compiled by the EU Platform on Food Losses and Food Waste to offer crucial guidance to all players in efforts to prevent food losses and waste, including EU citizens themselves [38]. The forward-looking recommendations include a set of horizontal or cross-cutting measures, as well as recommendations for action for primary production and food donation, for action at manufacturing, retail and consumer levels in hospitality/food services. The recommendations target both public and private entities, suggesting relevant actions for each group while emphasizing the need for cooperation among stakeholders. Despite existing EU guidelines, empirical research evaluating their effectiveness in reducing food waste remains scarce, underscoring the necessity for systematic assessment and standardized monitoring of member states' results.

To guide effective actions, the authors of [39] present a framework for the evaluation of food waste prevention actions and explain its use in assessing the effectiveness and efficiency of such actions. To facilitate the continuous sharing of resources, latest developments and best practices, the Commission has set up a digital platform called the EU Food Loss and Waste Prevention Hub. This serves as a comprehensive resource for all stakeholders involved in food waste prevention [40].

Research and innovation aimed at tackling the prevention of food waste are supported by the European Commission, including the development of blueprints for the establishment of national public–private partnerships (e.g., Voluntary Agreements) to reduce food waste throughout the entire food supply chain [41]. The EU Research and Innovation Framework Programme Horizon 2020 and Horizon Europe are offering new opportunities for research and innovation to address food loss and waste [26].

## 3. Materials and Methods

### 3.1. Data

Our research is based on the datasets of the Statistical Office of the European Union published on the Eurostat website. Table 1 contains the relevant tabular data about food waste datasets, while Table 2 summarizes the twenty-six selected structural indicators (environmental, social, economic and geographical parameters) that were collected for the EU-27 member states from 2020 to 2022. The downloaded data were directly used with no cleaning criteria. A rescaling to kg/year/capita units was applied before analysis. The total amount of food waste (TOTAL) is the sum of five distinct food waste parameters (Equation (1)): food waste from primary production of food (FP), manufacture of food products and beverages (MFP), retail and other distribution of food (RDF), restaurants and food services (RFS), and the total activity of households (HHA):

$$\text{TOTAL} = \text{FP} + \text{MFP} + \text{RDF} + \text{RFS} + \text{HHA} \tag{1}$$

**Table 1.** Food waste absolute parameters.

| Abbreviation | Indicators [kg/year/capita] |
|---|---|
| FP | Food waste from primary production of food—agriculture, fishing and aquaculture |
| MFP | Food waste from manufacture of food products and beverages |
| RDF | Food waste from retail and other distribution of food |
| RFS | Food waste from restaurants and food services |
| HHA | Food waste from total activities by households |
| TOTAL | Total food waste production, FP + MFP + RDF + RFS + HHA |

**Table 2.** Structural indicators (types of indicators: environmental (EN), economic (EC), social (S) and geographical (G).

| Abbreviation | Indicators | Unit | Type | Reference |
|---|---|---|---|---|
| AREA | Total area of the country | [km$^2$] | G | |
| CMR | Circular material use rate | [%] | EN | [42] |
| COFP | Consumption footprint index, year 2010 = 100 | [dimensionless] | EN | [43] |
| ELET | Early leavers from education and training by sex | [%] | S | [44] |
| FECH | Final energy consumption in households per capita | [kg of oil equivalent] | EC | [45] |
| GDP_PC | GDP per capita value | [EUR/capita] | EC | [46] |
| GVAEGS | Gross value added in environmental goods and services sector, reference year 2010, chain-linked volumes at 2010 exchange rates | [EUR/capita] | EN | [47] |

**Table 2.** *Cont.*

| Abbreviation | Indicators | Unit | Type | Reference |
| --- | --- | --- | --- | --- |
| IAF | Inability to afford a meal with meat, chicken or fish (or vegetarian equivalent) every second day—EU-SILC survey | [%] | S | [48] |
| MF | Material footprint | [tonne/capita] | EN | [49] |
| MSDR | Severe material and social deprivation rate | [%] | S | [50] |
| MWG_PC | Municipal waste by waste management operations | [kg/capita] | EN | [51] |
| NSFT | Nights spent in tourist accommodation establishments by residents/non-residents | [number] | EC | [52] |
| POP | Population on 01.01. | [capita] | EC | [53] |
| POPDENS | Population density of the country (POP/AREA) | [capita/km$^2$] | EC | [54] |
| PPGDP | Purchasing power adjusted GDP per capita, in purchasing power standards | [EUR/capita] | EC | [55] |
| PUKHW | Population unable to keep home adequately warm by poverty status | [%] | S | [56] |
| REG_BOR_PC | Border type area per capita | [km$^2$/capita] | G | [57] |
| REG_COA_PC | Coastal type area per capita | [km$^2$/capita] | G | [57] |
| REG_INT_PC | Intermediate type area per capita | [km$^2$/capita] | G | [57] |
| REG_ISL_PC | Island type area per capita | [km$^2$/capita] | G | [57] |
| REG_MOU_PC | Mountain type area per capita | [km$^2$/capita] | G | [57] |
| REG_RUR_PC | Rural type area per capita | [km$^2$/capita] | G | [57] |
| REG_URB_PC | Urban type area per capita | [km$^2$/capita] | G | [57] |
| RMRPG | Relative median at-risk-of-poverty gap | [%] | S | [57] |
| RPS_PC | People at risk of poverty or social exclusion | [%] | S | [58] |
| TEA | Tertiary educational attainment by sex | [%] | S | [59] |

Relative food waste parameters were calculated by dividing the absolute parameters by the TOTAL value (FP_REL = FP/TOTAL, MFP_REL = MFP/TOTAL, etc.). Absolute parameters are quantified in kg/year/capita, indicating the annual per person amount in kilograms. Relative parameters represent the percentage distribution of the specific waste types at the country level.

### 3.2. Methods

The correlation and cluster analyses were performed by a self-developed Python language program that used Scikit-Learn and Scientific Python modules. PCA for the dataset was carried out using SciKit-Learn software (version 1.4.1).

#### 3.2.1. Correlation Analysis

Linear (Pearson-type) and rank (Spearman-type) correlation tests were used to identify relationships between pairs of food waste parameters and structural indicators. The former are sensitive to approximately linear connection between the two variables, whereas the latter are capable of finding nonlinear but monotonic relationships. As the results did not reveal any non-trivial connections (see details in Section 4.1), Principal Component Analysis (PCA) and statistical tests for comparing clusters of countries were used to investigate the existence of multivariate and nonlinear connections.

#### 3.2.2. Principal Component Analysis

The food waste dataset includes five independent variables for each country per year, theoretically allowing data points to be visualized in a five-dimensional space. To overcome this difficulty, Principal Component Analysis method was applied to reduce the number of dimensions of the dataset [60]. PCA initiates by calculating the average vector of the dataset and then determining the difference between each vector and the mean vector.

The subsequent step involves identifying the direction in which the set of these difference vectors exhibits the greatest elongation—representing the "most important" direction—and describing it with a unit-length vector. Then, the subsequent (second, third, etc.) most important directions that are orthogonal to the previous ones are calculated.

In the case of food waste data, the original five dimensions were reduced to two dimensions using PCA, achieving a good approximation. Therefore, the principal component coefficients for each country and year were calculated and the temporal changes of the EU countries were visualized on a two-dimensional plot, with coordinates representing the PCA coefficients. This resulting plot was used to identify countries exhibiting similar behavioural patterns. In contrast to our prior similar PCA investigation [7], which was restricted to the year 2020 due to data limitations, the current research covers a three-year period from 2020 to 2022. All the three-year data (81 five-dimensional vectors) were used for a unique PCA-base vector calculation, and these were used to calculate the PCA components.

Given that the PCA method is based on linear operations, it was only applied to absolute food waste data, as absolute values in kg/capita/year units are additive, while relative components cannot be linearly combined.

#### 3.2.3. Cluster Analysis

Cluster analysis is a multivariate statistical method used to find groups of similar items based on multidimensional numerical data. Our research applied this method to find subsets of EU members with comparable food waste characteristics. Clustering calculations were performed for the five-dimensional vectors of absolute (FP, MFP, RDF, RFS and HHA) and relative (FP_REL, MFP_REL, RDF_REL, RFS_REL and HHA_REL) food waste parameters. Linear scaling for the parameters was applied to obtain the same standard deviation in every coordinate. This standardization step was necessary because the average values for different parameters vary significantly. For example, household food waste (HHA) typically ranges from 50 to 100 kg/capita/year, while food waste generated during primary food production (FP) is commonly under 20 kg/capita/year. Therefore, a 10 kg/capita/year change in HHA represents a small change, while the same difference in FP is a dramatic alteration.

The K-means clustering method [61] was applied with $N = 2$–$7$ clusters. The elbow method was utilized to determine the optimal cluster number. According to the results, $N = 5$ appears to be the best choice. Qualitative analysis of various cluster numbers also confirmed that $N = 5$ represents the optimal selection. The classification of a country seemed forced when $N$ was less than five; conversely, when $N$ exceeded five, it led to an unwarranted fragmentation of clusters. The cluster members were listed and stacked histograms of absolute and relative food waste parameters were generated, where the EU members were grouped based on the clusters. Similar calculations were performed in our previous study [6], but then food waste data were only available for the year 2020. The present investigation expands the temporal scope to 3 years.

Following the identification of the groups of countries with similar food waste characteristics, structural indicators for each group were collected. The Mann and Whitney test (U test) was used to find statistically significant differences between these parameter-sets. Significant differences suggest a connection between the examined structural indicator and the food waste values. The advantage of this complex approach over the simple correlation analysis is two-fold. First, the comparison in this case is based on more data points, thereby minimizing the impact of uncorrelated measurement errors. Second, comparing only specific subsets of EU members can reveal connections that are not valid for the whole EU, while correlation analysis tries to find globally true connections.

## 4. Results and Discussion

### 4.1. Results of Correlation Analysis

Pearson- and Spearman-type correlation coefficients were calculated between all food waste parameters and the examined structural indicators. In all cases, the correlation coefficients did not exceed 0.7, leading to the conclusion that there is no simple dependence between such pairs. However, in some cases the coefficients were between 0.5 and 0.7, which indicates a possible connection, but this is not accepted as statistically well established. The absence of statistically significant results does not mean that there is no relationship between the amount of food waste and other parameters. It is possible that such a relationship only exists between certain groups of countries or that the correlation coefficient is reduced by a large measurement error for the parameters.

### 4.2. Results of Principal Component Analysis

Reducing the number of dimensions with PCA is a significant help in overviewing multidimensional data, but it inherently implies an approximation. The required level of precision should be determined by considering both the data's accuracy and the research objectives. Table 3 presents the average difference of the original components and the reconstructed values, depending on the number of dimensions retained. Naturally, increasing the number of dimensions results in smaller reconstruction error.

**Table 3.** PCA reconstruction error in kg/year/capita units and relative explained variance values of percentages as a function of number of dimensions (D).

| D | 0 | 1 | 2 | 3 | 4 | 5 |
|---|---|---|---|---|---|---|
| Accuracy | 11.8 | 7.7 | 4.9 | 3.2 | 1.4 | 0.0 |
| Rel. expl. var. | | 60.4 | 23.8 | 7.7 | 6.4 | 1.7 |

D = 2 was selected for analysis. This case has an accuracy of 4.9 kg/capita/year, falling within the expected range of measurement error. Relative explained variance values show that the third PCA component gives only a 7.7% improvement. While increasing to D = 3 offers a marginal reduction in average deviation, it complicates visualization. Since the numerical values provided by PCA are not used for further analysis in this article, but rather to review the nature of the changes, the accuracy provided by the two components is considered to be sufficient.

The results of the PCA algorithm contain the "mean vector" (m) and two "base vectors" (s1 and s2). All data points can be approximated as a linear combination of them, as is written in Equation (2).

$$foodwastevector \approx m + s_1 \cdot P_1 + s_2 \cdot P_2 \tag{2}$$

For each data point, the PCA coefficients, represented by the dimensionless quantities $P_1$ and $P_2$, are computed and used to characterize the food waste values for a specific country in a given year. This approach reduces the dimensionality of the data from the original five components (FP, MFP, RDF, RFS and HHA) to just two.

The numerical results are summarized in Equation (3). Figure 1 illustrates the graphical representation of the mean and base vectors.

$$
\begin{aligned}
FP &\approx 13.2 + 0.148 \cdot P_1 + 0.134 \cdot P_2 \\
MFP &\approx 30.2 + 0.970 \cdot P_1 - 0.066 \cdot P_2 \\
RDF &\approx 11.6 + 0.186 \cdot P_1 + 0.134 \cdot P_2 \\
RFS &\approx 15.1 + 0.043 \cdot P_1 + 0.130 \cdot P_2 \\
HHA &\approx 69.4 + 0.014 \cdot P_1 + 0.971 \cdot P_2
\end{aligned} \tag{3}
$$

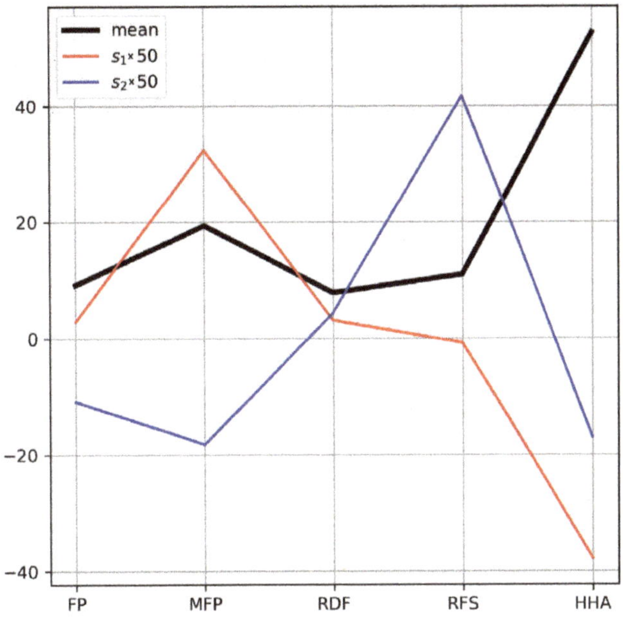


**Figure 1.** Mean vector and base vectors (scaled) of PCA in kg/capita/year units.

Numeric values in equation (3) indicate that the first primary component ($P_1$) is mainly composed of food waste generated from manufacturing for food production (MFP), which carries the highest weight, followed by primary food production (FP) and the distribution of food production (RDF) components, which have smaller, but still considerable weights. These activities exhibit less dependence on individual behaviours and instead reflect the specificity of the economic sector (like agriculture, industry and transportation). Conversely, the $P_2$ component encompasses waste generated in households (HHA) as the most predominant factor; alongside restaurants and other food services (RFS) with a higher weight than in $P_1$, the other three components have smaller coefficients. Consequently, the second principal component is more heavily influenced by individual consumption, eating and cooking habits than the first one. It is worth noting that this separation of economic system-level parameters and individual behaviour parameters is an automatic result of PCA. This demonstrates that the method was capable of separating parameters that behave differently. According to these, a positive value for $P_1$ or $P_2$ means above-the-average economic system-level or individual food waste production, respectively.

Figure 2 illustrates the countries in the $P_1$-$P_2$ plane. Symmetrical logarithmic scaling was applied on the axes to show the extremely large values and to resolve the high-density part of the diagram near the origin on the same plot.

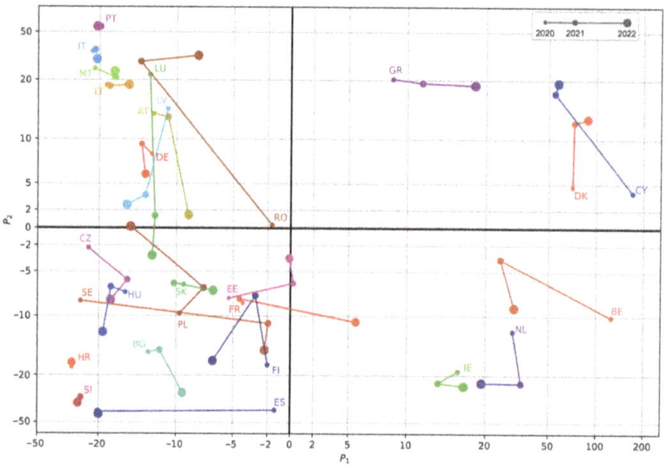


**Figure 2.** EU countries in $P_1$-$P_2$ plane [unit: dimensionless]. Symmetrical logarithmic scaling was applied on both axes. Small, medium and large dots correspond to years 2020, 2021 and 2022, respectively. (Abbreviations: AT, Austria; BE, Belgium; BG, Bulgaria; CY, Cyprus; CZ, Czech Republic; DE, Germany; DK, Denmark; EE, Estonia; ES, Spain; FI, Finland; FR, France; GR, Greece; HR, Croatia; HU, Hungary; IE, Ireland; IT, Italy; LT, Lithuania; LU, Luxembourg; LV, Latvia; MT, Malta; NL, the Netherlands; PL, Poland; PT, Portugal; RO, Romania; SE, Sweden; SI, Slovenia; SK, Slovakia).

Each country is represented by a distinct colour and characterized by three data points of different sizes, corresponding to the years 2020 (small), 2021 (medium) and 2022 (large). These data points are interconnected by straight lines to allow for the observation of value trends across the specified period.

According to their $P_1$ and $P_2$ values, the countries can be divided into three groups:

- $P_1 > 0$: Belgium, Cyprus, Denmark, Greece, Ireland and the Netherlands. Their differences from the average food waste characteristics are mainly in the s1 direction, which means high excess in food waste from manufacture of food products (MFP) and significant excess in food waste from primary production (FP) and food waste from distribution of food production (RDF) components.
- $P_1 < 0$ and $P_2 > 0$: Austria, Germany, Italy, Latvia, Lithuania, Luxembourg, Malta, Portugal and Romania. They have below-average values for FP, MFP and RDF, but excessive values for food waste from households (HHA) and food waste from restaurants and other food services (RFS).
- $P_1 < 0$ and $P_2 < 0$: Bulgaria, Croatia, Czech Republic, Estonia, Finland, France, Hungary, Poland, Slovakia, Slovenia, Spain and Sweden. All of their food waste components are smaller than or equal to the average.

The examined period from 2020 to 2022 reveals two principal trends in the movement of individual countries within the $P_1$-$P_2$ plane. For a few countries, the differences across the three years are negligible (Croatia and Portugal) or marginal (Germany, Greece, Italy, Portugal, Slovenia and Slovakia). Conversely, half of the countries exhibit a notable shift from 2020 to 2021, but a much smaller shift from 2021 to 2022 (Austria, Belgium, Bulgaria, Cyprus, Czech Republic, Estonia, Latvia, Lithuania, Luxembourg, Malta, Poland, Romania, Spain and Sweden).

### 4.3. Results of Cluster Analysis

Absolute and relative food waste datasets contain five-dimensional vectors for the 27 EU members for the 3 years. One objective of our research was to detect patterns in these datasets. The widely used K-means clustering method was applied to find the groups of similar countries with standard scaling, i.e., applying a linear transformation on the data to obtain a standard deviation of 1 for each component. Without this normalization, changes in the components with the highest average values (HHA and MFP) would dominate the clustering and the effect of the other components would be negligible.

Calculations were performed for cluster sizes of $N = 3, 4$ and 5. Several similar patterns can be observed across these versions: Cyprus is typically isolated as a separate cluster and the former socialist states tended to group together. Investigating the results, we determined that $N = 5$ provided the most informative clustering. Therefore, the results for five groups are presented in this article.

The results of the K-means clustering calculations consist of $N$ vectors: the so-called "cluster centres" and a labelling of each vector indicating its nearest cluster centres. This way the cluster centres represent the "typical" food waste parameter sets, while the cluster labels of the countries denote which cluster they most resemble.

4.3.1. Clusters of Absolute Food Waste Data

Clustering calculations were performed independently for each of the 81 food waste parameter vectors (27 countries for 3 years). Therefore, if the food waste profile of an EU member presents a significant shift during the 3-year period, data from different years for the same country may belong to different clusters.

The next list contains the members of the five clusters. When all three-year data for a specific country are grouped into the same cluster, only the two-letter abbreviation is displayed. Otherwise, the years are appended to the country code. Clusters are denoted by "ACL1", "ACL2", . . ., where "A" indicates that this clustering is based on the absolute food waste parameters. Figure 3 displays this clusterization on a color-coded map.

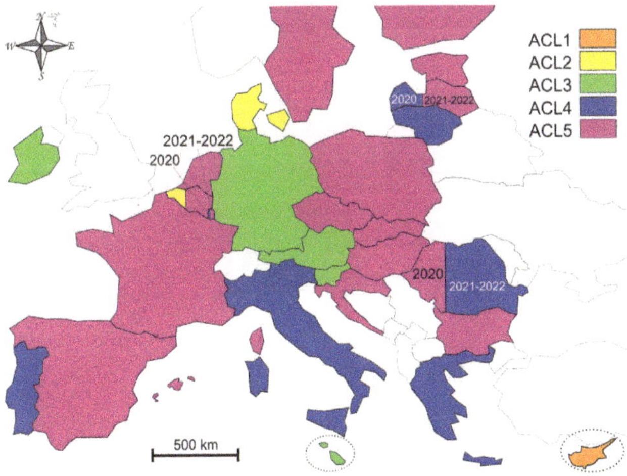


**Figure 3.** Location of countries belonging to specific clusters (food waste per capita, 2020–2022, for N = 5 clusters). In the case of cluster-changing countries, colours for both clusters appear.

- ACL1 (1 member): CY.
- ACL2 (2 members): BE (2020) and DK.
- ACL3 (5 members): AT, DE, IE, MT and SI.
- ACL4 (7 members): GR, IT, LT, LU (2020), LV (2020), PT and RO (2021, 2022).
- ACL5 (16 members): BE (2021, 2022), BG, CZ, EE, ES, FI, FR, HR, HU, LU (2021, 2022), LV (2021, 2022), NL, PL, RO (2020), SE and SK.

A graphical representation of this clustering is presented in Figure 4. The columns are sorted to reflect the results of the clusterization, and the blue vertical lines represent the barriers of clusters.

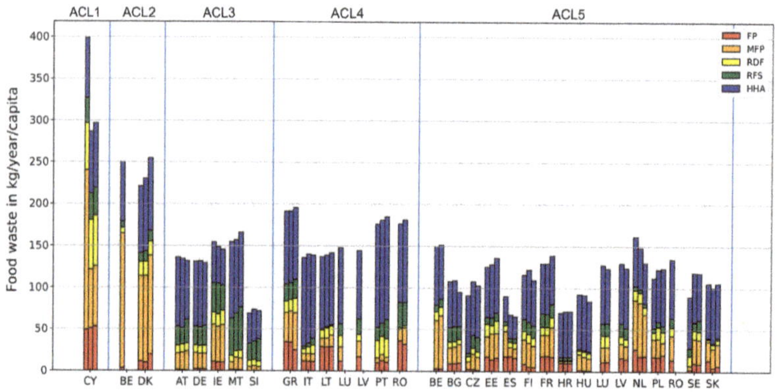


**Figure 4.** Sectoral distribution of food waste per capita, 2020–2022, for N = 5 clusters. First bar: year 2020, second bar: year 2021, third bar: year 2022. (Abbreviations: see Figure 2).

The analysis revealed that most countries preserved their respective cluster affiliations during the examined 3-year period. This indicates that the fundamental characteristics of these countries, as determined by the comprehensive and sectoral distribution of food waste, remained largely stable despite the COVID-19 pandemic. However, Belgium, Latvia, Luxembourg and Romania deviate from this pattern. In all these cases, the data from 2020 were grouped distinctly from the other two years. More specifically, the 2020 data for Luxembourg and Latvia are categorized in cluster ACL4, and for Belgium they are in ACL2, while their data from 2021 and 2022 are in ACL5. Conversely, Romania shifted from ACL5 in 2020 to ACL4 in 2021 and 2022.

Outlier data, such as those observed for Cyprus or Belgium in 2020, can be partly ascribed to methodological idiosyncrasies within specific reporting systems and highlight the need for increased harmonization in food waste data collection.

It is notable that the historical background aligns with the observed clustering patterns. For instance, almost all countries formerly belonging to the Eastern Bloc are in the ACL5 cluster (Figures 3 and 4). The prevalence of the former Eastern Bloc nations in the same cluster indicates that previous consumption patterns and economic histories continue to affect present-day food wastage trends.

Table 4 presents the cluster centre values, which serve as multivariate indicators representing the "typical" food waste characteristics of EU members. A yellow background highlights the maximum value in each column. In the line of ACL1, there are three of them, which indicates exceptionally large food waste production in three categories. On the other hand, ACL3 and ACL5 lack any yellow cells, indicating that these clusters comprise countries with low levels of food waste. Households are the primary contributors across all clusters, underscoring that changes in individual consumer habits are key to minimizing food waste in the EU.

The ACL5 nations have the lowest average total food waste quantity, with the lowest amounts of food waste originating from households (HHA), restaurants (RFS), retail and other food distribution channels (RDF). Cyprus, Belgium and Denmark, as members of ACL1 and ACL2, exhibit the highest total food waste levels, the waste being generated particularly in the manufacture of food products and beverages (MFP). The countries in ACL4 are characterized by large amounts of household food waste (HHA), while those in

ACL1 and ACL3 experience higher volumes of food waste originating from restaurants and related services (RFS).

**Table 4.** Sectoral distribution of cluster centre values based on food waste per capita data in kg/capita/year units. Yellow background indicates the column-wise maximum.

| | FP | MFP | RDF | RFS | HHA |
|---|---|---|---|---|---|
| ACL1 | 50.9 | 111.6 | 58.6 | 32.0 | 74.3 |
| ACL2 | 11.1 | 121.6 | 14.4 | 10.9 | 81.0 |
| ACL3 | 3.0 | 20.4 | 9.8 | 30.2 | 65.4 |
| ACL4 | 21.6 | 16.7 | 12.1 | 14.4 | 97.9 |
| ACL5 | 11.1 | 24.5 | 8.6 | 9.2 | 58.8 |

4.3.2. Clusters of Relative Food Waste Data

Cluster analysis was also performed on the relative food waste data. The next list presents the members of clusters in text, and in Figure 5 they are presented in a color-coded map, while Figure 6 shows the relative food waste parameters in a stacked histogram. The initial letter "R" of the cluster identifiers indicates the relative parameters.

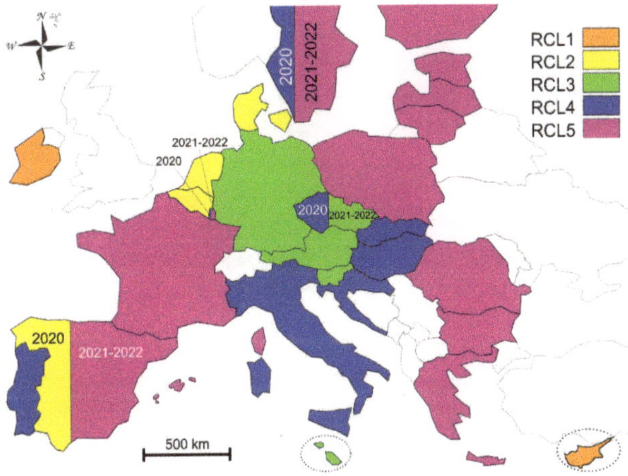


**Figure 5.** Location of countries belonging to specific clusters (relative food waste, 2020–2022, for $N = 5$ clusters). In the case of cluster-changing countries, the colours of both clusters appear.

- RCL1 (2 members): CY and IE.
- RCL2 (4 members): BE, DK, ES (2020) and NL.
- RCL3 (5 members): AT, CZ (2021, 2022), DE, MT and SI.
- RCL4 (8 members): CZ (2020), HR, HU, IT, LU (2020), PT, SE (2020) and SK.
- RCL5 (12 members): BG, EE, ES (2021, 2022), FI, FR, GR, LT, LU (2021, 2022), LV, PL, RO and SE (2021, 2022).

The analysis demonstrates that 23 out of 27 countries maintained their cluster affiliations over the three-year period. This significant stability confirms that the composition of food waste sources remains consistent despite short-term fluctuations. Moreover, similarly to the results of the absolute value-based clustering, the year 2020 is the standout year in all such cases (Spain, Czech Republic, Luxembourg and Sweden).

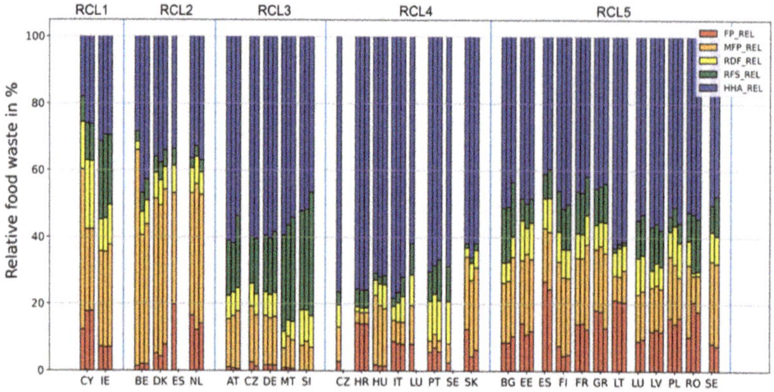


**Figure 6.** Sectoral distribution of relative food waste values, 2020–2022, for $N = 5$ clusters. First bar: year 2020, second bar: year 2021, third bar: year 2022. (Abbreviations: see Figure 2). The columns are sorted to reflect the results of clusterization, and the blue vertical lines represent the barriers of clusters.

Table 5 presents the cluster centre values for each cluster, representing the distribution of food waste relative data in different sectors. The maximum values within each column are highlighted with a yellow background. It is remarkable that each cluster features exactly one yellow cell, which means that all five clusters can be characterized by one of the relative food waste parameters. However, for all clusters except RCL1 and RCL2 the majority of waste is generated from household activities. Household food waste (HHA) alone represents 70% of the total in RCL4, showing that in these countries household activities are the dominant source of food waste production. In contrast, manufacture of food and beverages (MFP) makes up the largest portion in RCL2. RCL3 is primarily characterized by a higher proportion of food waste originating from restaurants and food services (RFS) compared to other clusters. Agricultural-origin food waste is almost 14% in RCL5, which is the highest ratio compared to the other clusters. This implies that policy interventions have to be tailored: for instance, restaurant-based interventions for RCL3 and household-based interventions for RCL4 nations.

**Table 5.** Sectoral distribution of cluster centre values based on food waste relative values in percentages. Yellow background indicates the column-wise maximum.

| | FP | MFP | RDF | RFS | HHA |
|---|---|---|---|---|---|
| RCL1 | 11.5 | 30.9 | 14.4 | 16.6 | 26.7 |
| RCL2 | 8.5 | 43.5 | 6.9 | 4.5 | 36.5 |
| RCL3 | 1.0 | 12.1 | 7.1 | 23.0 | 56.8 |
| RCL4 | 7.0 | 10.7 | 6.5 | 5.2 | 70.5 |
| RCL5 | 13.6 | 17.8 | 7.9 | 10.1 | 50.6 |

Clustering results based on absolute and relative food waste parameters differ significantly. This is understandable, considering their distinct approaches to defining "similarity". The purpose of research should guide the selection of the more relevant method. Absolute unit groups focus on the similarity of average food waste production among individual persons, whereas relative unit groups concentrate on the distribution of a country's food waste production generated by various sectors. For example, in terms of relative parameters, the food waste profile of Ireland is similar to that of Cyprus, while their absolute values differ significantly. This may indicate that food waste-generating processes have a similar structure in Ireland and Cyprus, but Ireland has a greater degree of economic efficiency overall. The similar behaviour of Ireland and Cyprus in terms of relative values, despite great differences in absolute values, also shows that patterns of food waste can coincide across economies of highly different sizes.

#### 4.3.3. Relationship Between Food Waste Clusters and the Structural Indicators

The structural (environmental, social, economic and geographical) indicators corresponding to the clusters identified in the preceding analysis were gathered, and a statistical analysis was conducted to determine which parameters present significant differences between the clusters. The Mann and Whitney U test was used to compare the distributions. Given the high number of possible relationships examined, strict significance levels were used to minimize the incidence of false-positive signals. Specifically, a *p*-value threshold of 0.01 was designated for "weak", and a threshold of 0.001 was designated for "strong" connections. Considering that there are ten possible pairs with the five group (clusters), 10 significance tests were performed per parameter. For the 26 parameters examined, this amounts to a total of 260 tests, meaning that on average there will be 2.6 cases where the test incorrectly indicates a weak correlation and 0.26 cases where a false strong correlation signal will be obtained. These numbers have been verified by trial calculations involving the random dividing of the countries' annual data into five groups.

The results of the 260 statistical tests are transparently presented in Tables 6 and 7. Columns correspond to cluster pairs, and each row is for a specific structural indicator. The cells are empty if no significant difference was found, but contain ">" or "<" signs for weak differences (0.001 < *p*-value <= 0.01) and ">>" or "<<" signs for strong differences (*p*-value <= 0.001). The direction of the relational signs indicates whether the first or the second cluster within a pair exhibits higher average values.

**Table 6.** Indicators with significant differences between clusters, based on food waste per capita data. (Abbreviations of indicators: see Table 2). Empty columns (ACL1—ACL2 and ACL2—ACL3) are not displayed.

| | ACL1—ACL3 | ACL1—ACL4 | ACL1—ACL5 | ACL2—ACL4 | ACL2—ACL5 | ACL3—ACL4 | ACL3—ACL5 | ACL4—ACL5 |
|---|---|---|---|---|---|---|---|---|
| AREA | | | < | < | < | | | |
| CMR | | | | | | | | < |
| COFP | | | | > | > | | > | |
| ELET | | | | | | | | |
| FECH | | | | > | | | | < |
| GDP_PC | | | | > | > | >> | > | |
| GVAEGS | | < | << | | | | | |
| IAF | < | < | < | < | | | | |
| MF | | | | | | | | |
| MSDR | | | | | | | | |
| MWG_PC | | > | > | >> | >> | > | > | |
| NSFT | | | | | | | | |
| POP | | < | < | | | | | |
| POPDENS | | | | | > | | | |
| PPGDP | | | | > | > | >> | > | |
| PUKHW | > | | > | < | | << | | >> |
| REG_BOR_PC | | < | < | | < | | < | |
| REG_COA_PC | | | | | | | | |
| REG_INT_PC | > | | | | | << | << | |
| REG_ISL_PC | | > | >> | | | | | > |
| REG_MOU_PC | | | | | | | | |
| REG_RUR_PC | | < | < | | | | | |
| REG_URB_PC | | < | < | | | | | |
| RMRPG | | < | | < | | << | | > |
| RPS_PC | | < | | < | | << | | >> |
| TEA | | | > | | | | | |

**Table 7.** Indicators with significant differences between clusters based on relative food waste data.

| | RCL1—RCL2 | RCL1—RCL3 | RCL1—RCL4 | RCL1—RCL5 | RCL2—RCL3 | RCL2—RCL4 | RCL2—RCL5 | RCL3—RCL4 | RCL3—RCL5 | RCL4—RCL5 |
|---|---|---|---|---|---|---|---|---|---|---|
| AREA | | | | | | < | << | | | |
| CMR | << | << | | | | > | >> | > | >> | |
| COFP | | | > | > | | | | | | |
| ELET | | | | | | | | | | |
| FECH | | | | | | | | | | |
| GDP_PC | | | > | > | > | >> | >> | > | | |
| GVAEGS | << | | | | | > | | | | |
| IAF | << | << | << | << | << | << | << | | | |
| MF | | | | | | | < | | | < |
| MSDR | | | | | | | | | | |
| MWG_PC | | | | | | > | > | >> | >> | |
| NSFT | | | | | | | | | | |
| POP | << | | < | < | | | | | | |
| POPDENS | << | << | | | | > | >> | | >> | > |
| PPGDP | | | > | | | >> | >> | >> | > | |
| PUKHW | | | | | | < | | | | |
| REG_BOR_PC | | | < | < | | < | << | | < | |
| REG_COA_PC | >> | >> | >> | | > | | | | << | |
| REG_INT_PC | | | | | | < | << | < | << | |
| REG_ISL_PC | >> | >> | >> | >> | | | | | | |
| REG_MOU_PC | | < | < | < | < | < | | | | |
| REG_RUR_PC | | | | | | < | << | | < | |
| REG_URB_PC | | < | | < | | | | | | |
| RMRPG | < | < | << | << | | < | < | < | < | |
| RPS_PC | | | | | | | < | | << | |
| TEA | >> | >> | >> | >> | >> | >> | | | | < |

Table 6 presents the results for absolute food waste parameters, while Table 7 shows the results for relative food waste parameters. For instance, in Table 6 the row corresponding to "GDP_PC" has ">" symbols in the ACL2—ACL4, ACL2—ACL5 and ACL3—ACL5 columns and a ">>" symbol in the ACL3—ACL4 column. This row indicates that the per capita GDPs of cluster 2 and cluster 3 are greater than those of cluster 4 and 5. This suggests a statistical connection between per capita GDP and absolute food waste parameters. However, this is not a simple, monotonic connection. For example, cluster 3 exhibits higher average per capita GDP values compared to cluster 4, yet their total food waste production is approximately the same, with lower average total food waste in the case of cluster 3 compared to cluster 4. The difference between these two clusters can be found in the characteristics of their food waste as represented by the five parameters.

Cluster analysis reveals significant differences in social–economic parameters among the five clusters. The results presented in Table 6 suggest that food waste levels per person are strongly linked to socio-economic inequalities. The ACL1–ACL5, ACL1–ACL4 and ACL2–ACL4 cluster pairs show the most significant differences in terms of the examined parameters, with distinctions found across 10–11 structural indicators. Additionally, there is a noticeable contrast between the ACL2–ACL5, ACL3–ACL4, ACL3–ACL5 and ACL4–ACL5 clusters in terms of 6–7 distinct indicators.

Countries belonging to ACL2 and ACL3 are richer than countries in ACL4 and ACL5, as they have a higher GDP per capita and a higher purchasing power adjusted GDP per capita. In parallel, they have a higher consumption footprint index (COFP) and higher municipal waste generation (MWG_PC), showing a connection between stronger economies and higher overall waste output (not limited to food).

The poorer countries of ACL4, in contrast, have higher PUKHW (population unable to keep home adequately warm), IAF (inability to afford a meal), RMRPG (at risk of poverty) and RPS_PC (people at risk of poverty or social exclusion) values than the other clusters have, indicating severer social vulnerabilities in the countries in this group.

According to the values, there is no evident relationship between educational status (like ELET and TEA values) and the food waste per capita characterization of a country. The effect of geographical factors (region types) can be observed mainly in the case of Cyprus (ACL1).

The results in Table 6 indicate that absolute food waste patterns are linked to macroeconomic strength and social vulnerability indicators. Wealthier countries, characterized by higher GDP per capita, higher municipal waste and lower deprivation indicators, form a distinct cluster. Conversely, poorer clusters exhibit greater social challenges and relatively lower waste per capita. This suggests that the extent of absolute food waste reflects broader socio-economic inequalities across the EU.

Table 7 compares the cluster results of the analysis performed using relative food waste values (% of total waste per sector). This shifts the focus from how much waste per capita is generated to which sectors dominate food waste in each country. This comparison reveals much greater differences between the individual clusters. The RCL2–RCL4 (15 differences) and RCL2–RCL5 (13 differences) cluster pairs present the most significant differences, but between the RCL1–RCL2, RCL1–RCL3, RCL1–RCL4, RCL1–RCL5 and RCL3–RCL5 cluster pairs there also high dissimilarities (9–11 cases).

Countries belong to RCL1, RCL2 and RCL3 have higher GDP per capita and purchasing power adjusted GDP per capita values than countries in RCL4 and RCL5. This suggests that sectoral distribution of food waste (e.g., manufacturing vs. households) is also influenced by national wealth. Meanwhile, RCL2 and RCL3 demonstrate a higher circular material use rate (CMR) compared to RCL1, RCL4 and RCL5. However, the amount of municipal waste (MWG_PC) is also higher in the countries of RCL2 and RCL3 than those of RCL4 and RCL5.

Significant differences exist in IAF (inability to afford a meal) and RMRPG (relative median at-risk-of-poverty-gap) values, with the best values observed in RCL1 and RCL2 and considerably higher values in RCL4 and RCL5. A significantly higher percentage of people in RCL5 countries face the risk of poverty or social exclusion (RPS_PC) compared to those in RCL2 and RCL3. RCL4 and RCL5 are predominantly characterized by household-level waste (70% in RCL4 and 50% in RCL5), indicating that food waste structure reflects social inequalities.

Tertiary educational attainment (TEA) is strongly connected with relative food waste patterns, with educational status being significantly better in countries within RCL1 and RCL2. People in rural and border regions dominate countries belonging to RCL4 and RCL5 (Eastern and Southern Europe).

To sum up, the results based on relative food waste data highlight structural and sectoral food waste differences. Wealthier clusters (RCL1, RCL2 and RCL3) tend to have higher food waste in manufacturing (RCL1 and RCL2) and restaurant services (RCL3), while in poorer clusters (RCL4 and RCL5) households are the primary source of food waste. This latter pattern may be attributed to weaker redistribution infrastructures, cultural habits and socio-economic vulnerabilities. Geographical context (e.g., rural vs. urban areas) further reinforces these patterns.

As was mentioned above, absolute and relative food waste parameters characterize the food waste production of the countries in different ways. Comparing Table 6 to Table 7 shows a significant difference: the former contains 13 strong and 47 weak connections, while in the latter there are 48 strong and 45 weak connections. This indicates that the clusterization based on relative data has a stronger relationship with the structural parameters than that based on absolute data. Uncovering the reason behind this difference can be a topic of further research, but the observation may be important in understanding the connection between food waste production and structural parameters.

## 5. Conclusions

The paper investigates patterns of food waste generation across the 27 EU member states between 2020 and 2022, with particular attention paid to the period of the COVID-19 pandemic. Eurostat data for five main food waste sources (primary production, food manufacturing, retail–distribution, food services and households) were analysed and compared with over twenty social, economic, environmental and geographical indicators. Correlation analysis, Principal Component Analysis (PCA) and cluster analysis were applied to identify multivariate relationships, to reveal similarities and differences and to classify countries according to their waste generation profiles.

The results indicate that the total volume of food waste in the EU remained relatively stable during the observed three years, but its composition changed significantly. Household food waste peaked in 2021 during lockdowns, while food waste originating from distribution and food services increased steadily. Principal Component Analysis revealed a clear distinction between system-level (linked to agriculture, industry and distribution) and individual-level (households and food services) food waste. Cluster analysis identified five groups of countries with largely stable food waste patterns, even under pandemic-related disruptions.

According to the results derived from absolute and relative food waste data, the level and structure of food waste are strongly linked to the economic strength and social inequalities of the EU countries. Higher-income Western and Northern European states tend to generate greater amounts of total food waste and more waste in manufacturing and food service sectors, while in Eastern and Southern Europe household food waste is the dominant factor.

Households represent the most important target for EU-wide food waste reduction policies. However, country-specific strategies are required to differentiate according to national and regional contexts. In higher-income states, efficiency improvements in the manufacturing and food service sectors should be prioritized, while in Eastern and Southern states household-level awareness campaigns and stronger social support systems should be prioritized. Future EU strategies must account for regional and social disparities in order to shift forward a more sustainable food system.

The key contribution of this paper is the demonstration that PCA facilitates obtaining a clean differentiation between system-level and individual-level drivers of food waste, which is a methodological contribution, in contrast to the descriptive analyses that are ubiquitously used in the area. Nevertheless, this approach is not without its limitations. First, relying on a three-year Eurostat dataset constrains detection capacity in capturing long-term structural changes, particularly in relation to residual effects from the COVID-19 pandemic. Second, the lack of harmonized datasets outside of the EU impedes a global comparative setting, which would situate EU trends within the wider context. Third, even though PCA and clustering were deemed useful, further statistical confirmation (e.g., robustness tests, regression or hierarchical models) could enhance the robustness of outcomes.

**Author Contributions:** Conceptualization, A.Z. and A.H.; methodology, A.H., A.Z. and G.Z.M.; software, A.H.; validation, G.Z.M. and A.H.; formal analysis, A.H. and A.Z.; investigation, A.Z., A.H. and G.Z.M.; resources, G.Z.M.; data curation, A.H. and G.Z.M.; writing—original draft preparation, A.Z., A.H. and G.Z.M.; writing—review and editing, G.Z.M., A.Z. and A.H.; visualization, A.H. and G.Z.M.; supervision, A.Z.; funding acquisition, A.Z. and G.Z.M. All authors have read and agreed to the published version of the manuscript.

**Funding:** This research received no external funding. The APC was funded by Széchenyi István University.

**Institutional Review Board Statement:** Not applicable.

**Informed Consent Statement:** Not applicable.

**Data Availability Statement:** The raw data used in this study are available in the Eurostat database, as referenced in Tables 1 and 2. The processed and derived data are available on request from the corresponding author.

**Conflicts of Interest:** The authors declare no conflicts of interest.

## Abbreviations

The following abbreviations are used in this manuscript:

| | |
|---|---|
| ACL | Clustering based on absolute food waste data |
| EU | European Union |
| FP | Food waste from primary production of food |
| GDP | Gross domestic product |
| HHA | Food waste from total activity of households |
| MFP | Food waste from manufacture of food products and beverages |
| PCA | Principal Component Analysis |
| RCL | Clustering based on relative food waste data |
| RDF | Food waste from retail and other distribution of food |
| RFS | Food waste from restaurants and food services |
| TOTAL | Total amount of food waste |

## References

1. Nemes, G.; Chiffoleau, Y.; Zollet, S.; Collison, M.; Benedek, Z.; Colantuono, F.; Dulsrud, A.; Fiore, M.; Holtkamp, C.; Kim, T.-Y.; et al. The impact of COVID-19 on alternative and local food systems and the potential for the sustainability transition: Insights from 13 countries. *Sustain. Prod. Consum.* **2021**, *28*, 591–599. [CrossRef] [PubMed]
2. Vărzaru, A.A.; Simion, D. Exploring the Drivers of Food Waste in the EU: A Multidimensional Analysis Using Cluster and Neural Network Models. *Foods* **2025**, *14*, 1358. [CrossRef]
3. Atalan-Helicke, N.; College Bürge Abiral, S.; Abiral, B.; Krieger, Z. Alternative food distribution networks, resilience, and urban food security in Turkey during the COVID-19 pandemic. *J. Agric. Food Syst. Community Dev.* **2021**, *10*, 89–104.
4. Kakaei, H.; Nourmoradi, H.; Bakhtiyari, S.; Jalilian, M.; Mirzaei, A. Effect of COVID-19 on food security, hunger, and food crisis. In *COVID-19 and the Sustainable Development Goals*; Elsevier: Amsterdam, The Netherlands, 2022; pp. 3–29.
5. Kumar, V.; Yetkin Ekren, B.; Wang, J.; Shah, B.; Frederico, G.F. Investigating the impact of COVID-19 on sustainable food supply chains. *J. Model. Manag.* **2023**, *18*, 1250–1273. [CrossRef]
6. Zseni, A.; Horváth, A.; Macher, G.Z.; Pécsinger, J.; Sipos, D. Exploring disparities in the generation of food waste from a spatial and sustainability perspective. *BIO Web Conf.* **2024**, *125*, 02002. [CrossRef]
7. Zseni, A.; Horváth, A.; Macher, G.Z.; Sipos, D.; Pécsinger, J. Using Multivariate Statistical Analysis for Examining the Relationship between Food Waste Generation and Socio-economic Factors. *J. Sustain. Dev. Energy Water Environ. Syst.* **2025**, *13*, 1–16. [CrossRef]
8. Paparella, A.; Purgatorio, C.; Chaves-López, C.; Rossi, C.; Serio, A. The Multifaceted Relationship between the COVID-19 Pandemic and the Food System. *Foods* **2022**, *11*, 2816. [CrossRef]
9. Borghesi, G.; Morone, P. A review of the effects of COVID-19 on food waste. *Food Secur.* **2023**, *15*, 261–280. [CrossRef]
10. Boyacı-Gündüz, C.P.; Ibrahim, S.A.; Wei, O.C.; Galanakis, C.M. Transformation of the Food Sector: Security and Resilience during the COVID-19 Pandemic. *Foods* **2021**, *10*, 497. [CrossRef]
11. Habib, M.; Singh, S.; Jan, S.; Jan, K.; Bashir, K. The future of the future foods: Understandings from the past towards SDG-2. *npj Sci. Food* **2025**, *9*, 1–27. [CrossRef]
12. Magableh, G.M. Supply Chains and the COVID-19 Pandemic: A Comprehensive Framework. *Eur. Manag. Rev.* **2021**, *18*, 363–382. [CrossRef]
13. Chen, Q.; Li, C.; Lu, L.; Ke, Y.; Kang, K.; Mao, S.; Liao, Z. Optimization of Fresh Produce Supply Chain Resilience Capacity: An Extension Strategy Generation Method. *Symmetry* **2025**, *17*, 272. [CrossRef]
14. Schudel, S.; Shoji, K.; Shrivastava, C.; Onwude, D.; Defraeye, T. Solution roadmap to reduce food loss along your postharvest supply chain from farm to retail. *Food Packag. Shelf Life* **2023**, *36*, 101057. [CrossRef]
15. Kumar, A.; Mangla, S.K.; Kumar, P.; Song, M. Mitigate risks in perishable food supply chains: Learning from COVID-19. *Technol. Forecast. Soc. Change* **2021**, *166*, 120643. [CrossRef]
16. Nguyen, T.T.T.; Hetherington, J.B.; O'Connor, P.J.; Malek, L. Sustainable food consumption: Sustainability-conscious consumers do not reduce food waste but nutrition-conscious consumers do. *Resour. Conserv. Recycl.* **2025**, *219*, 108296. [CrossRef]

17. Sjölund, A.; Malefors, C.; von Brömssen, C.; Svensson, E.; Brancoli, P.; Syed, S.; Rousta, K.; Eriksson, M. Unveiling the hidden patterns of household food waste. *Curr. Res. Environ. Sustain.* **2025**, *9*, 100292. [CrossRef]
18. Babbitt, C.W.; Babbitt, G.A.; Oehman, J.M. Behavioral impacts on residential food provisioning, use, and waste during the COVID-19 pandemic. *Sustain. Prod. Consum.* **2021**, *28*, 315–325. [CrossRef]
19. Cequea, M.M.; Vásquez Neyra, J.M.; Schmitt, V.G.H.; Ferasso, M. Household Food Consumption and Wastage during the COVID-19 Pandemic Outbreak: A Comparison between Peru and Brazil. *Sustainability* **2021**, *13*, 7583. [CrossRef]
20. Filimonau, V.; Vi, L.H.; Beer, S.; Ermolaev, V.A. The Covid-19 pandemic and food consumption at home and away: An exploratory study of English households. *Socio-Econ. Plan. Sci.* **2022**, *82*, 101125. [CrossRef]
21. Cavallo, C.; Sacchi, G.; Carfora, V. Resilience effects in food consumption behaviour at the time of Covid-19: Perspectives from Italy. *Heliyon* **2020**, *6*, e05676. [CrossRef]
22. Betz, A.; Buchli, J.; Göbel, C.; Müller, C. Food waste in the Swiss food service industry—Magnitude and potential for reduction. *Waste Manag.* **2015**, *35*, 218–226. [CrossRef] [PubMed]
23. Rao, M.; Bast, A.; de Boer, A. How COVID-19 impacted surplus food redistribution in the Netherlands: An explorative study. *Food Secur.* **2022**, *14*, 1377–1385. [CrossRef] [PubMed]
24. European Parliament; Council of the European Union. Directive 2008/98/EC of the European Parliament and of the Council of 19 November 2008 on waste and repealing certain Directives. *Off. J. Eur. Union* **2008**, *L 312*, 3–30.
25. European Parliament; Council of the European Union. Regulation (EC) No 178/2002 of the European Parliament and of the Council of 28 January 2002 laying down the general principles and requirements of food law, establishing the European Food Safety Authority and laying down procedures in matters of food safety. *Off. J. Eur. Union* **2002**, *L 31*, 1–24.
26. European Commission. *Commission Staff Working Document: Impact Assessment Report—Accompanying the Document Directive of the European Parliament and of the Council Amending Directive 2008/98/EC on Waste*; European Commission: Brussels, Belgium, 2023.
27. Parfitt, J.; Barthel, M.; MacNaughton, S. Food waste within food supply chains: Quantification and potential for change to 2050. *Philos. Trans. R. Soc. B Biol. Sci.* **2010**, *365*, 3065–3081. [CrossRef]
28. Stancu, V.; Haugaard, P.; Lähteenmäki, L. Determinants of consumer food waste behaviour: Two routes to food waste. *Appetite* **2016**, *96*, 7–17. [CrossRef]
29. Aschemann-Witzel, J.; de Hooge, I.; Amani, P.; Bech-Larsen, T.; Oostindjer, M. Consumer-Related Food Waste: Causes and Potential for Action. *Sustainability* **2015**, *7*, 6457–6477. [CrossRef]
30. Mena, C.; Adenso-Diaz, B.; Yurt, O. The causes of food waste in the supplier–retailer interface: Evidences from the UK and Spain. *Resour. Conserv. Recycl.* **2011**, *55*, 648–658. [CrossRef]
31. Parizeau, K.; von Massow, M.; Martin, R. Household-level dynamics of food waste production and related beliefs, attitudes, and behaviours in Guelph, Ontario. *Waste Manag.* **2015**, *35*, 207–217. [CrossRef]
32. Thyberg, K.L.; Tonjes, D.J. Drivers of food waste and their implications for sustainable policy development. *Resour. Conserv. Recycl.* **2016**, *106*, 110–123. [CrossRef]
33. Reisinger, H.; Monier, V. *Guidelines on the Preparation of Food Waste Prevention Programmes*; BIO Intelligence Service, European Commission, DG Environment: Brussels, Belgium, 2011.
34. European Commission. EU Platform on Food Losses and Food Waste—Food Safety. 2025. Available online: https://food.ec.europa.eu/food-safety/food-waste/eu-actions-against-food-waste/eu-platform-food-losses-and-food-waste_en (accessed on 9 October 2025).
35. European Commission. Delegated Decision—2019/1597—EN—EUR-Lex. 2019. Available online: https://eur-lex.europa.eu/eli/dec_del/2019/1597/oj/eng (accessed on 9 October 2025).
36. European Commission. *EU Guidelines on Food Donation (2017/C 361/01)*; European Commission: Brussels, Belgium, 2017.
37. European Commission. *Guidelines for the Feed Use of Food No Longer Intended for Human Consumption*; European Commission: Brussels, Belgium, 2018.
38. Jülicher, S. *Recommendations for Action in Food Waste Prevention*; EU Platform on Food Losses and Food Waste, European Commission: Brussels, Belgium, 2019.
39. Caldeira, C.; De Laurentiis, V.; Sala, S. *Assessment of Food Waste Prevention Actions. Development of an Evaluation Framework to Assess the Performance of Food Waste Prevention Actions*; Publications Office of the European Union: Luxembourg, 2019. [CrossRef]
40. European Commission. EU FOOD Loss and Prevention Hub, Official Website. 2025. Available online: https://ec.europa.eu/food/safety/food_waste/eu-food-loss-waste-prevention-hub/ (accessed on 9 October 2025).
41. REFRESH Wrap Global. Building partnerships, driving change—A voluntary approach to cutting food waste. 2019. Available online: http://tiny.cc/VABlueprint (accessed on 9 October 2025).
42. Eurostat. Circular Material Use Rate [cei_srm030]. 2025. Available online: https://ec.europa.eu/eurostat/databrowser/view/cei_srm030/default/table (accessed on 9 October 2025).
43. Eurostat. Consumption Footprint [cei_gsr010]. 2025. Available online: https://ec.europa.eu/eurostat/databrowser/view/CEI_GSR010/default/table?lang=en (accessed on 9 October 2025).

44. Eurostat. Early Leavers from Education and Training by Labour Status [edat_lfse_14]. 2025. Available online: https://ec.europa.eu/eurostat/databrowser/view/edat_lfse_14/default/table?lang=en (accessed on 9 October 2025).
45. Eurostat. Final Energy Consumption in Households per Capita [sdg_07_20]. 2025. Available online: https://ec.europa.eu/eurostat/databrowser/product/page/sdg_07_20 (accessed on 9 October 2025).
46. Eurostat. Purchasing Power Parities (PPPs), Price Level Indices and Real Expenditures for ESA 2010 Aggregates [prc_ppp_ind]. 2025. Available online: https://ec.europa.eu/eurostat/databrowser/view/prc_ppp_ind/default/table?lang=en (accessed on 9 October 2025).
47. Eurostat. Gross Value Added in Environmental Goods and Services Sector [sdg_12_61]. 2025. Available online: https://ec.europa.eu/eurostat/databrowser/view/sdg_12_61/default/table (accessed on 9 October 2025).
48. Eurostat. Inability to Afford a Meal with Meat, Chicken, Fish (or Vegetarian Equivalent) Every Second Day [ilc_mdes03]. 2025. Available online: https://ec.europa.eu/eurostat/databrowser/view/ilc_mdes03/default/table?lang=en (accessed on 9 October 2025).
49. Eurostat. Material Footprints—Main Indicators [env_ac_rme]. 2025. Available online: https://ec.europa.eu/eurostat/databrowser/product/page/env_ac_rme (accessed on 9 October 2025).
50. Eurostat. Severe Material and Social Deprivation Rate by Age and Sex [ilc_mdsd11]. 2025. Available online: https://ec.europa.eu/eurostat/databrowser/view/ILC_MDSD11/default/table?lang=en (accessed on 9 October 2025).
51. Eurostat. Municipal Waste by Waste Management Operations [env_wasmun]. 2025. Available online: https://ec.europa.eu/eurostat/databrowser/view/env_wasmun/default/table?lang=en (accessed on 9 October 2025).
52. Eurostat. Nights Spent at Tourist Accommodation Establishments [tour_occ_ninat]. 2025. Available online: https://ec.europa.eu/eurostat/databrowser/view/tour_occ_ninat/default/table?lang=en (accessed on 9 October 2025).
53. Eurostat. Population on 1 January [tps00001]. 2025. Available online: https://ec.europa.eu/eurostat/databrowser/product/page/TPS00001 (accessed on 9 October 2025).
54. Eurostat. Population Density [tps00003]. 2025. Available online: https://ec.europa.eu/eurostat/databrowser/view/tps00003/default/table (accessed on 9 October 2025).
55. Eurostat. Purchasing Power Adjusted GDP per Capita [sdg_10_10]. 2025. Available online: https://ec.europa.eu/eurostat/databrowser/view/sdg_10_10/default/bar?lang=en (accessed on 9 October 2025).
56. Eurostat. Population Unable to Keep Home Adequately Warm by Poverty Status [sdg_07_60]. 2025. Available online: https://ec.europa.eu/eurostat/databrowser/product/page/sdg_07_60 (accessed on 9 October 2025).
57. Eurostat. Area of the Regions by Other Typologies [urt_d3area]. 2025. Available online: https://ec.europa.eu/eurostat/databrowser/product/page/URT_D3AREA (accessed on 9 October 2025).
58. Eurostat. Persons at Risk of Poverty or Social Exclusion by Age and Sex [ilc_peps01n]. 2025. Available online: https://ec.europa.eu/eurostat/databrowser/view/ILC_PEPS01N/default/table (accessed on 9 October 2025).
59. Eurostat. Persons Aged 25–34 with Tertiary Educational Attainment Level [sdg_04_20]. 2025. Available online: https://ec.europa.eu/eurostat/databrowser/view/SDG_04_20/default/table?lang=en (accessed on 9 October 2025).
60. Jolliffe, I.T. PRINCIPAL COMPONENT ANALYSIS: A BEGINNER'S GUIDE—I. Introduction and application. *Weather* **1990**, *45*, 375–382. [CrossRef]
61. Scitovski, R.; Sabo, K.; Martínez-Álvarez, F.; Ungar, Š. *Cluster Analysis and Applications*; Springer: Cham, Switzerland, 2021; pp. 1–271.

*sustainability*

MDPI

*Article*

# An Empirical Comparison of Informational and Behavioral Food Waste Interventions in U.S. Households: The Advantage of Implementation Intentions

**John A. Aitken *, Grace Mika, Balca Alaybek, Laura Leets and Amber Sprenger**

The MITRE Corporation, 7515 Colshire Drive, McLean, VA 22102, USA; gmika@mitre.org (G.M.); balaybek@mitre.org (B.A.); lleets@mitre.org (L.L.); asprenger@mitre.org (A.S.)
* Correspondence: jaitken@mitre.org

**Abstract:** Household food waste is a major source of economic loss and environmental impact, but it is unclear whether interventions can successfully reduce food waste. Thus, the current study evaluated two interventions designed to change food waste-related behaviors and reduce food waste in the context of U.S. households. We tested these interventions via an online study of 1494 households, wherein individuals were randomized to three informational conditions (a "save money" message, a "holistic" message, or a "control" message) and two behavioral conditions (set implementation intentions or do not set implementation intentions). The results showed that informational interventions had minimal impact on food waste-related behaviors, only reducing the degree to which households wasted food due to date labels. Conversely, the behavioral intervention had a stronger impact on food waste-related behaviors: households enacted higher levels of efficient food acquisition behaviors, less frequently wasted food due to date labels, and enacted more efficient leftover usage behaviors. However, although these behaviors are related to food waste, neither intervention significantly reduced household food waste. Therefore, whereas behavioral interventions appear to be more promising than informational interventions alone, it appears that household food waste reduction is a more challenging endeavor than previously thought. We discuss implications of our results and a number of future research directions.

**Keywords:** food waste; food waste reduction behaviors; behavior change; messaging; implementation intentions

## 1. Introduction

As a major contributor to economic outcomes, food insecurity, and environmental strain, food waste is a far-reaching problem with national and global consequences. Worldwide, food waste amounts to over 1 billion tons and USD 940 billion in economic losses annually [1]. In the U.S., 40% of food is lost or wasted each year, costing an estimated USD 218 billion [2]. But, despite all this wasted food, food insecurity (i.e., uncertainty in having or inability to acquire adequate food for one's household due to insufficient resources) affected 17 million U.S. households at some time during 2022 [3]. This degree of food insecurity is in some ways a function of food prices, which can be reduced via reducing consumer food waste (i.e., lowering food demand and production [4]). Finally, beyond its economic and societal impacts, food waste is the largest source of human-related methane emissions in the U.S. and contributes to a wide range of environmental impacts including climate change, air pollutants, water scarcity, biodiversity loss, and soil and water quality degradation [2]. Thus, initiatives to reduce food waste have the potential to positively contribute to the economy, societal well-being, and the environment.

In June 2024, the U.S. White House, USDA, EPA, and Food and Drug Administration (FDA) announced the National Strategy for Reducing Food Loss and Waste and Recycling Organics as part of President Biden's whole-of-government approach to tackle climate change, feed people, address environmental justice, and promote a circular economy [5]. The strategy includes plans to develop, launch, and run a national food waste consumer education and behavior change campaign. Targeting consumers is a high priority for food waste reduction, given that roughly half of food waste occurs during the consumption phase by households and food service [6]. Moreover, food waste at the consumption phase carries larger environmental and economic cost than food losses upstream (e.g., farms or manufacturing [7]). As such, a consumer-facing behavior change campaign could have significant impacts, and its success depends on the effectiveness of its interventions.

To date, many consumer-facing food waste interventions have demonstrated mixed effectiveness for several reasons. First, there is a general lack of theoretical framing of intervention studies, such that the mechanisms through which interventions have their effects are not sufficiently articulated [8]. Thus, it is difficult to determine why an intervention did or did not demonstrate significant effects on food waste behaviors or amounts. Second, studies demonstrate significant methodological heterogeneity, which makes it difficult to come to an overall assessment of existing interventions in the field. Specifically, several review and meta-analysis papers have emphasized that there is significant heterogeneity in terms of experimental design (i.e., including "fair" control conditions (An example of a "fair" treatment-to-control comparison is Roe et al. [9]: the treatment group received a highly individualized food waste reduction intervention, and the control group received a highly individualized ("intensity-matched") stress management intervention. This study represents a more rigorous test of the intervention when comparing treatment and control groups and exemplifies the need for appropriate control groups.)), measurement of the outcome variable (i.e., measuring wasted food, including a pre-intervention measure), and using adequate statistical analyses [8,10–13]. Finally, many interventions utilize a one-size-fits-all approach and do not consider the need for tailoring interventions to particular households and lifestyles [9]. This is particularly worrisome due to the most effective intervention likely being one that targets individuals' specific interests and motivations—as has been demonstrated in several areas regarding behavioral change (e.g., health [14]; well-being [15]; workplace [16]).

Thus, the current study is designed to compare two different intervention approaches in an online experiment using random assignment and measuring self-reported food waste and behaviors before and after the intervention. Specifically, we compare an informational intervention approach and a behavioral intervention approach, testing (1) which informational conditions demonstrate a stronger effect on food waste and behaviors (i.e., "save money" versus "save money, reduce environmental impact, and help others") and (2) whether asking individuals to set if-then plans for themselves are incrementally effective relative to only providing food waste-related information. In doing so, we hope to extend the current literature by comparing these two common intervention paradigms.

## 2. Literature Review

The goal of the current study is to identify ways to help individuals reduce their household food waste. Thus, the amount of household food waste is the ultimate focus of this study. Borrowing language from intervention models in behavioral health and organizational science [17,18], food waste reduction is the distal outcome of the intervention. To impact distal outcomes, the intervention must first focus on impacting proximal outcomes:

the behaviors, thoughts, and patterns that determine whether a household wastes more or less food (distally).

Thus, the current intervention adopts an informational and behavioral approach to influencing these proximal outcomes so as to, ultimately, reduce food waste. To guide the intervention's design, we utilize the Theory of Planned Behavior [19] as a framework to determine how best to intervene in the process of attaining one's goal—which, in the present study, is food waste reduction.

In general, positive and cognitively accessible attitudes towards specific goal-relevant behaviors are more likely to give rise to those specific behaviors [20]. Thus, informational interventions can be delivered that bolster goal-relevant attitudes and, indirectly, promote the attainment of those goals [8]. However, there is often a gap between attitudes and behavior. In this regard, the Theory of Planned Behavior suggests that behavioral intentions are a more proximal determinant of behavior, but those intentions themselves are influenced by attitudes [21]. Thus, behavioral interventions that focus on this proximal link between intentions and behaviors stand to be especially effective in promoting goal attainment [22].

### 2.1. Informational Approach

In their review of food waste intervention studies, the Ref. [8] found that many studies focus on informational messaging: making individuals aware of the costs and environmental impact of food waste; posting signage in grocery stores; tips to prevent food waste and so forth. In general, such an approach to behavior change assumes that people will utilize relevant, useful information and adjust their behaviors accordingly.

An informational approach to food waste reduction may be especially useful to the extent that it targets issues relevant to people's lifestyles. Typically, messaging about food waste focuses on the potential to save money, to benefit the environment, and to feed families suffering from food insecurity [23]. For example, the goals of the recently announced National Strategy for Reducing Food Loss and Waste and Recycling Organics [5] are to "substantially reduce environmental impacts" and to "provide social and economic benefits that can help address the needs of underserved communities" (p. 4).

Thus, it seems critical to evaluate whether these messages can influence food waste behaviors and ultimately reduce food waste. However, it is possible that certain messages are more effective than others. In a nationally representative survey, saving money was found to be a key predictor of self-reported food waste amounts and was the most commonly cited concern related to food waste, whereas environmental concerns were seemingly less important—and even related to higher food waste [24].

Therefore, the present study evaluated whether a money-saving message alone was as impactful as a holistic message about saving money, reducing environmental impact, and helping others. These messages were delivered in narrative fashion, as some evidence suggests that stories create more openness to information than information presented as facts and statistics [25].

**Hypothesis 1.** *Participants who are assigned to an informational intervention (i.e., a "save money" message or a "holistic" message) will demonstrate (a) increased food waste reduction behaviors and (b) decreased food waste, relative to participants who are assigned to a control condition (i.e., a "sitting behavior" message).*

**Research Question 1.** *Is there a difference in (a) food waste reduction behaviors and (b) food waste between participants who receive the "save money" message versus participants who receive the "holistic" message?*

However, the provision of information alone may not be sufficient to change food waste behaviors. As stated by Hebrok and Heidenstrøm [26]: "Informing consumers about food waste as a societal problem is not sufficient enough to change how they handle food as part of their complex and interwoven everyday lives" (p. 1437). For example, general information about the benefits of reducing food waste may not be utilized if the information is not perceived as directly relevant to their lifestyles [27]. Thus, in the next section, we discuss a behavioral approach that may be impactful above and beyond an informational approach.

### 2.2. Behavioral Approach

As stated by the Theory of Planned Behavior, intentions are crucial predictors for behavior, but intentions alone are not sufficient to understand when a person will achieve their goals and execute the most appropriate behaviors [28]. This is likely to be especially the case for intentions to reduce food waste because there are many other ongoing goal pursuits that occupy people's daily lives and that may take priority, and, even more simply, reducing food waste is hard [26].

However, mirroring the gap between attitudes and behavior, intentions, too, do not always lead to the intended behavior [22]. One way to improve the effectiveness and situational relevance of intentions is via implementation intentions [29], which are if–then action plans through which a person specifies a situation ("If situation $x$ arises...") and an action to be performed ("... Then I will perform response $y$"). By setting such intentions, people are more likely to attain their goals because they have made a relevant situational cue accessible and tethered it to an appropriate goal-directed action: "People are in a good position both to see and to seize suitable opportunities to pursue their goals" (p. 300, [30]). Thus, when situational conditions match the cues formed while creating implementation intentions, people are more likely to recall and act upon their goals.

We reason that this behavioral approach to food waste reduction will be incrementally effective above and beyond a purely informational approach for several reasons. First, implementation intentions can effectively "shield" a person's goal to reduce food waste from becoming derailed by other concerns and goal pursuits (e.g., having sufficient food for the whole family during holidays), while also ensuring that such goal pursuit does not allow people to overextend themselves [31]. Thus, any food waste-related information may be better preserved by the act of setting intentions. Second, setting implementation intentions inherently involves tailoring one's action plans to one's household patterns, one's lifestyle, and so forth, rather than taking a "one size fits all" approach to food waste reduction [26]. Third, there is a significant amount of evidence that documents the effectiveness of implementation intentions on behavior change outcomes across a range of domains [32]. As such, a recent review paper recommended intention-setting as a framework for food waste reduction [33].

Therefore, the present study evaluated whether setting implementation intentions had a stronger impact on food waste relative to only reading informational food waste narrative messages.

**Hypothesis 2.** *Participants who are assigned to set implementation intentions will demonstrate (a) increased food waste reduction behaviors and (b) decreased food waste, relative to participants who only receive the informational intervention and who do not set implementation intentions.*

### 2.3. Current Study Design

The current study has been designed to compare informational versus behavioral approaches to household food waste interventions in the U.S. context. Due to the lightweight nature of these interventions, we have selected a relatively short time frame of two weeks to observe whether there are immediate, proximal changes in food waste behaviors and amounts, though it is possible that these interventions require longer periods of time for uptake and that their effects may regress over time. Additionally, we have opted to utilize self-reports for measures of key outcomes due to this method being previously validated [24], straightforward to implement, and also relatively convergent with more objective, intensive measurement methods [34,35]. However, these methods are unlikely to be as accurate as objective methods, and they may be subject to additional biases. Finally, though we have sought to make all conditions as similarly "strong" to allow for a fair comparison [36], it is possible that the purely informational approach is simply "weaker" than the behavioral approach [8]. However, we preferred this comparison, even despite the possible difference in strength, because it closely approximates how these approaches are utilized in the "real world".

## 3. Methods

### 3.1. Participants

Prior to recruitment, we conducted a one-tailed power analysis using G*Power (3.1) with alpha set at 0.05 and power at 0.80 for both an independent *t*-test (i.e., comparing participants assigned to different conditions) and a dependent *t*-test (i.e., comparing participants over time). The required sample size to detect small effects ($d = 0.20$) was 310 per condition for the independent *t*-test and 272 per condition for the dependent *t*-test. Thus, given these power considerations and our financial resources, we aimed to recruit at least 1500 participants (i.e., due to there being five study conditions).

We initially recruited 1999 adults from Prolific, an online participant panel platform, to complete Survey 1. However, we did not invite 31 participants to complete Survey 2 (two weeks later) due to failing the embedded attention check or for otherwise demonstrating low effort in responding to the survey. Additionally, of this number, we were only able to allow 1500 to complete Survey 2 due to financial resource limitations. Thus, 1494 successfully completed Survey 1 and Survey 2 and was the final sample size for this study. As compensation for their time, participants received USD 12 dollars for completion of the full study (USD 5.00 for Survey 1 and USD 7.00 for Survey 2).

As can be seen in Table 1 below, the sample consisted of a balanced representation of age, gender, household size, employment, and income generally consistent with national demographics. The sample comprised U.S. households. The sample reported an above-average level of education compared to the national average, with 86.9% of the sample having education beyond a high school graduate whereas only 63.3% of the nation is reported to have pursued education beyond high school [37].

**Table 1.** Participant demographics.

| Measure | N | Percent | M | SD | National Average |
|---|---|---|---|---|---|
| **Age** | 1492 | | 39.58 | 13.07 | 38.9 |
| **Household size** | 1488 | | 2.92 | 1.41 | 2.51 |
| **Number of children (aged < 18 years) in household** | 1489 | | 0.80 | 1.13 | 0.55 |
| **Gender** | | | | | |
| Female | 852 | 57.10% | | | 50.5% |
| Male | 616 | 41.29% | | | 49.5% |
| Non-binary | 19 | 1.27% | | | |
| Prefer not to answer | 5 | 0.34% | | | |

**Table 1.** *Cont.*

| Measure | N | Percent | M | SD | National Average |
| --- | --- | --- | --- | --- | --- |
| **Race** | | | | | |
| White | 989 | 66.20% | | | 75.3% |
| Black | 220 | 14.73% | | | 13.7% |
| Asian | 109 | 7.30% | | | 6.4% |
| Mixed/Other | 176 | 11.77% | | | 4.7% |
| **Employment** | | | | | |
| Employed full-time | 850 | 57.09% | | | 37.3% |
| Employed part-time | 275 | 18.47% | | | 10.4% |
| Not employed and not looking for work | 237 | 15.92% | | | 37.4% |
| Not employed but looking for work (unemployed) | 127 | 8.53% | | | 4.1% |
| **Annual Household Income** | | | | | |
| USD 12,000 to USD 23,999 | 86 | 5.77% | | | |
| USD 24,000 to USD 35,999 | 150 | 10.07% | | | |
| USD 36,000 to USD 47,999 | 121 | 8.12% | | | |
| USD 48,000 to USD 59,999 | 191 | 12.82% | | | |
| USD 60,000 to USD 89,999 | 281 | 18.86% | | | |
| USD 90,000 to USD 119,999 | 240 | 16.11% | | | |
| USD 120,000 to USD 179,999 | 226 | 15.17% | | | |
| USD 180,000 to USD 239,999 | 62 | 4.16% | | | |
| USD 240,000 and over | 53 | 3.56% | | | |
| **Education** | | | | | |
| Less than a high school diploma | 13 | 0.87% | | | 8.9% |
| High school graduate | 182 | 12.19% | | | 27.9% |
| Technical, trade, vocational, business school | 46 | 3.08% | | | N/A |
| Two-year associate degree | 135 | 9.04 | | | 10.5% |
| Some college, but no degree | 283 | 18.96% | | | 14.9% |
| Four-year bachelor's degree | 578 | 38.71% | | | 23.5% |
| Some postgrad. schooling, but no postgrad. degree | 32 | 2.14% | | | N/A |
| Postgraduate or professional degree | 224 | 15.00% | | | 14.4% |

*Note.* N = sample size; M = mean (the average value of a set of numbers); SD = standard deviation (a measure of how spread out the values are from the average). Responses to demographic questions were not required, and some respondents chose not to answer one or more questions. Therefore, counts of responses may sum to less than the full sample of 1494. Note that the median household income at the national level is USD 80,610.

### 3.2. Procedure and Measures

This study involved two surveys: Survey 1 was completed by participants on Monday, 19 August 2024, and Survey 2 was completed by participants two weeks later between Monday, September 2, and Tuesday, September 3. Each survey took participants approximately 10–12 min to complete. The research was approved by the institutional review board of The MITRE Corporation.

In Survey 1, participants first provided informed consent and then reported their household's food waste (in cups and tablespoons) and food waste-related behaviors over the past week. To measure household food waste, we adapted the self-report method used by [24]. Participants were told: "Please think about all the food that gets thrown away or disposed of at your home. By disposed food, we mean food that went in the trash bin, in the compost, or down the sink drain. This includes food wasted by all members currently residing in your household and any foods you purchased from a store, market, or a restaurant, obtained at a food bank or pantry, received as a gift, or grown or raised yourself." Then, participants were asked to estimate approximately how many cups (i.e., the size of a closed fist) and tablespoons (i.e., the size of a thumb) of food was wasted in their households "in the past 7 days" for each of the following categories: fruit, vegetables, grains, protein, dairy, mixed, scraps, and oils, fats, and sugars. Excluding scraps, these amounts were summed to form a total measure of household food waste. Of note, despite noted limitations of self-reported food waste measures, this method has demonstrated moderate convergent validity with a diary-based, objective measure of food waste [34].

To measure food waste-related behaviors, participants reported the frequency with which they enacted behaviors related to food acquisition (i.e., checking food at home before shopping, making and sticking to shopping lists), date label wasting (i.e., wasting food because it had passed its date label), and leftover usage (i.e., using leftovers as ingredients, eating leftovers as meals, prioritizing leftovers close to expiration, throwing away leftovers). These behaviors were measured by adapting items published in the MITRE Gallup State of Food Waste in America report [24].

Next, participants were randomly assigned to one of three narrative messages: (1) a "money" message wherein a narrator described their attempts to reduce food waste in order to save money; (2) a "holistic" message wherein a narrator described their attempts to reduce food waste in order to save money, reduce negative environmental consequences, and help the food-insecure; or (3) a "sitting" control message that was unrelated to food waste and included a narrator's description of their efforts to reduce sitting behavior to improve their health. The full text of these messages can be found in the online Supplementary Materials S1–S3.

Then, participants who read the "money" and "holistic" messages were randomly assigned to either set implementation intentions or were not prompted to set implementation intentions. Participants assigned to set intentions were first told: "Implementation intentions are simple IF-THEN statements that describe (1) a situation (e.g., point in time, place) relevant to your goal, and (2) an action to be performed relevant to your goal. The basic structure of these statements is: *"IF I encounter situation X, THEN I will perform action Y!"* Then, participants were instructed to make at least one or two implementation intentions to reduce food waste over the coming week in one of the following categories: (1) food acquisition (e.g., meal planning, grocery shopping), (2) food preparation (e.g., cooking, food storage), and (3) leftovers (e.g., using up leftovers, using leftovers in other meals). Participants were provided with examples for each (e.g., "IF I am going shopping, THEN I will make a shopping list beforehand."). The full text of these instructions can be found in the online Supplementary Materials S4.

One week after completing Survey 1, participants were sent a check-in message reminding them of the upcoming Survey 2 and also reminding them to keep track of their food waste. Participants who had set implementation intentions were reminded to enact their intentions.

Finally, two weeks following completion of Survey 1, participants were invited to Survey 2. Participants first completed the same food waste and food waste-related behavior measures in reference to the past week. Participants also completed a number of items regarding how they tracked food waste, their experiences with their implementation intentions, and demographic characteristics.

### 3.3. Study Design

The procedure described above resulted in a 2 (time: pre-vs. post-intervention) by 3 (informational condition: money vs. holistic vs. sitting) by 2 (behavioral condition: set intentions vs. did not set intentions) factorial design. The time factor was manipulated within participants, whereas the informational and behavioral conditions were manipulated between participants. Participants were randomly assigned to one of five study conditions, shown in Table 2 below.

**Table 2.** Study design overview and sample sizes.

| | Informational Condition | | |
|---|---|---|---|
| | *Money Message* | *Holistic Message* | *Sitting Message (Control)* |
| **Behavioral Condition** | | | |
| *Set intentions* | N = 311 | N = 288 | — |
| *Did not set intentions (control)* | N = 305 | N = 305 | N = 285 |

*Note.* N = sample size.

### 3.4. Data Analysis

Data was analyzed using R [38]. The first step was to conduct bivariate mean differences to test whether there were significant changes in food waste amounts before and after the intervention across all conditions (i.e., Survey 1 vs. Survey 2). Analyses involved paired-sample $t$-tests and calculation of Cohen's $d$ effect sizes. The second step was to conduct multiple regression analyses to compare the effects of the money message versus the control message (Model 1), the holistic message versus the control message (Model 2), the money message versus the holistic message (Model 3), and the implementation intention-setting versus no intention-setting conditions (Model 4). All analyses controlled for pre-intervention levels of the four outcome variables in their respective analyses [39]: food acquisition behaviors, date label wasting behavior, leftover usage behaviors, and food waste.

## 4. Results

Descriptive statistics are shown for focal variables in Table 3, and a visual of the mean differences in food waste and food behaviors is shown in Figures 1 and 2.

**Table 3.** Descriptive statistics for focal outcome variables.

| | Pre-Intervention | | | Post-Intervention | | |
|---|---|---|---|---|---|---|
| | M | SD | Median | M | SD | Median |
| Edible food waste (cups) | 9.01 | 8.02 | 7.00 | 7.97 | 7.35 | 6.00 |
| Food acquisition behaviors | 2.78 | 0.79 | 3.00 | 2.88 | 0.75 | 3.00 |
| Date label wasting behavior | 2.01 | 1.10 | 2.00 | 1.78 | 1.15 | 2.00 |
| Leftover usage behaviors | 2.35 | 0.71 | 2.50 | 2.47 | 0.72 | 2.50 |

*Note.* M = mean (the average value of a set of numbers); SD = standard deviation (a measure of how spread out the values are from the average). Edible food waste is the sum of all food waste reported per category (fruit, vegetables, grains, protein, dairy, mixed, oils) but excludes food waste reported as scraps.

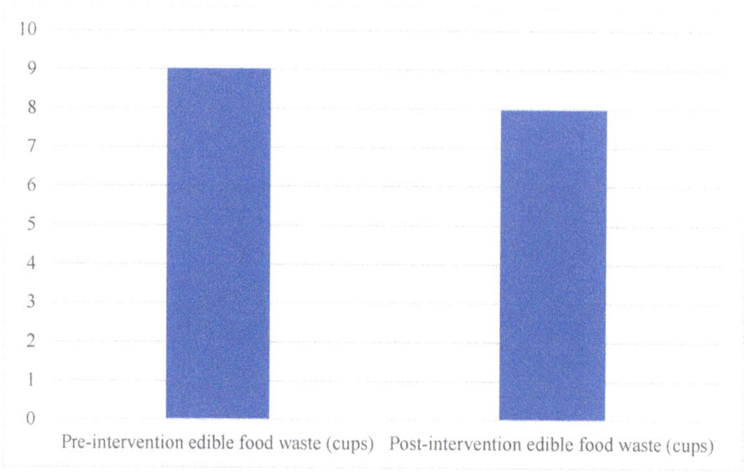


**Figure 1.** Mean differences in pre- and post-intervention food waste across study conditions.

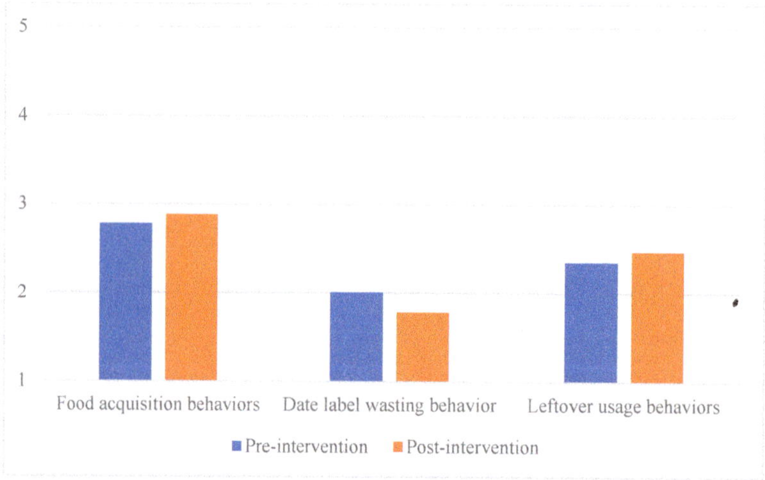


**Figure 2.** Mean differences in pre- and post-intervention food behaviors across study conditions.

Prior to testing this study's hypotheses and research question, we began by computing bivariate mean differences in food waste amounts across study conditions, as food waste reduction is the ultimate goal of these interventions. As can be seen in Table 4 and Figure 3, there was a significant reduction in post-intervention food waste across all study conditions. Furthermore, the largest effect sizes are associated with the combined condition where participants read the holistic message and then set implementation intentions ($d$ = 0.26), the implementation intention-setting condition ($d$ = 0.21), and the holistic message informational condition ($d$ = 0.19). Figure 4 shows the distribution of these effect sizes for each condition, given that each is relatively small and may be more easily distinguished visually. Of note, there was a significant reduction in food waste in the control condition as well, which was comparable to the reductions in the message-only conditions with no prompts for implementation setting.

**Table 4.** Bivariate mean differences comparing pre- and post-intervention waste.

| | N | Pre-Intervention Food Waste | Post-Intervention Food Waste | *t* | *d* |
|---|---|---|---|---|---|
| **Message** | | | | | |
| *Money* | 616 | 9.18 (8.14) | 8.21 (7.41) | 3.59 *** | 0.14 |
| *Holistic* | 593 | 8.70 (7.68) | 7.50 (7.10) | 4.55 *** | 0.19 |
| *Sitting (Control)* | 285 | 9.32 (8.47) | 8.45 (7.68) | 2.28 * | 0.13 |
| **Intention** | | | | | |
| *Set intentions* | 599 | 9.13 (7.50) | 7.82 (7.18) | 5.06 *** | 0.21 |
| *No intentions (Control)* | 896 | 8.94 (8.36) | 8.08 (7.46) | 3.87 *** | 0.13 |
| **Message * Intention** | | | | | |
| *Money * Set intentions* | 311 | 9.42 (7.75) | 8.37 (7.70) | 2.82 ** | 0.16 |
| *Holistic * Set intentions* | 288 | 8.81 (7.21) | 7.22 (6.53) | 4.43 *** | 0.26 |
| *Money * No prompt* | 305 | 8.93 (8.51) | 8.05 (7.12) | 2.25 * | 0.13 |
| *Holistic * No prompt* | 305 | 8.59 (8.11) | 7.76 (7.60) | 2.17 * | 0.12 |
| *Sitting * No prompt* | 285 | 9.32 (8.47) | 8.45 (7.68) | 2.28 * | 0.13 |

*Note.* N = sample size; $t$ = $t$-value (a statistic that shows how much a result differs from what is expected by chance); $d$ = Cohen's standardized mean difference effect size (a measure of how large the difference is between two groups, expressed in standard deviation units). As described in the Method section, the "Set intentions" and the "No intentions (Control)" intention conditions include participants who received a holistic message and who received a money message. Similarly, the "Money" and "Holistic" message conditions include participants who set intentions and participants who did not, whereas participants in the "Sitting (Control)" condition did not set intentions. *** $p < 0.001$. ** $p < 0.01$. * $p < 0.05$.

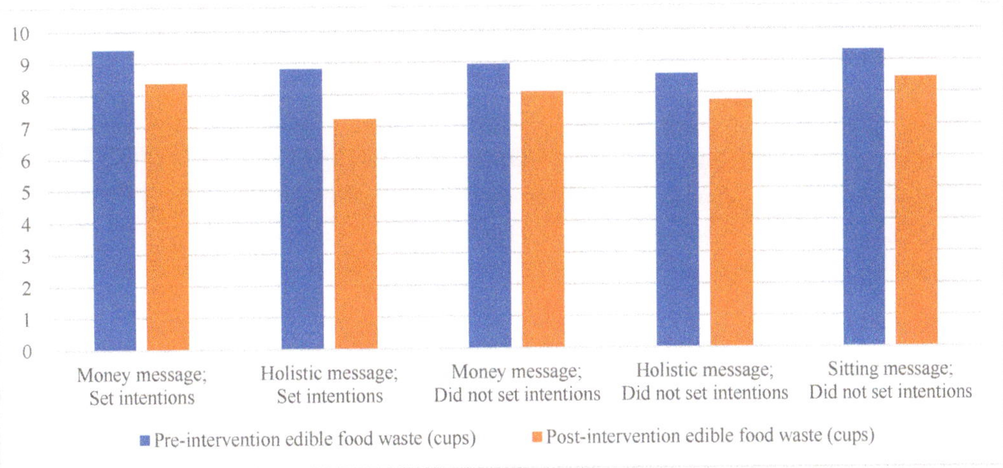


**Figure 3.** Mean differences in pre- and post-intervention food waste in each study condition.

To test study hypotheses and research question, we conducted a series of multivariate tests using multiple linear regression for four separate outcomes: post-intervention food acquisition behaviors, date label wasting behavior, leftover usage behaviors, and food waste. In each model, we controlled for the pre-intervention level of each outcome. These results can be seen in Table 5.

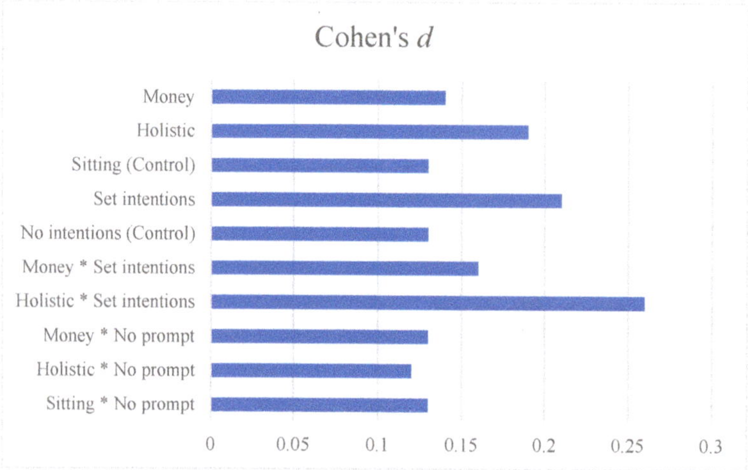


**Figure 4.** Comparison of effect sizes in each study condition. Note that the first five values are for individual conditions, whereas the conditions with an asterisk are the combined conditions (e.g., "Money * Set intentions" = individuals who received the money message and were assigned to set implementation intentions).

**Table 5.** Regression analysis of post-intervention outcomes.

| | Food Acquisition | | Date Label Wasting | | Leftover Usage | | Food Waste | |
|---|---|---|---|---|---|---|---|---|
| | *B* | *SE* | *B* | *SE* | *B* | *SE* | *B* | *SE* |
| **Model 1: H1 (N = 901)** | | | | | | | | |
| Message: Money vs. Sitting | 0.04 | 0.04 | −0.17 * | 0.07 | 0.04 | 0.04 | −0.15 | 0.41 |
| Control: Pre-intervention level | 0.63 *** | 0.03 | 0.51 *** | 0.03 | 0.68 *** | 0.03 | 0.59 *** | 0.02 |
| **Model 2: H1 (N = 878)** | | | | | | | | |
| Message: Holistic vs. Sitting | 0.06 | 0.04 | −0.24 ** | 0.07 | 0.06 | 0.04 | −0.58 | 0.40 |
| Control: Pre-intervention level | 0.64 *** | 0.02 | 0.52 *** | 0.03 | 0.67 *** | 0.03 | 0.59 *** | 0.02 |
| **Model 3: RQ1 (N = 1209)** | | | | | | | | |
| Message: Money vs. Holistic | −0.02 | 0.03 | 0.07 | 0.06 | −0.02 | 0.03 | 0.44 | 0.33 |
| Control: Pre-intervention level | 0.62 *** | 0.02 | 0.49 *** | 0.03 | 0.63 *** | 0.02 | 0.58 *** | 0.02 |
| **Model 4: H2 (N = 1209)** | | | | | | | | |
| Intentions: Set intentions vs. No intentions | 0.09 ** | 0.03 | −0.26 *** | 0.06 | 0.14 *** | 0.03 | −0.30 | 0.33 |
| Control: Pre-intervention level | 0.62 *** | 0.02 | 0.50 *** | 0.03 | 0.63 *** | 0.02 | 0.58 *** | 0.02 |

*Note. B* = unstandardized regression coefficient (shows how much the outcome changes for each one-unit change in a predictor variable); *SE* = standard error (an estimate of how much a measured value is likely to vary due to random chance). Model 1 excludes participants who received the holistic message, and "Money vs. Sitting" is a dummy-coded variable such that 0 = money message and 1 = sitting message. Model 2 excludes participants who received the money message, and "Holistic vs. Sitting" is a dummy-coded variable such that 0 = holistic message and 1 = sitting message. Model 3 excludes participants who received the sitting message, and "Money vs. Holistic" is a dummy-coded variable such that 0 = money message and 1 = holistic message. Model 1 excludes participants who received the sitting message, and "Set intentions vs. No intentions" is a dummy-coded variable such that 0 = did not set implementation intentions and 1 = set implementation intentions. Pre-intervention level is the amount of food wasted reported in Survey 1 (i.e., pre-intervention) for the food waste amount analysis, the level of food acquisition behaviors reported in Survey 1 for the food acquisition behaviors analysis, and so forth. *** $p < 0.001$. ** $p < 0.01$. * $p < 0.05$.

Beginning with Hypothesis 1a (Model 1 and Model 2 in Table 5), we found that the "save money" message and the "holistic" message were associated with significant reductions in post-intervention date label wasting (relative to the control condition of the "reduce sitting" message), but otherwise neither message was associated with other post-intervention food waste reduction behaviors. Turning to Hypothesis 1b, neither message was associated with significant reductions in post-intervention food waste. Thus, Hypothesis 1a was partially supported, and Hypothesis 1b was not supported. Furthermore, in answer to Research Question 1, inspection of Model 3 results in Table 5 reveals that there was no difference in effect on date label wasting between the money and holistic messages.

In summary, it appears that there may be some effects of messaging on a subset of food waste reduction behaviors, but purely informational approaches to behavior change may be limited in effectiveness. Moreover, although these messages were associated with significant reductions in the bivariate analysis (Table 4), controlling for the effects of the pre-intervention level of the outcome may have washed out those effects in these analyses. Finally, we conducted a secondary analysis of Model 1 and Model 2 for post-intervention date label wasting including only individuals who did not set implementation intentions, and we found that the effects of the money message (vs. sitting; $B = -0.05$, $SE = 0.08$, $p = 0.56$) and the holistic message (vs. sitting; $B = -0.09$, $SE = 0.08$, $p = 0.25$) became non-significant. Therefore, messaging appears to have little practical utility as a food waste reduction effort.

Finally, turning to Hypothesis 2 and Model 4 results in Table 5, we found that participants assigned to set implementation intentions demonstrated superior food waste reduction behaviors (i.e., higher food acquisition, less likely to waste food based on its date label, and higher leftover usage) relative to participants who did not set implementation intentions (i.e., information-only conditions). However, there was no significant difference in post-intervention food waste. Thus, Hypothesis 2a was supported, but Hypothesis 2b was not supported.

In summary, implementation intentions seemed to have enabled individuals to enact higher levels of behaviors that are associated with lower food waste, although there was no direct effect of the intentions on food waste itself. In an exploratory analysis, we tested whether post-intervention food waste reduction behaviors themselves predicted post-intervention food waste (while controlling for pre-intervention levels of all variables). We found that food acquisition behaviors were unrelated to food waste ($B = -0.45$, $SE = 0.26$, $p = 0.08$), but date label wasting behavior ($B = 0.88$, $SE = 0.14$, $p < 0.001$) and leftover usage behaviors ($B = -1.78$, $SE = 0.27$, $p < 0.001$) were significantly related to food waste. Therefore, the effect of the implementation intentions may be indirect, rather than direct, but this explanation is tentative and requires further exploration in future research. Messages alone, however, are clearly insufficient to promoting behavior change and furthering goals of food waste reduction in the household.

## 5. Discussion

The purpose of this study was to develop a consumer-facing food waste behavior change intervention that could be scaled up to a national food waste reduction campaign. To that end, using a two-time survey with pre- and post-intervention food waste and behavior measurements, we tested the effectiveness of a message-based, informational intervention approach and an action-based, behavioral intervention approach.

Our findings indicate that a message-based approach alone may have little practical utility as a food waste reduction tool. Relative to a control message unrelated to food waste, holistic and money-focused messages had significant impacts only on date label related behaviors, but not on food acquisition behaviors, leftover usage behaviors, or food waste itself. This is in line with Simões et al.'s [40] assessment from a recent review paper: "But knowing the impacts is not as effective as knowing specific actions to reduce food waste, which suggests that information and awareness raising on the impacts of consumer food waste alone arenot enough to change consumer behavior" (p. 8). However, as noted by one reviewer, future research may test a stronger message-based approach by utilizing an in situ delivery method for the information, possibly relying on technological aids (e.g., mobile applications, reminders) that can ensure that the information is provided in a timely, relevant fashion [17]. Additionally, as discussed by Castro et al. [41], smartphone applications can provide a full suite of information within a person's daily life, such as recipe ideas, sustainability information, and even a platform for sharing information and requests with other users. Thus, future research may consider testing this approach.

In contrast, an action-based approach involving setting implementation intentions did increase food waste reduction behaviors across food acquisition, leftover usage, and date label domains. While these effects on behavior did not translate into decreased food waste amounts relative to the control group, it is possible that such effects may take longer than one week to manifest. Indeed, Roe et al. [9] tested a tailored food waste reduction intervention and similarly found promising results with regard to antecedent behaviors but a non-significant reduction in plate waste, which they attributed in part to the short duration of the intervention. Future studies examining intervention effects across longer durations with multiple measurement points are needed. Of note, this sample was more highly educated and thus not fully representative of national demographics, and so future studies may also test these interventions in samples of varying educational levels as well as levels that are nationally representative.

Interestingly, we observed small, statistically significant decreases in reported food waste between pre- and post-test across all participants, including those in the control group who did not receive food waste messages or set implementation intentions. Although the control group did not participate in the interventions, they did report their respective household food waste levels and behaviors during the pretest session. In line with the notion that "what gets measured gets managed," it is possible that simply measuring food waste at pretest increased awareness and motivation to reduce food waste [42]. It is also possible that the control group's decrease in food waste was due to demand effects. Participants may have inferred the study's focus was on food waste and responded with lower food waste levels at posttest in line with intuition of study aims. The finding of significant reductions in food waste amounts in the control group highlights the need for careful consideration of control group design in food waste intervention research. Simply observing pre- post decreases in food waste amounts in an intervention condition could be a measurement artifact, rather than an indication that the intervention itself caused the food waste change.

Future research investigating whether and how food waste measurement alone impacts behavior and waste outcomes is needed. The current study utilized self-report methods, which, though generally convergent with other methods [34,35], is likely to have produced an underestimate of actual food waste amounts and may have exerted a more obvious demand effect. Therefore, one possible approach to study measurement impacts includes having others report on a person's food waste amounts, to reduce demand effects and biases that may motivate participants to report lower food waste at posttest. Another possible approach is to measure food waste amounts via curbside waste audits, rather than self-reported food waste. With this measurement approach, researchers collect, sort, and weigh food waste components of household trash streams, providing an objective estimate of food waste over time [43]. If future studies with such methods find that measuring food waste itself is a viable, effective food waste reduction intervention, behavior change campaigns could consider implementing measurements with feedback (e.g., via trend line charts) as a potential tool to change behavior.

An additional direction for future research is to test for whom either informational or behavioral intervention approaches are more or less effective. As raised by one reviewer, demographics can influence outcomes, and in the current study we conducted an exploratory analysis of results by education level, income, and household size. Specifically, for participants who were assigned to set implementation intentions, we examined correlations between change in food waste pre-/post-intervention and education, income, and household size. We found only weak relationships: education was unrelated to change in food waste ($r = -0.04$, $p = 0.36$), income was negatively related to change in food waste ($r = -0.09$, $p = 0.04$), and household size was unrelated to change in food waste ($r = -0.06$, $p = 0.14$). Thus, it may be that behavioral approaches are slightly more effective for households with higher income levels, but future research may continue to explore this and other moderating demographic and lifestyle variables.

Finally, future research may consider testing alternative control conditions. Specifically, as raised by one reviewer, it is possible that the various intervention conditions were not matched in intensity and thus may have exerted a demand effect. We would suggest that asking participants to set implementation intentions to reduce their sitting behavior (as a control) is similar in strength to asking participants to set implementation intentions to reduce their food waste (as a treatment), in that both require participants to take steps to change aspects of their daily lifestyle. Indeed, past research has found that implementation intentions are effective for increasing physical activity [44], meaning that this is a relatively like-for-like comparison. However, future research should test alternative control conditions to determine whether different results may appear.

## 6. Conclusions

Changing consumer food behaviors to improve climate, food insecurity, and financial outcomes is of critical importance, yet it is a difficult endeavor. The present research yielded three key insights for researchers and practitioners in this realm. First, simply focusing on educational/informational messaging alone may not yield impacts of the size needed to halve food waste by 2030. Second, setting implementation intentions has greater promise for effecting behavior change, though impacts on food waste one week after setting intentions were null. Third, asking households to simply track their weekly food waste may itself act as an intervention.

**Supplementary Materials:** The following supporting information can be downloaded at: https://www.mdpi.com/article/10.3390/su172310752/s1, S1: Money Message; S2: Holistic Message; S3: Sitting Message (Control); S4: Implementation Intentions.

**Author Contributions:** J.A.A.: Conceptualization, Methodology, Formal analysis, Investigation, Data curation, Writing—original draft, Writing—review and editing. G.M.: Conceptualization, Methodology, Formal analysis, Investigation, Data curation, Writing—original draft, Writing—review and editing. B.A.: Conceptualization, Supervision, Project Administration, Funding acquisition, Investigation, Resources, Writing—original draft, Writing—review and editing. L.L.: Supervision, Funding acquisition. A.S.: Conceptualization, Methodology, Supervision, Project Administration, Funding acquisition, Investigation, Resources, Writing—original draft, Writing—review and editing. All authors have read and agreed to the published version of the manuscript.

**Funding:** This research was funded by The MITRE Corporation Independent Research and Development Program.

**Institutional Review Board Statement:** The study was conducted in accordance with the Declaration of Helsinki, and approved by the Institutional Review Board of The MITRE Corporation (protocol code 2024-032 and 21 June 2024).

**Informed Consent Statement:** Informed consent was obtained from all subjects involved in the study.

**Data Availability Statement:** The data for this research is publicly available at the following link: https://osf.io/y5zma/?view_only=95004d74675e486296f219057f470bd5, accessed on 10 November 2025.

**Acknowledgments:** The authors thank Mark Roberts, Erin McCluskey, Valerie Leney, David Rogers, Leah Karrer, Michael Jones, and Brian Roe for their feedback on study methodology.

**Conflicts of Interest:** Authors John A. Aitken, Grace Mika, Balca Alaybek, Laura Leets and Amber Sprenger were employed by the company The MITRE Corporation.

## References

1. United Nations Environment Programme. Food Waste Index Report 2024. 2024. Available online: https://wedocs.unep.org/20.500.11822/45230 (accessed on 23 October 2024).
2. U.S. Environmental Protection Agency. International Efforts on Wasted Food Recovery. 2024. Available online: https://www.epa.gov/international-cooperation/international-efforts-wasted-food-recovery#:~:text=In%20the%20United%20States,%2040,billion%20or%201.3%25%20of%20GDP (accessed on 23 October 2024).
3. U.S. Department of Agriculture Economic Research Service. Food Security in the U.S. 2023. Available online: https://www.ers.usda.gov/topics/food-nutrition-assistance/food-security-in-the-u-s/key-statistics-graphics/ (accessed on 23 October 2024).
4. Hegwood, M.; Burgess, M.G.; Costigliolo, E.M.; Smith, P.; Bajželj, B.; Saunders, H.; Davis, S.J. Rebound effects could offset more than half of avoided food loss and waste. *Nat. Food* **2023**, *4*, 585–595. [CrossRef]
5. The White House. National Strategy for Reducing Food Loss and Waste and Recycling Organics. 2024. Available online: https://www.usda.gov/sites/default/files/documents/NATIONAL-STRATEGY-FOR-REDUCING-FOOD-LOSS-AND-WASTE-AND-RECYCLING-ORGANICS.pdf (accessed on 23 October 2024).
6. U. S. Environmental Protection Agency. From Farm to Kitchen: The Environmental Impacts of Food Waste (Part 1). 2021. Available online: https://www.epa.gov/system/files/documents/2021-11/from-farm-to-kitchen-the-environmental-impacts-of-u.s.-food-waste_508-tagged.pdf (accessed on 23 October 2024).
7. U. S. Environmental Protection Agency. From Field to Bin: The Environmental Impacts of U. S. Food Waste Management Pathways (Part 2). 2023. Available online: https://www.epa.gov/land-research/field-bin-environmental-impacts-us-food-waste-management-pathways (accessed on 23 October 2024).
8. National Academies of Sciences, Engineering, and Medicine. *A National Strategy to Reduce Food Waste at the Consumer Level*; The National Academy Press: Cambridge, MA, USA, 2020. [CrossRef]
9. Roe, B.E.; Qi, D.; Beyl, R.A.; Neubig, K.E.; Apolzan, J.W.; Martin, C.K. A randomized controlled trial to address consumer food waste with a technology-aided tailored sustainability intervention. *Resour. Conserv. Recycl.* **2022**, *179*, 106121. [CrossRef]
10. Casonato, C.; Garcia-Herrero, L.; Caldeira, C.; Sala, S. What a waste! Evidence of consumer food waste prevention and its effectiveness. *Sustain. Prod. Consum.* **2023**, *41*, 305–319. [CrossRef]
11. Jenkins, E.L.; Brennan, L.; Molenaar, A.; McCaffrey, T.A. Exploring the application of social media in food waste campaigns and interventions: A systematic scoping review of the academic and grey literature. *J. Clean. Prod.* **2022**, *360*, 132068. [CrossRef]
12. Tian, X.; Xia, Z.; Xie, J.; Zhang, C.; Liu, Y.; Xu, M. A meta-analytical review of intervention experiments to reduce food waste. *Environ. Res. Lett.* **2022**, *17*, 064041. [CrossRef]
13. Zhang, J.; Huang, Y.; Zhu, J.; Zhao, L. A meta-analysis on the effectiveness of food-waste reducing nudges. *Food Policy* **2023**, *120*, 102480. [CrossRef]
14. Di Clemente, C.C. Motivation for change: Implications for substance abuse treatment. *Psychol. Sci.* **1999**, *10*, 209–213. [CrossRef]
15. Lyubomirsky, S.; Layous, K. How do simple positive activities increase well-being? *Curr. Dir. Psychol. Sci.* **2013**, *22*, 57–62. [CrossRef]
16. Person, A.L.; Colby, S.E.; Bulova, J.A.; Eubanks, J.W. Barriers to participation in a worksite wellness program. *Nutr. Res. Pract.* **2010**, *4*, 149–154. [CrossRef] [PubMed]
17. Nahum-Shani, I.; Smith, S.N.; Spring, B.J.; Collins, L.M.; Witkiewitz, K.; Tewari, A.; Murphy, S.A. Just-in-time adaptive interventions (JITAIs) in mobile health: Key components and design principles for ongoing health behavior support. *Ann. Behav. Med.* **2018**, *52*, 446–462. [CrossRef]
18. Zhu, Z.; Aitken, J.A.; Dalal, R.S.; Kaplan, S.A. The promise of just-in-time adaptive interventions for organizational scholarship and practice: Conceptual development and research agenda. *Organ. Res. Methods* **2023**, *27*, 543–587. [CrossRef]
19. Ajzen, I. The theory of planned behavior. *Organ. Behav. Hum. Decis. Process.* **1991**, *50*, 179–211. [CrossRef]
20. Ajzen, I.; Cote, N.G. Attitudes and the prediction of behavior. In *Attitudes and Attitude Change*; Crano, W.D., Prislin, R., Eds.; Psychology Press: London, UK, 2008; pp. 289–311.
21. Ajzen, I. Nature and operation of attitudes. *Annu. Rev. Psychol.* **2001**, *52*, 27–58. [CrossRef]

22. Sheeran, P.; Webb, T.L. The intention–behavior gap. *Soc. Personal. Psychol. Compass* **2016**, *10*, 503–518. [CrossRef]
23. ReFED. Food Waste Problem. Available online: https://refed.org/food-waste/the-problem/ (accessed on 23 October 2024).
24. Aitken, J.A.; Alaybek, B.; Hartman, H.; Leets, L.; Mika, G.; Rao, D.; Sprenger, A.; Viani, D.; Archer, K.; Davoodi, T.; et al. The State of Food Waste in America. 2023. Available online: https://sites.mitre.org/household-food-waste/ (accessed on 23 October 2024).
25. Dudley, M.Z.; Squires, G.K.; Petroske, T.M.; Dawson, S.; Brewer, J. The use of narrative in science and health communication: A scoping review. *Patient Educ. Couns.* **2023**, *112*, 107752. [CrossRef]
26. Hebrok, M.; Heidenstrøm, N. Contextualising food waste prevention—Decisive moments within everyday practices. *J. Clean. Prod.* **2019**, *210*, 1435–1448. [CrossRef]
27. Richetin, J.; Perugini, M.; Conner, M.; Adjali, I.; Hurling, R.; Sengupta, A.; Greetham, D. To reduce and not to reduce resource consumption? That is two questions. *J. Environ. Psychol.* **2012**, *32*, 112–122. [CrossRef]
28. Ajzen, I. The theory of planned behavior: Frequently asked questions. *Hum. Behav. Emerg. Technol.* **2020**, *2*, 314–324. [CrossRef]
29. Gollwitzer, P.M. Implementation intentions: Strong effects of simple plans. *Am. Psychol.* **1999**, *54*, 493–503. [CrossRef]
30. Webb, T.L.; Sheeran, P. How do implementation intentions promote goal attainment? A test of component processes. *J. Exp. Soc. Psychol.* **2007**, *43*, 295–302. [CrossRef]
31. Gollwitzer, P.M.; Sheeran, P. Implementation intentions and goal achievement: A meta-analysis of effects and processes. In *Advances in Experimental Social Psychology*; Zanna, M.P., Ed.; Elsevier Academic Press: Amsterdam, The Netherlands, 2006; Volume 38, pp. 69–119. [CrossRef]
32. Webb, T.L.; Sheeran, P. Does changing behavioral intentions engender behavior change? A meta-analysis of the experimental evidence. *Psychol. Bull.* **2006**, *132*, 249–268. [CrossRef]
33. van Geffen, L.; van Herpen, E.; van Trijp, H. Household food waste—How to avoid it? An integrative review. In *Food Waste Management*; Närvänen, E., Mesiranta, N., Mattila, M., Heikkinen, A., Eds.; Palgrave Macmillan: New York, NY, USA, 2020. [CrossRef]
34. Aitken, J.A.; Sprenger, A.; Alaybek, B.; Mika, G.; Hartman, H.; Leets, L.; Maese, E.; Davoodi, T. Surveys and diaries and scales, oh my! A critical analysis of household food waste measurement. *Sustainability* **2024**, *16*, 968. [CrossRef]
35. van Herpen, E.; van der Lans, I.A.; Holthuysen, N.; Nijenhuis-de Vries, M.; Quested, T.E. Comparing wasted apples and oranges: An assessment of methods to measure household food waste. *Waste Manag.* **2019**, *88*, 71–84. [CrossRef]
36. Cooper, W.H.; Richardson, A.J. Unfair comparisons. *J. Appl. Psychol.* **1986**, *71*, 179–184. [CrossRef]
37. U.S. Census Bureau. Educational Attainment in the United States: 2021. 6 September 2022. Available online: https://www.census.gov/newsroom/press-releases/2022/educational-attainment.html#:~:text=8.9%25%20had%20less%20than%20a,Sex (accessed on 23 October 2024).
38. R Core Team. *R: A Language and Environment for Statistical Computing*; R Foundation for Statistical Computing: Vienna, Austria, 2024. Available online: https://www.R-project.org/ (accessed on 23 October 2024).
39. Vickers, A.J.; Altman, D.G. Analysing controlled trials with baseline and follow up measurements. *BMJ* **2001**, *323*, 1123–1124. [CrossRef] [PubMed]
40. Simões, J.; Carvalho, A.; de Matos, M.G. How to influence consumer food waste behavior with interventions? A systematic review. *J. Clean. Prod.* **2022**, *373*, 133866. [CrossRef]
41. Castro, C.; Chitikova, E.; Magnani, G.; Merkle, J.; Heitmayer, M. Less is more: Preventing household food waste through an integrated mobile application. *Sustainability* **2023**, *15*, 10597. [CrossRef]
42. Ramos, G.J.; Borges, J.A.R.; Domingues, C.H.D.F.; van Herpen, E. Reducing food waste by simply measuring it: Insights from interventions to reduce household food waste. *Br. Food J.* **2024**, *126*, 812–833. [CrossRef]
43. van der Werf, P.; Seabrook, J.A.; Gilliland, J.A. Food for thought: Comparing self-reported versus curbside measurements of household food wasting behavior and the predictive capacity of behavioral determinants. *Waste Manag.* **2020**, *101*, 18–27. [CrossRef]
44. Silva, M.A.V.d.; São-João, T.M.; Brizon, V.C.; Franco, D.H.; Mialhe, F.L. Impact of implementation intentions on physical activity practice in adults: A systematic review and meta-analysis of randomized clinical trials. *PLoS ONE* **2018**, *13*, e0206294. [CrossRef]

sustainability

MDPI

*Article*

# Knowledge-Driven Food Waste Reduction: A Mediation Analysis of Behavioral Determinants in Saudi Arabian Households

**Othman Mohammed Al-Tuwaijri [1], Fahd Owais Aldosari [1], Maged Ail Algannbi [1], Mohamed I. Motawei [2], Hassan M. Mousa [3] and Hassan Barakat [3,*]**

1 Department of Agricultural Extension and Rural Society, College of Food and Agriculture Sciences, King Saud University, Riyadh 11451, Saudi Arabia; othman050017@gmail.com (O.M.A.-T.); fadosri@ksu.edu.sa (F.O.A.); magedalgannbi@gmail.com (M.A.A.)

2 Department of Plant Production, College of Agriculture and Food, Qassim University, Buraydah 51452, Saudi Arabia; mtaoa@qu.edu.sa

3 Department of Food Science and Human Nutrition, College of Agriculture & Food, Qassim University, Buraydah 51452, Saudi Arabia

* Correspondence: haa.mohamed@qu.edu.sa

**Abstract**

Food waste undermines the four dimensions of food security, availability, accessibility, utilization, and stability, while imposing adverse economic, social, and environmental impacts on sustainable food systems. Understanding the behavioral determinants of food consumption rationalization is essential for addressing this challenge in the Kingdom of Saudi Arabia. This study examines household food waste behaviors within a knowledge-based framework that integrates three interconnected constructs: awareness of food waste consequences, behavioral knowledge of waste-reduction practices, and actual engagement in conservation strategies. Data were collected from 255 households (response rate: 66%) in Buraydah City through an electronic questionnaire administered in shopping malls. Using Baron and Kenny mediation analysis and multiple linear regression, awareness of waste consequences influences conservation practices both directly ($\beta = 0.132$, $p < 0.001$) and indirectly through behavioral knowledge (accounting for 68.6% of the total effect), explaining 74.9% of the variance in household conservation behaviors ($R^2 = 0.749$). The analysis reveals that awareness of waste consequences influences conservation practices both directly and indirectly through behavioral knowledge, establishing a mediation pathway. Together, these knowledge dimensions significantly explain variations in household conservation behaviors. The findings highlight the critical interplay between awareness and practical behavioral knowledge in driving sustainable food consumption practices. These insights provide empirical guidance for policymakers and agencies seeking to develop targeted interventions that integrate consequence messaging with practical behavioral training to effectively reduce household food waste and promote food security in Saudi Arabia.

**Keywords:** Saudi Arabia; food security; food waste; sustainability; food supply

## 1. Introduction

Minimizing food waste is a critical strategy for addressing global challenges, including climate instability, food insecurity, and resource scarcity. This multifaceted problem not only degrades the natural environment but also poses significant risks to public health [1]. Research shows that reducing food waste improves food quality and sustainability [2].

Food waste has serious consequences for food security, the environment, and the economy, as discussed globally, with 79 kg of food waste per capita and 1 billion meals lost daily [1,3]. Economic losses from global food waste total USD 2.6 trillion. This includes USD 700 billion in environmental effects, USD 1 trillion in production and waste losses, and USD 900 billion in quality-of-life losses [4]. Food waste accounts for approximately 28% of the world's agricultural land, resulting in 1.03 billion tons of waste annually. Currently, 29.6% of the world's population is moderately to severely food insecure, yet consumer behavior results in the wastage of 60% of food [1]. These facts show that achieving Sustainable Development Goal targets, particularly reducing food waste by 50% by 2030, is difficult [5].

Food waste has far-reaching consequences beyond food security. They deplete resources and reduce future productivity in the food system. Water, land, energy, labor, and capital are needed to produce and distribute food waste. Long-term agricultural sustainability is compromised [6]. Air, water, and soil pollution degrade the ecosystem while depleting resources [7]. Over time, agricultural systems become increasingly responsive to production. In vulnerable agrarian systems, water shortages and nitrogen loss-induced soil erosion increase the costs of resources [8]. Food waste accounts for 8% of global greenhouse gas emissions, third after the US and China [1]. Climate change-causing emissions make climate stabilization and environmental protection harder [9]. Food waste emits greenhouse gases, which can contribute to changes in weather patterns, including increased precipitation, droughts, and poorer agricultural production due to fluctuations in temperature and water availability. Environmental impacts on complex food systems create interconnected vulnerabilities [10]. Many processes affect food security and nutritional adequacy for current and future populations. Food waste reduces US and global food availability. Commodity pricing raises food prices, making it harder for people to obtain. Physical waste depletes nutrition. Unsustainable resource use harms food and nutritional security. Food waste affects all sectors of the global food system; therefore, we must address production, distribution, consumption, and waste management [11].

Food waste represents a critical sustainability issue that directly undermines the dimensions of food security, including availability, access, utilization, and stability [12]. Reducing food waste is a crucial priority for sustainable development, as it provides cost-effective alternatives to increased production and improves food availability [13]. Achieving the 2050 population projections requires a 60% increase in food production, provided that food waste reduction is also implemented simultaneously [1,14].

In Saudi Arabia, food waste represents a critical sustainability challenge. Current estimates indicate an annual waste generation of approximately 2330 million tons (105 kg per capita), accounting for 18.9% of the national waste index and valued at USD 40.48 billion annually. These losses generate annual economic impacts equivalent to 1.9% of gross domestic product [15,16]. Given that Saudi Arabia imports 80% of its food commodities at a yearly cost exceeding USD 12 billion, reducing food waste has a direct impact on national economic sustainability and import dependency. The significance of this challenge is evident in the Kingdom's strategic prioritization, with food waste identified as a critical risk to food security in national food security strategy documentation [17,18]. Research indicates that food waste behavior functions as a conscious phenomenon, primarily linked to cognitive factors rather than habitual patterns, suggesting substantial potential for intervention through enhanced awareness initiatives [19–21]. Consumer-level interventions are crucial, as production-focused solutions are limited without simultaneous changes in consumer behavior [22,23]. Addressing food waste requires comprehensive, consumer-focused measures, particularly as 8.1% of the Saudi population faces severe food insecurity [24].

Consumer engagement occurs when people acknowledge the consequences of food waste and are willing to change their behavior [25]. Awareness campaigns about the

https://doi.org/10.3390/su18020686

environmental, economic, and moral impacts of certain actions can affect behavior, such as shopping with a purpose and reusing leftovers. However, a lack of knowledge and practical assistance makes waste-reduction strategies difficult to implement [26]. Changing habits requires adopting ethical attitudes and acquiring knowledge about food storage and purchasing [27]. Knowledge of environmental challenges, notably landfills and water scarcity, increases responsible consumerism and sustainability [28]. Focusing on food storage and shopping habits reduces waste by 10% in households that shop once a week [29]. This indicates that practical planning abilities significantly reduce waste [30,31]. Cognitive variables, such as food management knowledge and awareness, have the most significant impact on food waste. Some cultural factors, including religion, can prevent food waste [31]. Since it changes many people's behavior, consumer-level prevention is the most effective intervention. Educational initiatives for young adults, combined with digital communication tools and social media platforms, may be beneficial [23]. Understanding how food waste affects knowledge, behavioral knowledge, and conservation practices is crucial to reducing food waste at all stages of consumption and encouraging sustainable food consumption [24,32,33]. Finding the best means to distribute knowledge ensures that everyone can access it, thereby bridging the gap between knowledge and practice. This supports Saudi Arabia's Vision 2030 initiatives, which aim to reduce food waste [10,27]. This study proposes a paradigm that examines the interaction between knowledge of food waste, knowledge of food conservation, and food conservation practices. This study extends traditional KAP by: (1) distinguishing consequences knowledge (theoretical knowledge of impacts) from behavioral knowledge (practical knowledge of implementation), and (2) testing whether behavioral knowledge mediates the relationship between consequence-awareness and practice, thereby identifying distinct motivational and capability mechanisms. The framework integrates the Knowledge-Attitude-Practice (KAP) model's sequential progression with NAT's mechanism and SCT's emphasis on capability development. Behavioral knowledge serves as a mediator through which consequence-awareness translates into practice, as illustrated in Figure 1.

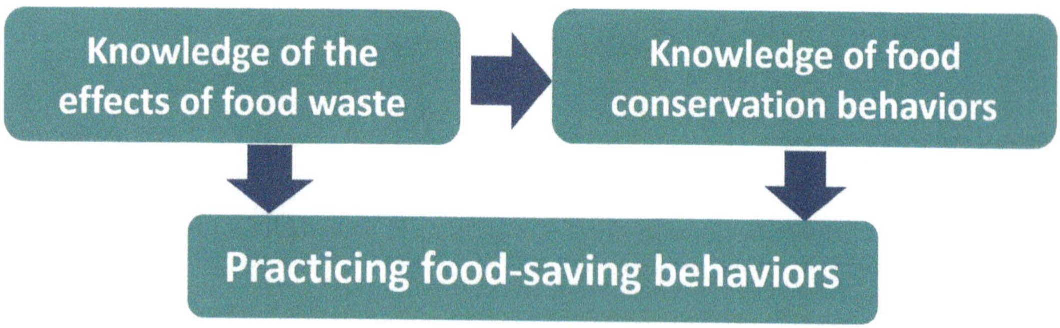


**Figure 1.** Knowledge-practice integration in household food waste reduction: a mediation framework indicating the behavioral knowledge mediates food waste awareness and conservation practice.

Understanding household food waste requires theoretical frameworks integrating knowledge, attitudes, and practices. While Theory of Planned Behavior (TPB) emphasizes intention, Social Practice Theory (SPT) reframes consumption as socially situated, and Value-Belief-Norm (VBN) Theory establishes pathways from values through awareness to personal norms, this study employs a KAP framework distinguishing between consequence knowledge (understanding why waste reduction matters) and behavioral knowledge (understanding how to implement practices), operating as motivational drivers and capability enablers, respectively. This dual-knowledge model clarifies why awareness campaigns

https://doi.org/10.3390/su18020686

alone fail to produce sustained behavior change, provides cultural resonance within the Saudi Arabian context where food waste is psychologically driven, and enables precision in intervention design. The mediation model quantifies direct versus indirect effects, enabling evidence-based resource allocation, with approximately 40% of intervention resources allocated to consequence messaging and 60% to practical training. Food waste in Saudi Arabia amounts to 2330 million tons annually, valued at USD 40.48 billion, equivalent to 18.9% of the country's total waste. This figure is particularly urgent, given the 80% dependency on food imports, which incurs yearly costs exceeding USD 12 billion. Consumer-level interventions that integrate consequence messaging with practical behavioral training demonstrate significantly greater effectiveness than either approach independently, making the dual-knowledge mediation model essential for advancing intervention effectiveness and supporting Saudi Vision 2030's food security objectives [10,27].

*Research Hypotheses and Theoretical Rationale*

This study proposes that knowledge of consequences translates into food waste reduction practices through a dual-mechanism mediation framework that integrates Norm Activation Theory (NAT) and Social Cognitive Theory (SCT). H1: posits that consequence knowledge has a statistically significant total effect on conservation practices through both direct motivational pathways ($\beta = 0.132$, $p < 0.001$) and indirect capability pathways. Grounded in NAT, consequence awareness across four domains—religious-ethical (M = 4.73, highest), economic (M = 3.82), food security (M = 3.58), and environmental (M = 3.15, lowest), activates personal moral responsibility. The religious-ethical domain dominates (83.1% report "completely know" Islamic prohibition of wastefulness), reflecting the cultural embeddedness of Islamic stewardship principles. However, the modest direct effect (31.4% of total impact) reveals that motivation alone produces limited behavior change, requiring capability enhancement for sustained practice [22]. H2: specifies that motivated individuals actively seek behavioral knowledge, operating through Social Cognitive Theory's knowledge-seeking mechanism. Individuals who understand the consequences experience psychological dissonance between their values (reducing waste and honoring Islamic teachings) and their capabilities (a lack of practical knowledge), motivating them to seek information and acquire skills. This is evidenced empirically: 47.8% of respondents engage with food preservation organizations explicitly teaching behavioral knowledge, with 86.9% reporting high satisfaction. Communication preferences reveal that motivated individuals prioritize efficient knowledge-delivery channels (WhatsApp, M = 3.70; Snapchat, M = 3.53; SMS, M = 3.38) over passive formats (Facebook, M = 2.22; Zoom, M = 2.49), demonstrating that motivation drives active knowledge acquisition rather than passive exposure [34]. H3: proposes that behavioral knowledge mediates the consequence-practice relationship through partial mediation, with capability (not motivation) as the dominant mechanism. The bivariate correlations reveal this pattern: behavioral knowledge predicts practice strongly (r = 0.86), while consequence knowledge predicts practice moderately (r = 0.42), indicating that knowing how (capability) substantially outweighs knowing why (motivation). The indirect pathway through behavioral knowledge accounts for 68.6% of the total effect, while direct motivational effects account for 31.4%. This decomposition has critical intervention implications: approximately 68% of resources should target capability development (practical training, skill-building), while 32% should target motivation (awareness messaging). Partial (not complete) mediation persists because activated personal moral norms maintain independent effects: individuals with strong awareness of consequences implement practices even with imperfect knowledge, driven by an intrinsic sense of responsibility. This integrated model explains why awareness campaigns alone achieve modest behavior change (10–20% variance), while dual-component interventions

https://doi.org/10.3390/su18020686

combining consequence messaging with behavioral training achieve substantial change (74.9% variance explanation), a 4-fold improvement, supporting evidence-based intervention design for household food waste reduction in Saudi Arabia [35]. The integrated model predicts that knowledge of consequences is necessary but insufficient for behavior change. Direct effects reflect motivational activation; indirect effects reflect capability enhancement. Mediation analysis enables the decomposition of mechanisms and informs the design of evidence-based interventions, balancing consequence messaging with practical behavioral training.

## 2. Materials and Methods

### *2.1. Description of the Study Area*

The city of Buraydah is situated in the north-central part of the Kingdom of Saudi Arabia, spanning an area of approximately 1300 square kilometers. Its population is estimated, according to the 2022 Saudi population census, to be approximately 677,647 people. The city of Buraydah is considered the agricultural hub of the Qassim region and the entire Kingdom. Agriculture represents the foundation of the regional economy, characterized by the availability of groundwater and fertile lands suitable for agriculture, and is rich in the quality of agricultural production of various types of crops, including wheat, vegetables, fruits, and dates [36].

The region features the world's largest date palm endowment farm. The project features approximately 200,000 diverse palm trees, with a production quantity exceeding 5000 tons, and has obtained multiple international certifications, including recognition from the Guinness Book of World Records and the European Organization for Organic Agriculture (Ecocert) certificate for the production of organic dates. The study area also encompasses the world's largest date market, which produces 390,000 tons annually and exports to over 20 countries worldwide [37].

### *2.2. Research Design and Data Collection Method*

The survey was conducted using a convenience, non-probability sample of household shoppers at various locations in major grocery store outlets. This method was used because a complete database of the study population, including contact information for the study population units, was not available. Therefore, this method is suitable for ensuring access to respondents in shopping malls, as grocery stores are a key stage in the food consumption process and are relevant to the study's topic and objectives. It is also effective for practical communication with shoppers who play a direct role in purchasing and managing food within the household. An electronic questionnaire was presented to shoppers via a link containing a QR code (See Supplementary Materials, Annex S1). Data collection began on 1 June 2025, and continued for up to 2 successive months. Respondents were given the freedom to complete the questionnaire at any location, whether in the shopping mall, upon returning home, or through a family member in their home. The total number of families was determined to be 163,904, and the average number of family members in the study area is four individuals, according to the General Authority for Statistics [38]. The sample size was determined to be 383 according to Stephen Thompson's equation [39]. A total of 255 Participants completed the questionnaire, which represents a response rate of 66%.

$$N = \frac{N\,x\,p\,(1-P)}{\left[N - 1\,x\left(d^2 \div Z^2\right)\right] + p\,(1-p)}$$

https://doi.org/10.3390/su18020686

where N is the community size, Z, the standard score corresponding to the significance level of 0.95, is equal to 1.96, D, the error rate is equal to 0.05, and *P*, Property availability and neutrality ratio = 0.50

*2.3. Questionnaire as a Survey Tool*

A comprehensive, structured questionnaire was developed to systematically evaluate participants' knowledge, attitudes, and behavioral practices concerning food waste management. The instrument was organized into four distinct sections, each designed to assess specific dimensions of food waste awareness and behavior. The initial section assessed participants' foundational knowledge of the food waste concept utilizing a three-point ordinal scale, with response categories denoted as follows: 1 (Don't know), 2 (Somewhat know), and 3 (Know). This measurement approach provided a preliminary evaluation of conceptual understanding across the respondent population. The three-point scales were selected because they are easier for participants to understand and conceptually suitable for measuring binary capability in knowledge or behavioral items. Statistical limitations due to reduced variance were addressed by increasing the sample size (N = 255, exceeding the minimum required N = 217 at $\alpha = 0.05$, power = 0.80) and employing nonparametric tests (Kruskal–Wallis or Mann–Whitney) for ordinal data analysis. The subsequent section examined participants' knowledge of the multifaceted consequences of food waste, encompassing ethical and religious dimensions, as well as economic, environmental, and food security implications. Responses to this section were elicited using a five-point Likert scale ranging from 1 (not completely known) to 5 (completely known). The operational definitions of food waste and the constituent knowledge components were derived from and aligned with established literature in the field [11,15,16,40]. The third section comprised a dual assessment of food-saving behaviors, distinguishing between knowledge and actual practice. Knowledge of food-saving behaviors was measured on a three-point scale (1 = "Don't know," 2 = "Somewhat know," 3 = "Completely know").

In contrast, the frequency of behavioral practice was evaluated using a parallel three-point scale (1 = "Don't practice," 2 = "Sometimes practice," 3 = "Always practice"). Item development adhered to guidelines established by the Food and Drug Administration (FDA) [41], while the design of practice-oriented items was informed by contemporary empirical research on food conservation behaviors [42,43]. Notably, only respondents who demonstrated at least partial knowledge of food-saving practices (response categories 2 or 3) were subsequently queried regarding their behavioral engagement; respondents indicating no knowledge were systematically excluded from this component of the survey. The final section investigated participants' management strategies for surplus food, presenting four mutually exclusive behavioral responses: 4 (applying appropriate refrigeration or freezing for subsequent consumption), 3 (donating to individuals in need), 2 (providing to animals), and 1 (disposing of via trash). Complementary items assessed respondents' satisfaction with their experiences utilizing food preservation organizations on a five-point scale (1 = not at all satisfied to 5 = very satisfied). They evaluated their preferences for information dissemination channels regarding food waste using an analogous five-point Likert scale (1 = not strongly preferred to 5 = strongly preferred).

Regarding the rationale for selecting the 5-point Likert scale, it captures the intensity and certainty of understanding regarding food waste impacts, as consequence awareness operates on a continuum from unawareness to absolute certainty. The 3-point scales for behavioral knowledge and practice measure binary capability, whether individuals possess knowledge of or engage in specific conservation behaviors, reflecting the categorical, implementation-oriented nature of these constructs. Our sample size (N = 255) exceeds the minimum required by power analysis (N = 217, $\alpha = 0.05$, power = 0.80), ensuring adequate

https://doi.org/10.3390/su18020686

statistical power despite the more parsimonious 3-point scales. Pilot testing demonstrated adequate variance in 3-point scales (SD = 0.32–0.43), confirming sufficient discriminative ability. Sensitivity analysis confirmed that regression coefficients remained stable (r > 0.95) across alternative scale treatments, validating the robustness of our findings regardless of scale specification. This differential scaling approach optimizes measurement precision while maintaining statistical validity and practical interpretability.

## 2.4. Data Analysis

A descriptive analysis, utilizing percentages, frequency distributions, arithmetic means, and standard deviations (SD), was employed to achieve the research objectives. The Statistical Package for the Social Sciences (SPSS) version 30 was used. Inferential statistical analyses, including independent samples t-tests, one-way analysis of variance (ANOVA), and multiple linear regression, were used to interpret the relationships between variables. Pearson's correlation coefficient was used to determine the strength and direction of the relationship between the independent variable and the dependent variable. The Baron and Kenny methodology was also followed to analyze the mediation between the components of the paper's conceptual framework [44]. The mediation model was selected to address a critical gap in the food waste literature: understanding not only whether knowledge influences behavior, but also how this influence occurs. By testing behavioral knowledge as a mediator, this study identifies two pathways: a direct impact of consequence-awareness on behavior (through motivation and responsibility), and an indirect influence through practical skill development. This distinction has important implications for intervention design. Bootstrapped confidence intervals were calculated for indirect effects to provide robust estimates of the significance of the mediation pathway. Sobel tests were also conducted to confirm the significance of mediated effects. All statistical analyses employed two-tailed significance tests with an alpha level of 0.05.

## 2.5. Validity and Reliability

The questionnaire was evaluated to achieve the study's objectives and then presented to the scientific committees in the Department of Agricultural Extension and the College of Food and Agricultural Sciences at King Saud University. A pilot test of the questionnaire was conducted with 40 participants in the study area prior to data collection, and reliability was assessed using Cronbach's alpha reliability coefficient, as shown in Table 1. However, the full reliability and validity assessment of the measurement model, including consequences, knowledge, behavioral knowledge, and behavioral practice, is provided in Table S1.

**Table 1.** Reliability coefficients (Cronbach's alpha) for the dimensions and axes of the study tool.

| Dimensions/Axes | | Number of Phrases | Cronbach's Alpha Reliability Coefficient |
|---|---|---|---|
| Dimensions of the axis of food waste effects | Knowledge of the economic impacts of food waste | 5 | 0.824 |
| | Knowledge of the environmental impacts of food waste | 4 | 0.848 |
| | Knowledge of the impact of food waste on food security and society | 6 | 0.888 |

https://doi.org/10.3390/su18020686

**Table 1.** *Cont.*

| Dimensions/Axes | | Number of Phrases | Cronbach's Alpha Reliability Coefficient |
|---|---|---|---|
| Axes | The overall axis of knowledge on the effects of food waste | 16 | 0.930 |
| | Knowledge axis on food rationalization behaviors | 12 | 0.826 |
| | Axis of practicing food rationalization behaviors | 12 | 0.793 |
| | Knowledge Transfer Axis to Reduce Food Waste | 10 | 0.861 |

## 3. Results

### *3.1. Socio-Economic Characteristics of the Respondents (Participants)*

The demographic profile of the study group exhibited significant characteristics across various dimensions. The gender distribution indicated that females comprised the majority, at 62.7%, in contrast to males, who made up 37.3%. The participants' ages were widespread, with the biggest group being those between the ages of 35 and 49 (42.7%), followed by those under 34 (32.2%), and those over 50 (25.1%). The participants had high levels of education, with 69.8% having completed university studies and 6.7% having pursued postgraduate studies (Master's or PhD). Fifteen percent of the sample had completed secondary school, while 5% had completed intermediate school, and 2% had completed elementary school. Analysis of family composition revealed that most participants resided in medium-sized homes. The largest group (45.5%) consisted of families with 5–7 persons. There were 32.5% of the sample families with 2–4 members, and 22.0% of the sample families with 8–10 members. The monthly family income exhibited considerable economic variability within the sample. The distribution was pretty even across income groups: 15.3% earned < 5000 riyals, 25.1% earned between 5000 and 10,000 riyals, 18.8% earned between 10,000 and 15,000 riyals, and 18.8% earned 15,000 riyals or more. A significant 22.0% of participants chose not to reveal their income information, suggesting a possible sensitivity towards financial disclosure (Table 2).

**Table 2.** Distribution of study sample members according to their demographic characteristics.

| Demographic Characteristics | | Number | Percentage % |
|---|---|---|---|
| Gender | Male | 95 | 37.3 |
| | Female | 160 | 62.7 |
| Age | Under 34 years old | 82 | 32.2 |
| | 35–49 years old | 109 | 42.7 |
| | 50 years and older | 64 | 25.1 |
| Educational level | Elementary | 6 | 2.4 |
| | Intermediate | 15 | 5.9 |
| | Secondary | 39 | 15.3 |
| | University | 178 | 69.8 |
| | Postgraduate studies (Master's-PhD) | 17 | 6.7 |

https://doi.org/10.3390/su18020686

**Table 2.** *Cont.*

| Demographic Characteristics | | Number | Percentage % |
|---|---|---|---|
| Number of family members | 2–4 | 83 | 32.5 |
| | 5–7 | 116 | 45.5 |
| | 8–10 | 56 | 22.0 |
| monthly family income level | <5000 riyals | 39 | 15.3 |
| | From 5000 riyals to <10,000 riyals | 64 | 25.1 |
| | From 10,000 riyals to <15,000 riyals | 48 | 18.8 |
| | riyals or more 15,000 | 48 | 18.8 |
| | Better not to disclose | 56 | 22.0 |

1 Saudi Riyal = USD 0.2665 (approximately).

### 3.2. *Knowledge of the Concept of Food Waste*

The vast majority of participants demonstrated a deep awareness of food waste, with 86.3% correctly stating that it refers to food that is intended to be eaten but is not. Another 11.4% were somewhat aware of the concept, and only 2.4% stated they knew nothing about food waste. The results show that the people in the research had a clear idea of what food waste is. Most of the people who answered demonstrated responsible ways to dispose of extra food. Most of them (64.7%) used the best storage methods, putting extra food in the fridge or freezer for later use. In total, 20.4% of people gave items to people in need, while 13.3% of people used them as animal feed. A tiny percentage (1.6%) threw away extra food in the trash, which suggests that most people know how to handle food properly and reduce waste. Approximately half of the participants (53.3%) remained in contact with food preservation groups, and 47.8% reported using their services. People who used these groups were generally satisfied with the services they received: 86.9% reported being delighted with the services overall (62.3% stated they were entirely happy, and 24.6% said they were glad to some extent). There was very little dissatisfaction, with only 1.6% of people being unhappy and none reporting depression. On the other hand, 11.5% reported being somewhat satisfied. This indicates that most people had positive experiences with the Food Preservation Society's services, as shown in Table 3.

**Table 3.** Frequency distributions and percentages of knowledge, behavior, and coping with food preservation societies.

| Awareness and Management of Surplus Food and Dealing with Food Preservation Associations | | Number | Percentage % |
|---|---|---|---|
| Food waste is defined as food intended for consumption but not used. | I don't know | 6 | 2.4 |
| | I don't know to some extent | 29 | 11.4 |
| | I know | 22 | 86.3 |
| What do you do if there is a surplus of food? | Store it properly in the refrigerator or freezer for later consumption. | 165 | 64.7 |
| | I donate it to the needy | 52 | 20.4 |
| | I offer it as animal food | 34 | 13.3 |
| | We put it in the waste container | 4 | 1.6 |

https://doi.org/10.3390/su18020686

**Table 3.** *Cont.*

| Awareness and Management of Surplus Food and Dealing with Food Preservation Associations | | Number | Percentage % |
|---|---|---|---|
| Do you have contact information for food preservation associations? | Yes | 136 | 53.3 |
| | No | 119 | 46.7 |
| Have you ever dealt with the services of food preservation societies? | Yes | 122 | 47.8 |
| | No | 133 | 52.2 |
| How satisfied are you with the services provided by food preservation societies? | Not completely satisfied | 0 | 0.0 |
| | Not satisfied | 2 | 1.6 |
| | Satisfied to some extent | 14 | 11.5 |
| | Satisfied | 30 | 24.6 |
| | Completely satisfied | 76 | 62.3 |

### 3.3. *Level of Knowledge of the Effects of Food Waste*

In Table 4, participants understood the impact of food waste on various aspects, including religious, economic, environmental, and food security. Table 4 demonstrates knowledge levels by effect, with mean scores from 4.73 to 3.15. The Islamic religious principle against wastefulness received the highest mean score (4.73), indicating it was the most significant influence. The participants' notion of resource responsibility matches Islamic values. In economics, the awareness that wasted food leads to decreased income rated second (M = 4.29), while the comprehension that wasted food costs manufacturing, transportation, and storage ranked third (M = 4.16). That throwing away food makes things more expensive was fifth (M = 3.87), and that Kingdom people waste a lot of food was sixth (M = 3.77). Food waste was recognized as a significant issue, with a mean of 3.61. Participants knew that food waste reduces food availability (M = 4.13, rank 1) and raises food prices (M = 4.00, rank 2). Throwing away healthy food makes it harder to find it in third place (M = 3.81), and it wastes natural resources in fourth place (M = 3.72). Environmental comprehension was moderate to good. People initially learned that food waste wastes production resources (M = 3.82). Food waste contributes to ecological contamination through excessive resource use, a phenomenon that individuals are aware of (M = 3.64). Third place went to awareness of pesticide and fertilizer loss environmental issues (M = 3.52). Knowledge that the National Program aims to halve food waste by 2030 ranked fourth (M = 3.40). The least known information was that food waste contributes to increased greenhouse gas emissions (M = 3.15), scoring lowest in all categories. The consequences of food waste on all areas were moderately well-known, as the average knowledge score was 3.80.

https://doi.org/10.3390/su18020686

**Table 4.** Knowledge of the Effects of Food Waste across Religious, Economic, Environmental, and Food Security Dimensions. A: I don't know exactly, B: don't know, C: I know to some extent, D: I know, F: I know exactly.

| Statement | A | B | C | D | F | Mean | SD | Rank |
|---|---|---|---|---|---|---|---|---|
| Dimension Religious | | | | | | | | |
| Extravagance is prohibited in the teachings of Islam | 0.8 | 2.0 | 3.9 | 10.2 | 83.1 | 4.73 | 0.70 | 1 |
| Dimension Economic | | | | | | | | |
| Lost spending on wasted food leads to reduced income. | 1.6 | 4.7 | 11.0 | 28.2 | 54.5 | 4.29 | 0.95 | 2 |
| Wasted food represents a loss in the cost of inputs, production, transportation, and storage. | 2.0 | 5.9 | 14.9 | 28.2 | 49.0 | 4.16 | 1.01 | 3 |
| Wasted food represents a loss of financial investment in agriculture | 2.0 | 4.3 | 18.0 | 29.4 | 46.3 | 4.14 | 0.99 | 4 |
| Food waste leads to higher food prices. | 3.1 | 13.3 | 18.4 | 23.9 | 41.2 | 3.87 | 1.18 | 5 |
| The amount of food waste per person in the Kingdom is considerable. | 2.4 | 12.5 | 26.3 | 23.1 | 35.7 | 3.77 | 1.13 | 6 |
| The cost of food waste in the Kingdom is a critical priority risk. | 3.1 | 17.3 | 24.3 | 26.3 | 29.0 | 3.61 | 1.17 | 7 |
| Dimension Food Security | | | | | | | | |
| Food waste leads to reduced food availability. | 1.2 | 7.8 | 14.9 | 29.0 | 47.1 | 4.13 | 1.01 | 1 |
| Food waste contributes to higher food prices, which in turn affects access to food. | 2.0 | 10.2 | 14.9 | 31.4 | 41.6 | 4.00 | 1.07 | 2 |
| Wasting nutritious food reduces the availability of healthy food. | 3.5 | 12.2 | 18.4 | 31.8 | 34.1 | 3.81 | 1.14 | 3 |
| Food waste leads to unsustainable use of natural resources. | 3.5 | 14.1 | 20.4 | 31.0 | 31.0 | 3.72 | 1.15 | 4 |
| Dimension Environmental | | | | | | | | |
| Food waste wastes the natural resources used in production. | 2.7 | 13.7 | 18.4 | 29.4 | 35.7 | 3.82 | 1.14 | 1 |
| A source of environmental pollution due to the excessive use of landfills. | 6.7 | 13.7 | 20.8 | 26.7 | 32.2 | 3.64 | 1.25 | 2 |
| Food waste leads to environmental risks from lost materials (pesticides, fertilizers). | 5.1 | 19.2 | 22.4 | 25.5 | 27.8 | 3.52 | 1.23 | 3 |
| The National Program aims to reduce food waste by half by 2030. | 8.2 | 18.0 | 24.3 | 23.9 | 25.5 | 3.40 | 1.27 | 4 |
| Food waste leads to the emission of greenhouse gases. | 10.6 | 29.4 | 18.0 | 18.0 | 23.9 | 3.15 | 1.36 | 5 |
| General Average | | | | | | 3.80 | 0.82 | |

### 3.4. Knowing the Behaviors of Rationalizing Food Consumption

Data in Table 5 illustrated that participants demonstrated significant comprehension of food conservation methods and initiatives aimed at reducing food waste. The average scores for these behaviors, ranging from 2.85 to 2.51, are presented in Table 5. The average score was 2.67. The fundamental aspect of water conservation is the efficient utilization of this resource to reduce waste. The mean score was 2.85, indicating a considerable awareness of this behavior among individuals. The second most critical factor was ensuring that food was devoid of stains, decay, stickiness, and discoloration (M = 2.82). The third most critical aspect was ensuring that canned items were not bulging, rusted, cracked, or leaking (M = 2.79). Consuming items with the shortest shelf life ranked fourth (M = 2.74), while refraining from providing excess food at meals ranked fifth (M = 2.73). Monitoring expiration dates during purchasing ranked sixth (M = 2.69), indicating a moderate level of comprehension. Documenting a list of dietary requirements prior to shopping ranked ninth (M = 2.63). Refraining from allowing prepared food to remain at room temperature for over


https://doi.org/10.3390/su18020686

two hours ranked as the eighth most significant factor (M = 2.62), while avoiding grocery shopping while hungry was the ninth most considerable factor (M = 2.60). Refraining from purchasing unnecessary food items due to sales ranked tenth (M = 2.54), while adhering to the food requirements list during shopping ranked eleventh (M = 2.52). The evaluation of storage instructions on products yielded the lowest mean score (M = 2.51), indicating that respondents deemed this task the least significant in relation to their comprehension of conservation techniques. The mean knowledge score of 2.67 suggests that individuals demonstrated a moderate to high level of awareness regarding the food conservation activities examined.

**Table 5.** Knowledge of Food Conservation Behaviors and Preventive Measures. A: I don't know exactly, B: don't know, C: I know to some extent.

| Behavior | A | B | C | Mean | SD | Rank |
|---|---|---|---|---|---|---|
| Consume water wisely to avoid Waste | 0.4 | 14.1 | 85.5 | 2.85 | 0.37 | 1 |
| Ensure food is free of stains, rot, stickiness, and discoloration | 2.0 | 14.5 | 83.5 | 2.82 | 0.44 | 2 |
| Ensure canned foods are not bulging, rusty, cracked, or leaking | 2.0 | 17.3 | 80.8 | 2.79 | 0.46 | 3 |
| Start consuming foods with the shortest shelf life first | 3.9 | 18.4 | 77.7 | 2.74 | 0.52 | 4 |
| Do not serve more food than necessary at meals | 2.7 | 21.2 | 76.1 | 2.73 | 0.50 | 5 |
| Check the expiration date while shopping | 5.1 | 20.4 | 74.5 | 2.69 | 0.56 | 6 |
| Write down a list of food needs before shopping | 4.7 | 27.5 | 67.8 | 2.63 | 0.57 | 7 |
| Do not leave cooked food at room temperature > 2 h | 7.5 | 23.1 | 69.4 | 2.62 | 0.62 | 8 |
| Avoid shopping when you feel hungry | 6.3 | 27.1 | 66.7 | 2.60 | 0.61 | 9 |
| Avoid sales that lead to buying unneeded food | 5.5 | 35.3 | 59.2 | 2.54 | 0.60 | 10 |
| Adhere to the food requirements list while shopping | 5.9 | 36.1 | 58.0 | 2.52 | 0.61 | 11 |
| Read the storage instructions written on the products | 7.5 | 34.1 | 58.4 | 2.51 | 0.63 | 12 |
| General Average | | | | 2.67 | 0.32 | |

### 3.5. Practice and Implementation of Food Conservation Behaviors

Participants conserved food moderately to highly in their daily activities. Table 6 ranks these activities by mean score, which varied from 2.77 to 2.23 with an average of 2.47 (SD = 0.43). The majority of practices were constant. Buying food without stains, rot, stickiness, or discoloration was of high importance. Its mean score of 2.77 indicates quality-consciousness. The second most essential thing was checking canned goods for bulging, rust, damage, or leaking (M = 2.76). Drinking water sensibly to avoid waste was the third most important (M = 2.75). Fourth place (M = 2.71) was achieved by eating the shortest-lasting items first. Shoppers checked expiration dates fifth (M = 2.67). The sixth most important thing was not leaving prepared food at room temperature for more than two hours (M = 2.63). Not overserving meals took seventh place (M = 2.58) with mediocre implementation. Making a food list before shopping was tenth (M = 2.55), and sticking to it while shopping was ninth (M = 2.51). Reading product storage instructions was rated eighth (M = 2.50), with the least utilization. Avoiding sales that led to buying undesirable food was rated tenth (M = 2.31), with 10.4% and 48.5% unsure due to limited practice. Avoiding shopping when hungry yielded the lowest mean score (M = 2.23), indicating the poorest execution, with 13.4% of participants being uninformed and 50.2% being confused. The knowledge score of 2.67 was substantially higher than the practice score of 2.47. Participants were more aware of food conservation practices in theory than in practice, suggesting a gap between knowledge and actual behavior (Table 6).

https://doi.org/10.3390/su18020686

**Table 6.** Practice and Implementation of Food Conservation Behaviors. A: I don't know exactly, B: don't know, C: I know to some extent.

| Behavior | A% | B% | C% | Mean | SD |
|---|---|---|---|---|---|
| Ensure food is free of stains, rot, stickiness, and discoloration. | 3.6 | 15.6 | 80.8 | 2.77 | 0.50 |
| Ensure canned foods are not bulging, rusty, cracked, or leaking. | 4.0 | 16.0 | 80.0 | 2.76 | 0.51 |
| Consume water wisely to avoid Waste. | 1.6 | 22.0 | 76.4 | 2.75 | 0.47 |
| Start consuming foods with the shortest shelf life first. | 5.7 | 17.6 | 76.7 | 2.71 | 0.57 |
| Check the expiration date while shopping. | 2.9 | 26.9 | 70.2 | 2.67 | 0.53 |
| Do not leave cooked food at room temperature for more than 2 h. | 4.2 | 28.8 | 67.0 | 2.63 | 0.57 |
| Do not serve more food than necessary at meals. | 3.6 | 34.3 | 62.1 | 2.58 | 0.56 |
| Write down a list of food needs before shopping. | 3.3 | 38.7 | 58.0 | 2.55 | 0.56 |
| Adhere to the food requirements list while shopping. | 2.9 | 43.3 | 53.8 | 2.51 | 0.56 |
| Read the storage instructions written on the products. | 5.5 | 39.0 | 55.5 | 2.50 | 0.60 |
| Avoid sales that lead to buying unneeded food. | 10.4 | 48.5 | 41.1 | 2.31 | 0.65 |
| Avoid shopping when you feel hungry. | 13.4 | 50.2 | 36.4 | 2.23 | 0.67 |
| General Average | | | | 2.47 | 0.43 |

### 3.6. Respondents' Preference for Means of Knowledge Transfer on the Subject of Food Waste

Respondents had several options for learning about reducing food waste and conserving resources. Table 7 ranks ten communication platforms by mean preference scores from 3.70 to 2.22. All techniques were "Average" preferred. The most popular method to communicate knowledge was WhatsApp, with a mean score of 3.70. A 38.4% considered it highly recommended, and 23.5% agreed. Snapchat ranked second (M = 3.53), with 32.2% highly recommending it and 27.8% favoring it. The X platform placed fourth (M = 3.35), while mobile text messaging placed third (M = 3.38). Lectures ranked sixth (M = 3.32), indicating that traditional teaching methods were still popular. TikTok placed sixth (M = 3.29), suggesting that people are becoming increasingly interested in short-form video, despite 22% not liking it. Instagram was ninth (M = 3.01), and YouTube was seventh (M = 3.26). The Zoom app. ranked tenth (M = 2.49), with 31.8% intensely hating it and 10.2% recommending it remarkably. This indicates that synchronous virtual meetings are less popular. Facebook scored lowest (M = 2.22). Only 10.2% strongly approved it, while 44.3% strongly disliked it. This suggests that people are reluctant to share knowledge on this platform. WhatsApp, Snapchat, and mobile text messages were the most popular platforms for food waste education. Facebook and Zoom were the least popular.

**Table 7.** Level of preference for knowledge transfer methods.

| Means | Strongly Unfavorable | Unfavorable | Neutral | Favorable | Highly Recommended | Mean | SD | Rank | Preference Level |
|---|---|---|---|---|---|---|---|---|---|
| | | | No. (%) | | | | | | |
| Whatsap | 25 (9.8) | 27 (10.6) | 45 (17.7) | 60 (23.5) | 98 (38.4) | 3.70 | 1.34 | 1 | Average |
| Snapchat | 41 (16.1) | 17 (6.7) | 44 (17.3) | 71 (27.8) | 82 (32.2) | 3.53 | 1.41 | 2 | Average |
| Mobile text messages | 29 (11.4) | 39 (15.3) | 52 (20.4) | 76 (29.8) | 59 (23.1) | 3.38 | 1.30 | 3 | Average |
| X platformx | 26 (10.2) | 42 (16.5) | 62 (24.3) | 67 (26.3) | 58 (22.8) | 3.35 | 1.28 | 4 | Average |


https://doi.org/10.3390/su18020686

**Table 7.** *Cont.*

| Means | Strongly Unfavorable | Unfavorable | Neutral | Favorable | Highly Recommended | Mean | SD | Rank | Preference Level |
|---|---|---|---|---|---|---|---|---|---|
| | | | No. (%) | | | | | | |
| Lectures | 40 (15.7) | 26 (10.2) | 55 (21.6) | 80 (31.4) | 54 (21.2) | 3.32 | 1.34 | 5 | Average |
| TikTok | 56 (22.0) | 21 (8.2) | 44 (17.3) | 62 (24.3) | 72 (28.2) | 3.29 | 1.50 | 6 | Average |
| YouTube | 34 (13.3) | 37 (14.5) | 66 (25.9) | 65 (25.5) | 53 (20.8) | 3.26 | 1.31 | 7 | Average |
| Instagram | 58 (22.8) | 34 (13.3) | 59 (23.1) | 56 (22.0) | 48 (18.8) | 3.01 | 1.42 | 8 | Average |
| Zoom application | 81 (31.8) | 56 (22.0) | 56 (22.0) | 36 (14.1) | 26 (10.2) | 2.49 | 1.34 | 9 | Average |
| Facebook | 113 (44.3) | 54 (21.2) | 34 (13.3) | 28 (11.0) | 26 (10.2) | 2.22 | 1.38 | 10 | Average |

### 3.7. Examining the Correlation Between Awareness of Food Waste Consequences and Rationalization of Food Consumption: A Mediation Analysis

A mediation model was employed to investigate the interrelations among three principal variables: awareness of the consequences of food waste (theoretical knowledge), understanding of food conservation practices (behavioral knowledge), and the implementation of food rationalization activities. This investigation aimed to determine whether behavioral knowledge acts as a mediator in the relationship between theoretical knowledge and actual behavioral practices, using the Baron and Kenny technique. Initial analysis indicated statistically substantial relationships among all three variables. A considerable link existed between knowing how to conserve food and actually doing so (r = 0.86, $p < 0.001$). This means that being aware of the right actions is closely linked to doing them. Understanding the repercussions of food waste was positively correlated with actual activity (r = 0.42, $p < 0.001$), indicating a direct link between awareness of the consequences of food waste and subsequent action. Moreover, awareness of the repercussions of food waste showed a strong correlation with awareness of food conservation practices (r = 0.36, $p < 0.001$), thereby affirming that theoretical comprehension enhances the development of behavioral knowledge (Table 8).

**Table 8.** Correlation Coefficients Between Knowledge of Food Waste, Knowledge of Food Conservation Behaviors, and Actual Practice of Food Rationalization Behaviors.

| Variables | Level of Practice of Food Rationalization Behaviors | |
|---|---|---|
| | Relation | Significance |
| Knowledge of food conservation behaviors | 0.86 | 0.001 |
| Knowledge of the effects of food waste | 0.42 | 0.001 |
| | Knowledge of food conservation behaviors | |
| Knowledge of the effects of food waste | 0.36 | 0.001 |

https://doi.org/10.3390/su18020686

*Step 1: The Immediate Impact of Food Waste Awareness on Behavioral Practice*

Regression analysis investigated the direct influence of awareness on the effects of food waste on food rationalization habits (Table 9). Awareness of food waste impacts has shown a favorable and statistically significant impact on behavioral practice (B = 0.221, β = 0.421, $p < 0.001$), signifying a robust link. This model elucidated 17.7% of the variance in food conservation practices ($R^2 = 0.177$, $F(1, 253) = 54.59$, $p < 0.001$), demonstrating that theoretical knowledge alone yields significant predictive value for behavioral results.

**Table 9.** Regression results for predicting behavioral practices through knowledge of the effects of food waste.

| Independent Variable | B | SE | *β* | *t*-Value | *p*-Value |
|---|---|---|---|---|---|
| (The constant) | 1.628 | 0.116 | - | 14.001 | <0.001 |
| Knowledge of the effects of food waste | 0.221 | 0.03 | 0.421 | 7.389 | <0.001 |

*Step 2: The effect of knowing about food waste on knowing how to behave*

The correlation between awareness of food waste consequences and understanding of food conservation practices was further analyzed (Table 10). Theoretical knowledge was a strong predictor of behavioral knowledge (B = 0.139, β = 0.357, $p < 0.001$), and knowledge of the impacts of food waste explained 12.7% of the variance in behavioral knowledge ($R^2 = 0.127$). This phase confirmed that moving from knowing what happens to food waste to having practical behavioral information made sense.

**Table 10.** Regression results for predicting behavioral knowledge through knowledge of the effects of food waste.

| Independent Variable | B | SE | *β* | *t*-Value | *p*-Value |
|---|---|---|---|---|---|
| (The constant) | 2.141 | 0.089 | | 24.028 | <0.001 |
| Knowledge of the effects of food waste | 0.139 | 0.023 | 0.357 | 6.078 | <0.001 |

*Step 3: Complete Mediation Model Incorporating Behavioral Knowledge*

The full mediation model was evaluated by concurrently inputting both the independent variable (awareness of food waste consequences) and the mediating variable (awareness of food conservation methods) to predict actual behaviors (Table 11). The standardized effect of understanding the impacts of food waste decreased from β = 0.421 to β = 0.132; however, it remained statistically significant ($p < 0.001$), indicating partial mediation. Behavioral knowledge had a significantly greater impact on practice (B = 1.088, β = 0.809, $p < 0.001$), substantially exceeding the direct influence of theoretical information. The complete model accounted for 74.9% of the variance in food rationalization behaviors ($R^2 = 0.749$), indicating a markedly enhanced predictive capability relative to the simple model, hence illustrating the approach's considerable efficacy.

**Table 11.** Regression results for predicting the practice of behaviors through knowledge of effects and knowledge of behaviors.

| Independent Variable | B | SE | *β* | *t*-Value | *p*-Value |
|---|---|---|---|---|---|
| (The constant) | −0.70 | 0.117 | | −6.001 | <0.001 |
| Knowledge of the effects of food waste | 0.069 | 0.018 | 0.132 | 3.919 | <0.001 |
| Knowledge of food conservation behaviors | 1.088 | 0.045 | 0.809 | 23.93 | <0.001 |

https://doi.org/10.3390/su18020686

These findings validate partial mediation, indicating that awareness of food waste impacts behavior both directly and indirectly, through shaping behavioral knowledge. The significant drop in the direct effect coefficient, while still statistically substantial, indicates that behavioral knowledge is a crucial link between theoretical understanding and practical behavior. These findings highlight the need for formulating comprehensive awareness campaigns that integrate theoretical insights into the ramifications of food waste with practical behavioral directives, thereby enhancing the effectiveness of translating awareness into sustainable food conservation practices. The direct effect of knowledge on conservation practice ($R^2 = 0.177$) accounts for 17.7% of the variance, which may appear modest. However, this is consistent with social psychological research on behavioral prediction [cite: relevant studies]. When behavioral knowledge is included as a mediator, the model explains 74.9% of the variance in conservation practice (combined $R^2$), demonstrating substantial improvement. This suggests that the indirect pathway through behavioral knowledge accounts for a considerable amount of behavioral variance. Additionally, conservation behavior in real-world household contexts is influenced by numerous factors beyond knowledge (financial constraints, household structure, storage availability, time availability, habit persistence), indicating that explaining 74.9% of variance through knowledge-based mechanisms is substantial. The direct pathway captures intrinsic motivation effects (understanding consequences directly activates norms), while the indirect pathway captures capability effects (motivation drives knowledge-seeking, enabling behavior). Both mechanisms make a meaningful contribution to understanding behavior.

Knowledge of food conservation behaviors served as a critical mechanism for translating consequence awareness into food-saving practices, with the model explaining 74.9% of the variance in behavioral outcomes ($R^2 = 0.749$), as shown in Figure 2. Knowledge of the effects of food waste demonstrated a substantial total effect on practice ($\beta = 0.421$, $p < 0.001$), decomposing into direct ($\beta = 0.132$, $p < 0.001$) and indirect pathways through conservation knowledge ($\beta = 0.357$, $p < 0.001$). The strongest relationship was between conservation knowledge and the implementation of practice ($\beta = 0.809$, $p < 0.001$), with the indirect effect accounting for 68.6% of the total influence. These results support partial mediation, wherein behavioral knowledge, understanding concrete, actionable strategies for food waste reduction, represents the proximal determinant of food-saving practice. The findings align with Social Cognitive Theory's emphasis on behavioral capability as an essential mediator that translates motivational awareness into action, while supporting Norm Activation Theory by demonstrating that consequence awareness activates motivation for behavioral knowledge acquisition. Intervention implications are significant: awareness campaigns establishing the consequences of food waste are necessary but insufficient; sustained behavior change requires complementary investment in behavioral knowledge development through practical instruction in meal planning, food storage, and waste reduction. The high explanatory power of the knowledge-based model positions behavioral knowledge acquisition as a high-leverage intervention target for promoting household food waste reduction aligned with Saudi Vision 2030 objectives.

https://doi.org/10.3390/su18020686

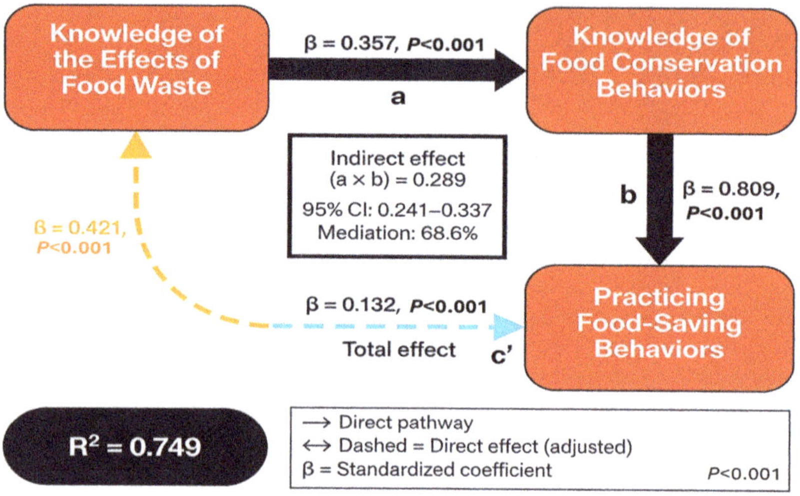


**Figure 2.** Knowledge of food conservation behaviors mediates the relationship between awareness of the consequences of food waste and food-saving practices. Path a represents the effect of consequence awareness on conservation knowledge ($\beta = 0.357$, $p < 0.001$); path b represents the effect of conservation knowledge on food-saving behavior ($\beta = 0.809$, $p < 0.001$); path c′ represents the direct effect adjusted for mediation ($\beta = 0.132$, $p < 0.001$). The indirect effect ($a \times b = 0.289$, 95% CI: 0.241–0.337) represents 68.6% of the total effect (0.421), indicating strong mediation. The model explains 74.9% of the variance in practicing food-saving behaviors ($R^2 = 0.749$). All pathways are statistically significant at $p < 0.001$.

## 4. Discussion

### *4.1. General Discussion*

This study shows that household food waste reduction is driven by a tripartite knowledge pathway linking awareness of consequences, behavioral knowledge of conservation practices, and their enactment in daily life. A dual-component intervention that combines consequence awareness (moral motivation) with behavioral knowledge (practical capability) explains 74.9% of practice variance, versus 17.7% for consequence awareness alone, confirming the central mediating role of behavioral knowledge. Effective programs must integrate "why" and "how" elements by clarifying environmental (e.g., climate change), economic, and food security impacts, while strengthening key behaviors such as reading storage instructions, sticking to shopping lists, avoiding shopping when hungry, and resisting unnecessary promotions. Allocating approximately 40% of communication resources to consequence-focused messaging and 60% to practical training provides concrete guidance for actors working toward Saudi Vision 2030's goal of halving food waste, especially when framed through Islamic stewardship and economic responsibility, and targeted at behaviors such as impulse buying and poor label use. Preferences for WhatsApp, Snapchat, and SMS among female household decision-makers highlight promising channels for integrated outreach that blends consequence messaging with structured skills training, supported by partnerships with food preservation and food security associations. At the same time, future work should test optimal message ratios and identify barriers that hinder the translation of knowledge into sustained conservation behavior.

### *4.2. Theoretical Contributions and Framework Alignment*

This study advances the KAP model through methodological innovation, distinguishing between consequence knowledge (theoretical understanding of impacts) and behavioral

https://doi.org/10.3390/su18020686

knowledge (practical understanding of implementation), with behavioral knowledge functioning as a mediator [32]. This refinement addresses a critical gap in the environmental behavior literature, where traditional KAP models assume a linear progression [25–27]. Our findings demonstrate that knowledge operates through dual pathways, aligning with both Norm Activation Theory (NAT) and Social Cognitive Theory (SCT) [35,45,46].

The mediation model's empirical findings offer theoretical clarity lacking in prior KAP research. The direct pathway ($\beta = 0.132$, $p < 0.001$) reflects NAT's mechanism whereby consequence awareness activates personal moral responsibility, maintaining independent effects on behavior [47]. The indirect pathway (accounting for 68.6% of total influence) reflects SCT's emphasis on behavioral capability as a prerequisite for translating motivation into action [48,49]. This dual-mechanism model explains why awareness campaigns alone produce modest behavior change (17.7% variance explained) while integrated consequence-plus-capability interventions achieve substantial improvement (74.9% variance explained), resulting in a 4.2-fold effectiveness gain [22,35]. The partial mediation pattern (not complete mediation) confirms that consequence knowledge maintains direct effects while facilitating knowledge-seeking, demonstrating theoretically distinct motivational and capability pathways [44,50].

### 4.3. Contextualized Findings: Cultural and Regional Significance

The finding that religious-ethical consequence awareness dominates (M = 4.73, with 83.1% reporting "completely know" Islamic prohibitions) compared to environmental awareness (M = 3.15, in the lowest category) reflects Saudi Arabia's unique cultural context, where Islamic stewardship principles provide powerful motivational frames [51,52]. This pattern contradicts Western literature, which emphasizes environmental appeals, and validates the theoretical argument that knowledge effectiveness is culturally contingent [23,53]. Environmental knowledge deficiency (greenhouse gas emissions M = 3.15) identifies a specific intervention target: connecting abstract climate impacts to concrete regional concerns (water scarcity, Islamic stewardship principles) rather than relying on global climate messaging [9,54].

The 98.4% surplus food conservation rate (storage, donation, or animal feed) versus 1.6% disposal reflects conscious, deliberate food management behavior driven by cognitive factors rather than habitual patterns [19,20]. This finding empirically supports the behavioral nature hypothesis that food waste in Saudi Arabia is amenable to behavioral intervention through knowledge enhancement [24,32].

The knowledge-practice gap (knowledge M = 2.67 vs. practice M = 2.47; a 7.5-percentage-point difference) reveals implementation barriers that operate independently of awareness [26,31]. Low scores for impulse-control behaviors (avoiding hungry shopping, M = 2.23, and avoiding sales-driven purchases, M = 2.31) and list adherence (M = 2.52) suggest that competing immediate priorities and habit persistence override awareness [27,55]. This pattern suggests that interventions must address implementation barriers through environmental design, habit formation support, and concrete feedback mechanisms, rather than merely transferring knowledge [56,57].

### 4.4. Mediation Model: Mechanism and Intervention Implications

The mediation analysis reveals that behavioral knowledge serves as the proximal determinant of food-saving practice, with substantially greater effect ($\beta = 0.809$) than theoretical knowledge alone [48]. This finding operationalizes SCT's argument that knowing "how to reduce waste" (behavioral knowledge) is more influential than knowing "why to reduce waste" (consequence knowledge) [35,49]. However, the presence of direct effects

https://doi.org/10.3390/su18020686

($\beta = 0.132$) indicates that motivated individuals implement practices even with imperfect practical knowledge, driven by intrinsic moral responsibility [25,47].

The empirical pathway decomposition provides actionable resource allocation: consequence messaging should represent approximately 40% of intervention resources (direct effect pathway with 31.4% of total influence). In comparison, behavioral training should account for 60% (an indirect effect pathway with a 68.6% total influence). This evidence-based allocation differs substantially from typical awareness-focused campaigns that emphasize consequences without equivalent investment in practical skill-building [22,58].

*4.5. Implications for Intervention Design and Policy*

Integration of Dual-Component Messaging: Effective interventions require simultaneous emphasis on consequence awareness ("Why: Food waste depletes resources, violates Islamic principles, threatens food security") and behavioral capability ("How: List-making before shopping, checking expiration dates, proper storage methods") [26,28,31]. Current separation of these components in awareness campaigns explains their limited effectiveness [23]. Integrated messaging campaigns should present consequences and practical solutions as complementary elements within single communications [59,60].

Channel Selection and Digital Strategy: WhatsApp (M = 3.70) and Snapchat (M = 3.53) emerge as high-efficacy channels, reflecting Saudi demographic preferences for direct, rapid messaging over passive platforms [61,62]. Facebook (M = 2.22) and Zoom (M = 2.49) demonstrate limited utility, suggesting that synchronous/passive platforms conflict with knowledge-acquisition preferences [63]. Given that 62.7% of household food managers are female, WhatsApp-based interventions targeting female household decision-makers represent an optimal allocation of resources [64,65].

Food Preservation Association Expansion: Current usage (47.8%) coupled with exceptionally high satisfaction (86.9% completely satisfied) indicates proven institutional effectiveness with substantial expansion potential [30,32]. Organizations should: (1) shift from in-person services to WhatsApp-based behavioral training modules, increasing accessibility and reach [61,66]; (2) develop modular training addressing low-performance domains (list-making, label-reading, impulse shopping avoidance) [29,67]; (3) establish retail partnerships positioning behavioral support at critical decision points (grocery store information kiosks) [68]; and (4) target female household managers through gender-sensitive program design acknowledging competing time demands [64].

Environmental Knowledge Enhancement: The systematic deficiency in environmental awareness, particularly regarding greenhouse gas emissions (M = 3.15), represents both a limitation and an opportunity for intervention [9,28]. Rather than relying on abstract global climate messaging, effective campaigns should localize environmental impacts through water scarcity frames that align with Saudi regional concerns and Islamic stewardship principles, connecting food waste to resource depletion that contradicts Islamic teachings on stewardship [52,54,69]. Targeted Behavior Modification: Specific behaviors demonstrate intervention resistance: avoiding hungry shopping (M = 2.23 practice score, 13.4% uninformed) and avoiding impulse purchases (M = 2.31). These low scores suggest that point-of-purchase psychological interventions (e.g., smartphone shopping list notifications, contactless payment reminders) may prove more effective than knowledge transfer alone [55,67,68].

Future research priorities include: (1) longitudinal studies tracking behavior change persistence following dual-component interventions using objective waste audits rather than self-report; (2) randomized controlled trials manipulating consequence-versus-behavioral messaging ratios to establish optimal allocation; (3) qualitative research exploring knowledge-practice gap barriers (time constraints vs. competing priorities vs.

https://doi.org/10.3390/su18020686

active non-compliance); (4) systematic moderation analysis across demographic strata; (5) comparative regional research testing generalizability; and (6) mechanistic research explicitly testing Norm Activation versus Self-Efficacy pathways through moderated mediation analyses.

## 5. Study Limits

Convenience sampling at shopping mall locations introduces selection bias; however, the findings remain descriptive of urban, regularly shopping households in Buraydah, and shopping malls provide strategic access to primary food managers. Resource constraints precluded the use of stratified random sampling, and geographic confinement limits the generalizability to other Saudi regions. Self-reported behavioral data is prone to social desirability bias, female overrepresentation (62.7%), and university education overrepresentation (76.5%), which may inflate knowledge levels. Income non-disclosure (22%) can mask economic barriers, and a lack of seasonal assessment provides only a temporal snapshot. The cross-sectional design precludes causal inference; however, mediation analysis suggests the theorized mechanisms. Future research employing longitudinal designs with stratified sampling by region, income, and education level, as well as objective behavioral measures and complete demographic representation, would strengthen the evidence. Despite these limitations, the findings provide robust mechanistic evidence for knowledge-practice mediation and establish guidance on resource allocation for dual-component interventions that support Saudi Vision 2030′s food security objectives.

**Supplementary Materials:** The following supporting information can be downloaded at: https://www.mdpi.com/article/10.3390/su18020686/s1, the questionnaire is attached as Annex S1 and Table S1 for the reliability and validity assessment of the measurement model, including consequences, knowledge, behavioral knowledge, and behavioral practice.

**Author Contributions:** Methodology, Preparation, and Formal analysis, O.M.A.-T.; Data collection and organization, O.M.A.-T., M.A.A. and F.O.A.; Review and editing, Preparation of original draft, O.M.A.-T., F.O.A. and H.B. Conceptualization and methodology, O.M.A.-T. and F.O.A.; Formal analysis, M.I.M.; Writing—original draft preparation, H.M.M. and H.B.; Writing—review and editing, H.B. All authors have read and agreed to the published version of the manuscript.

**Funding:** The Researchers would like to thank the Deanship of Graduate Studies and Scientific Research at Qassim University for financial support (QU-APC-2026).

**Institutional Review Board Statement:** The study was conducted in accordance with the Declaration of Helsinki and approved by the Research Ethics Committee of the Deanship of Graduate Studies, King Saud University, REF NO. KSU–HE-25 (protocol code 693 and date of approval 19 May 2025).

**Informed Consent Statement:** Informed consent was obtained from all study participants.

**Data Availability Statement:** The original contributions presented in this study are included in the article/Supplementary Material. Further inquiries can be directed to the corresponding author.

**Conflicts of Interest:** The authors declare no conflicts of interest.

## References

1. United Nations Environment Programme. *Food Waste Index Report 2024*; UNEP: Nairobi, Kenya, 2024. Available online: https://www.unep.org/resources/publication/food-waste-index-report-2024 (accessed on 12 August 2025).
2. Brennan, A.; Browne, S. Food waste and nutrition quality in the context of public health: A scoping review. *Int. J. Environ. Res. Public Health* **2021**, *18*, 5379. [CrossRef]
3. Abiad, M.G.; Meho, L.I. Food loss and food waste research in the Arab world: A systematic review. *Food Secur.* **2018**, *10*, 311–322. [CrossRef]
4. Food and Agriculture Organization (FAO). *Food Wastage Footprint: Full Cost Accounting*; FAO: Rome, Italy, 2014.

https://doi.org/10.3390/su18020686

5. Food and Agriculture Organization (FAO). The State of Food and Agriculture 2019. In *Moving Forward on Food Loss and Waste Reduction*; FAO: Rome, Italy, 2019. Available online: https://www.fao.org/interactive/state-of-food-agriculture/2019/en/ (accessed on 22 May 2025).
6. Bisangwa, E. The Effects of Halving Lost and Wasted Food on Food Security, Water, Land and Fertilizer Resources—A Quantitative Comparison Between Developed and Less Developed Regions. Ph.D. Thesis, University of Tennessee, Knoxville, TN, USA, 2019. Available online: https://trace.tennessee.edu/utk_graddiss (accessed on 22 May 2025).
7. Baig, M.B.; Gorski, I.; Neff, R.A. Understanding and addressing waste of food in the Kingdom of Saudi Arabia. *Saudi J. Biol. Sci.* **2019**, *26*, 1633–1648. [CrossRef]
8. Kummu, M.; De Moel, H.; Porkka, M.; Siebert, S.; Varis, O.; Ward, P.J. Lost food wasted resources: Global food supply chain losses and their impacts on freshwater, cropland, and fertiliser use. *Sci. Total Environ.* **2012**, *438*, 477–489. [CrossRef]
9. Murdock, A. *What You Need to Know About Food Waste and Climate Change*; University of California News: Oakland, CA, USA, 2017. Available online: https://www.universityofcalifornia.edu/news/what-you-need-know-about-food-waste-and-climate-change (accessed on 22 May 2025).
10. Baig, M.B.; Al-Zahrani, K.H.; Schneider, F.; Straquadine, G.S.; Mourad, M. Food waste posing a serious threat to sustainability in the Kingdom of Saudi Arabia–A systematic review. *Saudi J. Biol. Sci.* **2019**, *26*, 1743–1752. [CrossRef]
11. High Level Panel of Experts on Food Security and Nutrition (HLPE). Food Losses and Waste in the Context of Sustainable Food Systems. In *A Report by the High Level Panel of Experts on Food Security and Nutrition of the Committee on World Food Security*; FAO: Rome, Italy, 2014. Available online: www.fao.org/cfs/cfs-hlpe (accessed on 20 May 2025).
12. Ishangulyyev, R.; Kim, S.; Lee, S.H. Understanding food loss and waste—Why are we losing and wasting food? *Foods* **2019**, *8*, 297. [CrossRef]
13. KC, K.B.; Haque, I.; Legwegoh, A.F.; Fraser, E.D.G. Strategies to Reduce Food Loss in the Global South. *Sustainability* **2016**, *8*, 595. [CrossRef]
14. Ministry of Environment, Water and Agriculture. *Food Security Strategy and Implementation Plan: Executive Summary*; Ministry of Environment, Water and Agriculture: Riyadh, Saudi Arabia, 2018. Available online: https://www.mewa.gov.sa/en/Ministry/initiatives/SectorStratigy/Pages/default.aspx (accessed on 22 May 2025).
15. Saudi Grains Organization (SAGO). *Baseline: Food Loss and Waste Index in the Kingdom of Saudi Arabia*; King Fahd National Library: Riyadh, Saudi Arabia, 2021. Available online: https://www.academia.edu/51050246/Quantifying_Food_Loss_and_Waste_in_Saudi_Arabia (accessed on 18 May 2025).
16. Baig, M.B.; Alotaibi, B.A.; Alshammari, G.M.; Shah, A.A. Food waste in Saudi Arabia: Causes, consequences, and combating measures. *Sustainability* **2022**, *14*, 10362. [CrossRef]
17. Mir, M.A.; Chang, S.K.; Hefni, D. A comprehensive review on challenges and choices of food waste in Saudi Arabia: Exploring environmental and economic impacts. *Environ. Syst. Res.* **2024**, *13*, 40. [CrossRef]
18. Mu'azu, N.D.; Blaisi, N.I.; Naji, A.A.; Abdel-Magid, I.M.; AlQahtany, A. Food waste management current practices and sustainable future approaches: A Saudi Arabian perspective. *J. Mater. Cycles Waste Manag.* **2019**, *21*, 678–690. [CrossRef]
19. Food and Agriculture Organization (FAO). Rural Transformation—Key for Sustainable Development in the Near East and North Africa. In *Overview of Food Security and Nutrition 2018*; FAO: Cairo, Egypt, 2019.
20. Althumiri, N.A.; Basyouni, M.H.; Duhaim, A.F.; AlMousa, N.; AlJuwaysim, M.F.; BinDhim, N.F. Understanding food waste, food insecurity, and the gap between the two: A nationwide cross-sectional study in Saudi Arabia. *Foods* **2021**, *10*, 681. [CrossRef] [PubMed]
21. The General Grain Organization (SAGO). *Results and Initiatives of the Food Loss and Waste Study in the Kingdom of Saudi Arabia*; King Fahd National Library: Riyadh, Saudi Arabia, 2019. Available online: https://www.scribd.com/document/545730749/Baseline-230719 (accessed on 18 May 2025).
22. Stancu, V.; Haugaard, P.; Lähteenmäki, L. Determinants of consumer food waste behaviour: Two routes to food waste. *Appetite* **2016**, *96*, 7–17. [CrossRef]
23. Food and Agriculture Organization (FAO). *Global Food Losses and Food Waste—Extent, Causes and Prevention*; FAO: Rome, Italy, 2011; ISBN 978-92-5-107205-9.
24. Jagau, H.L.; Vyrastekova, J. Behavioral approach to food waste: An experiment. *Br. Food J.* **2017**, *119*, 882–894. [CrossRef]
25. Keller, V.; Gombos, S. Household food waste reduction determinants in Hungary: Towards understanding responsibility, awareness, norms, and barriers. *Foods* **2025**, *14*, 728. [CrossRef]
26. Aydin, A.E.; Yildirim, P. Understanding food waste behavior: The role of morals, habits, and knowledge. *J. Clean. Prod.* **2021**, *280*, 124250. [CrossRef]
27. Rasool, S.; Cerchione, R.; Salo, J.; Ferraris, A.; Abbate, S. Measurement of consumer awareness of food waste: Construct development with a confirmatory factor analysis. *Br. Food J.* **2021**, *123*, 337–361. [CrossRef]
28. Ananda, J.; Karunasena, G.G.; Mitsis, A.; Kansal, M.; Pearson, D. Analysing behavioural and socio-demographic factors and practices influencing Australian household food waste. *J. Clean. Prod.* **2021**, *306*, 127280. [CrossRef]


https://doi.org/10.3390/su18020686

29. Jahan, I.; Kamal, K.T.; Bhattacharjee, P.; Taqi, H.M.M.; Ali, S.M. Improving consumer awareness for reducing food waste using partial least squares structural equation modelling (PLS-SEM) approach. *Clean. Responsible Consum.* **2025**, *100*, 282. [CrossRef]
30. Qian, L.; Li, F.; Zhao, X.; Liu, H.; Liu, X. The association between religious beliefs and food waste: Evidence from Chinese rural households. *Sustainability* **2022**, *14*, 8555. [CrossRef]
31. Tsai, W.C.; Chen, X.; Yang, C. Consumer food waste behavior among emerging adults: Evidence from China. *Foods* **2020**, *9*, 961. [CrossRef] [PubMed]
32. Szakos, D.; Szabó-Bódi, B.; Kasza, G. Consumer awareness campaign to reduce household food waste based on structural equation behavior modeling in Hungary. *Environ. Sci. Pollut. Res.* **2021**, *28*, 24580–24589. [CrossRef]
33. Zarei, F.; Dehghani, A.; Ratansiri, A.; Ghaffari, M.; Raina, S.K.; Halimi, A.; Rakhshanderou, S.; Isamel, S.A.; Amiri, P.; Aminafshar, A.; et al. ChecKAP: A Checklist for Reporting a Knowledge, Attitude, and Practice (KAP) Study. *Asian Pac. J. Cancer Prev.* **2024**, *25*, 2573–2577. [CrossRef] [PubMed] [PubMed Central]
34. Alzghoul, B.I.; Seedahmed, H.M.; Ibraheem, K.M. Developing and Validating a Knowledge, Attitude, and Practice (KAP) Questionnaire for Pain Management. *Open Public Health J.* **2025**, *18*, e18749445181057. [CrossRef]
35. Ministry of Municipal and Rural Affairs. *Comprehensive Urban Vision Report for the City of Buraydah (Future Saudi Cities Programme)*; Ministry of Municipal and Rural Affairs: Riyadh, Saudi Arabia, 2019. Available online: https://www.momra.gov.sa (accessed on 18 May 2025).
36. Ministry of Media, Buraydah. Saudi Digital Encyclopedia (Saudipedia). 2025. Available online: https://saudipedia.com/en/article/271/government-and-politics/ministries/ministry-of-media (accessed on 18 May 2025).
37. General Authority for Statistics. *Saudi Census 2022 Statistics: Household Statistics (Buraydah Region)*; General Authority for Statistics: Riyadh, Saudi Arabia, 2022. Available online: https://www.stats.gov.sa/en/ (accessed on 13 May 2025).
38. Thompson, S.K. *Sampling*; John Wiley & Sons: New York, NY, USA, 2012; Volume 755. [CrossRef]
39. Boiteau, J.M. Food Loss and Waste in the United States and Worldwide. World Hunger News. Available online: https://www.worldhunger.org/food-loss-and-waste-in-the-united-states-and-worldwide/ (accessed on 13 June 2018).
40. General Food and Drug Authority. Reducing Food Waste. Awareness Campaigns. Riyadh, Kingdom of Saudi Arabia. 2018. Available online: https://www.sfda.gov.sa/ar/awarenesscampaign/74336 (accessed on 10 May 2025).
41. Ali, I.A.S. Rural women's behavior related to reducing household food waste in some sugar beet villages in Alexandria Governorate. *J. Adv. Agric. Res.* **2020**, *25*, 140–161. [CrossRef]
42. Al-Nasser, M.K.; Al-Tarawneh, M.S. Analyzing the Behavior of Jordanian Families Towards Reducing Household Food Waste: A Case Study of Families in Amman. Master's Thesis, Jerash University, Jerash, Jordan, 2022. Available online: https://search.emarefa.net/detail/BIM-1454006 (accessed on 6 May 2025).
43. Baron, R.M.; Kenny, D.A. The moderator-mediator variable distinction in social psychological research: Conceptual, strategic, and statistical considerations. *J. Pers. Soc. Psychol.* **1986**, *51*, 1173–1182. [CrossRef] [PubMed]
44. Neff, R.A.; Spiker, M.L.; Truant, P.L. Wasted food: US consumers' reported awareness, attitudes, and behavior. *PLoS ONE* **2015**, *10*, e0127881. [CrossRef]
45. Attiq, S.; Habib, M.D.; Kaur, P.; Hasni, M.J.S.; Dhir, A. Drivers of food waste reduction behaviour in the household context. *Food Qual. Prefer.* **2021**, *94*, 104300. [CrossRef]
46. Robelia, B.; Murphy, T. What do people know about key environmental issues? A review of environmental knowledge surveys. *Environ. Educ. Res.* **2012**, *18*, 299–321. [CrossRef]
47. Schanes, K.; Dobernig, K.; Gözet, B. Food waste matters—A systematic review of household food waste practices and their policy implications. *J. Clean. Prod.* **2018**, *182*, 978–991. [CrossRef]
48. Rani, Z.M.; Ismail, A.; Rahim, N.; Apandi, S.R.M.; Farook, F. The Impact of Environmental Knowledge on Food Waste Reduction and Sustainability Practices among Hospitality Students in Malaysia. *Inf. Manag. Bus. Rev.* **2024**, *16*, 51–58. [CrossRef]
49. Van der Werf, P.; Seabrook, J.A.; Gilliland, J.A. The quantity of food waste in the garbage stream of southern Ontario, Canada, in households. *PLoS ONE* **2018**, *13*, e0198470. [CrossRef]
50. Ministry of Commerce and Industry. *Initiative for Rationalizing Food Consumption and Preserving It from Waste*; The National Committee for the Campaign on Rationalizing Food Consumption, Ministry of Commerce and Industry: Kuwait City, Kuwait, 2023. Available online: https://pafn.gov.kw/assets/images/Uploads/18022024.pdf (accessed on 2 May 2025).
51. Etim, E.; Choedron, K.T.; Ajai, O.; Duke, O.; Jijingi, H.E. Systematic review of factors influencing household food waste behaviour: Applying the theory of planned behaviour. *Waste Manag. Res.* **2024**, *43*, 803–827. [CrossRef]
52. Schanes, K.; Stagl, S. Food waste fighters: What motivates people to engage in food sharing? *J. Clean. Prod.* **2019**, *211*, 1491–1501. [CrossRef]
53. Papargyropoulou, E.; Lozano, R.; Steinberger, J.K.; Wright, N.; bin Ujang, Z. The food waste hierarchy as a framework for the management of food surplus and food waste. *J. Clean. Prod.* **2014**, *76*, 106–115. [CrossRef]
54. Azazz, A.M.; Elshaer, I.A. Amid the COVID-19 pandemic, social media usage and food waste intention: The role of excessive buying behavior and religiosity. *Sustainability* **2022**, *14*, 6786. [CrossRef]


https://doi.org/10.3390/su18020686

55. Sobaih, A.E.E. Excessive food buying in Saudi Arabia amid COVID-19: Examining the effects of perceived severity, religiosity, consumption culture and attitude toward behavior. *Int. J. Environ. Res. Public Health* **2023**, *20*, 3126. [CrossRef] [PubMed]
56. Alotaibi, B.A.; Kassem, H.S. Adoption of sustainable water management practices among farmers in Saudi Arabia. *Sustainability* **2021**, *13*, 11260. [CrossRef]
57. Kollmuss, A.; Agyeman, J. Mind the Gap: Why Do People Act Environmentally and What Are the Barriers to Pro-Environmental Behavior? *Environ. Educ. Res.* **2002**, *8*, 239–260. [CrossRef]
58. Stern, P.C. New Environmental Theories: Toward a Coherent Theory of Environmentally Significant Behavior. *J. Soc. Issues* **2000**, *56*, 407–424. [CrossRef]
59. Young, W.; Hwang, K.; McDonald, S.; Oates, C.J. Sustainable Consumption: Green Consumer Behaviour When Purchasing Products. *Sustain. Dev.* **2010**, *18*, 20–31. [CrossRef]
60. Vermeir, I.; Verbeke, W. Sustainable Food Consumption: Exploring the Consumer "Attitude–Behavioral Intention" Gap. *J. Agric. Environ. Ethics* **2006**, *19*, 169–194. [CrossRef]
61. Bamberg, S.; Möser, G. Twenty Years after Hines, Hungerford, and Tomera: A New Meta-Analysis of Psycho-Social Determinants of Pro-Environmental Behaviour. *J. Environ. Psychol.* **2007**, *27*, 14–25. [CrossRef]
62. Denton, A. Voices for Environmental Action? Analyzing Narrative in Environmental Governance Networks in the Pacific Islands. *Glob. Environ. Change* **2017**, *43*, 62–71. [CrossRef]
63. Dunlap, R.E.; Van Liere, K.D.; Mertig, A.G.; Jones, R.E. New Trends in Measuring Environmental Attitudes: Measuring Endorsement of the New Ecological Paradigm: A Revised Nep Scale. *J. Soc. Issues* **2000**, *56*, 425–442. [CrossRef]
64. Aldagari, S.; Kabir, S.F.; Fini, E.H. Investigating Aging Properties of Bitumen Modified with Polyethylene-Terephthalate Waste Plastic. *Resour. Conserv. Recycl.* **2021**, *173*, 105687. [CrossRef]
65. Alandia, G.; Rodriguez, J.P.; Jacobsen, S.E.; Bazile, D.; Condori, B. Global Expansion of Quinoa and Challenges for the Andean Region. *Glob. Food Secur.* **2020**, *26*, 100429. [CrossRef]
66. Canova, L.; Bobbio, A.; Benincà, A.; Manganelli, A.M. Predicting Food Waste Avoidance: Analysis of an Extended Model of the Theory of Planned Behavior and of Relevant Beliefs. *Ecol Food Nutr.* **2024**, *63*, 539–563. [CrossRef]
67. Giordono, L.S.; Flora, J.; Zanocco, C.; Boudet, H. Food Practice Lifestyles: Identification and Implications for Energy Sustainability. *Int. J. Environ. Res. Public Health* **2022**, *19*, 5638. [CrossRef]
68. Fujiwara, N. The Approach of Social Practice Theory Towards Sustainable Food Consumption: An Analysis of a Questionnaire Survey on Purchasing Behavior. *J. Environ. Sociol.* **2020**, *26*, 95–110.
69. Jiang, Z.; Xu, S.; Hua, Y.; Li, M.; Tang, L.; Lin, Z. Impact of Knowledge-Attitude-Practice Model Nursing Based on Reproductive Health Education on Patients with Reproductive Tract Infections. *Eur. J. Obstet. Gynecol. Reprod. Biol.* **2025**, *313*, 114593. [CrossRef]



https://doi.org/10.3390/su18020686

sustainability

MDPI

*Article*

# Decoding Food Waste: Sociodemographic Determinants in Polish Households for Achieving Sustainable Consumption Goals

Agnieszka Bem [1], Paulina Ucieklak-Jeż [2], Marek Szajt [3,*] and Paweł Prędkiewicz [4]

1 Department of Corporate and Public Finance, Wroclaw University of Economics and Business, ul. Komandorska 118/120, 53-345 Wroclaw, Poland; agnieszka.bem@ue.wroc.pl

2 Department of Economics and Finance, Faculty of Law and Economics, Jan Długosz University in Częstochowa, Waszyngtona 4/8, 42-217 Częstochowa, Poland; p.ucieklak-jez@ujd.edu.pl

3 Department of Econometrics and Statistics, Faculty of Management, Czestochowa University of Technology, Dąbrowskiego 69, 42-201 Częstochowa, Poland

4 Department of Finance, Faculty of Economics and Finance, Wroclaw University of Economics and Business, ul. Komandorska 118/120, 53-345 Wroclaw, Poland; pawel.predkiewicz@ue.wroc.pl

* Correspondence: marek.szajt@pcz.pl

**Abstract**

Household food waste remains a significant barrier to sustainable consumption goals. This study investigates the impact of sociodemographic characteristics influence self-reported food waste levels in Poland. A cross-sectional CAWI survey (N = 1000), based on the HFSSM, was conducted among adults responsible for purchasing household food. Associations between food waste and structural factors were analysed using $\chi^2$ tests, Spearman rank correlations, and Mann–Whitney U tests. Age appears as the strongest determinant: younger respondents consistently report higher food waste, while older adults indicate markedly lower levels. Household composition is equally important—the presence and number of children significantly increase waste. Economic status is also relevant: pensioners and disability-benefit recipients report substantially less waste than employed and self-employed individuals. Net household income shows no significant effect. Education does not necessarily reduce food waste; in some comparisons, higher-educated respondents report slightly higher levels, suggesting that formal education does not automatically translate into effective food management routines. The findings highlight that food waste is primarily influenced by life stage, household structure, and daily habits rather than income or education. Interventions should focus on younger adults and families with children, emphasising practical skills such as meal planning, inventory management, appropriate storage, and the use of leftovers.

**Keywords:** household food waste; sociodemographic determinants; sustainable consumption; SDG 12.3; HFSSM survey

## 1. Introduction

Food waste remains one of the most pressing challenges for sustainable food systems, as it depletes limited resources (land, water, and energy), causes avoidable greenhouse gas emissions and biodiversity loss, and increases the social and economic burdens on households. SDG 12.3 aims for countries to halve per-capita food waste at the retail and consumer levels by 2030. According to the latest Food Waste Index 2024, around 1.05 billion tonnes of food were wasted in 2022 (roughly 19% of food available at the consumption stage), with households responsible for about 60% of this total [1,2].

Research in behavioural economics and sustainability science shows that the amount of food waste is not simply related to prices or income. Instead, it is shaped by bounded rationality [3], loss–gain asymmetry [4], and the attitudinal, normative, and control-related elements of the Theory of Planned Behaviour [5]. On a practical level, the main factors include habits around meal and shopping planning, stock management, proper storage, the use of leftovers, individual skills, perceived behavioural control, and social norms—both descriptive and injunctive. Numerous survey-based studies from various countries consistently demonstrate that practices such as planning purchases, monitoring household inventories, and ensuring adequate storage are strongly linked to lower levels of food waste [6–8].

While behavioural theories are commonly used to explain pro-environmental behaviours, including household food waste, this study does not aim to test specific behavioural models. Instead, concepts such as everyday routines, life stage, and social norms are used as interpretative lenses to contextualise sociodemographic patterns without overstating causal or theoretical claims.

This research contributes to the sustainable consumption and production (SCP) agenda by systematically distinguishing between (a) structural determinants—objective household characteristics and their context (e.g., household size and composition, age, education, income, place of residence) [9].

We pose the following research question: Which sociodemographic and household-structure characteristics meaningfully differentiate self-reported household food waste in Poland, and which commonly assumed determinants—such as income and education—do not?

The specific objectives are to: (1) determine the extent to which household characteristics (age, structure, income sources, economic status) differentiate declared levels of food waste; (2) assess which structural and economic factors show stable associations with food-waste intensity; (3) verify hypotheses concerning the relationships between age, education, economic status and household composition and the level of food waste; (4) provide an empirical basis for designing precisely targeted consumer-level interventions aimed at reducing food waste.

The survey instrument was based on the U.S. Household Food Security Survey Module (HFSSM), expanded to include a socio-economic questionnaire and a self-assessment module on food waste, thereby enhancing the comparability and replicability of the measurement. The level of food waste was assessed subjectively, and the analysis employed the $\chi^2$ test of independence with Yates' correction, Yule's $\varphi$ coefficient, Spearman's rank correlation coefficient, and the Mann-Whitney U test.

At the same time, we recognise the inherent limitations of self-reported measures: survey responses are vulnerable to recall bias and social desirability effects. For this reason, we depend on validated instruments and established measurement guidelines. The HFWQ (Household Food Waste Questionnaire) allows for comparability across households and reduces some measurement error; however, the literature shows systematic differences between methods (diaries, direct weighing, waste-composition analysis), with diary- and survey-based approaches usually underestimating waste compared to weighing and WCA [3–5].

This study contributes to science in three areas:

(1) Precise verification of sociodemographic determinants—while the literature increasingly stresses that habits and competencies drive food-waste behaviours, the findings demonstrate that certain structural characteristics—especially age and household composition—remain statistically robust predictors even under conservative non-parametric tests.
(2) Reassessment of the role of education—the study challenges a widely held assumption that higher education reduces food waste. The results show no protective effect of

https://doi.org/10.3390/su18020880

educational attainment; in some comparisons, more highly educated respondents report slightly higher waste levels. This shifts the emphasis from formal education to practical skills and everyday routines.

(3) Clear evidence for targeted interventions—the analysis identifies high-risk groups—young adults, working households and families with children. These findings provide a strong empirical basis for designing tailored, non-generic interventions within policies aimed at limiting household food waste in line with SDG 12.3.

The novelty of this study lies not in proposing new behavioural mechanisms, but in embedding robust sociodemographic evidence on household food waste within the specific institutional and policy context of Poland—still weakly represented in Central and Eastern European research—using a large, nationally and quota-based sample of 1000 adults that enables systematic comparison with findings from countries such as the Czech Republic, Romania, and Lithuania.

By situating sociodemographic determinants of household food waste within the Polish institutional context, the study also provides policy-relevant insights that may support the design and targeting of food-waste reduction measures, particularly those addressing households at specific life stages.

## 2. Literature Review

### 2.1. Scale and Consequences of the Phenomenon

The literature consistently emphasises the significant role played by consumers in the overall flow of wasted food, along with its economic, environmental, and social impacts. Aschemann-Witzel et al. [6] highlight various demand-side drivers and suggest potential strategies for reducing household food waste. In contrast, the review by Dou and Toth [7] emphasises the considerable variability of measurement indicators and the methodological difficulties that hinder cross-study comparisons.

The Sustainable Development Goals reaffirm the need to halve food waste by 2030 [10]. According to the latest Food Waste Index, around 19% of food available at the consumer level (retail, food service, households) was discarded in 2022, amounting to roughly 79 kg per person annually [1]. Simultaneously, approximately 733 million people suffered from hunger in 2023, vividly highlighting the social cost of coexisting with undernutrition and high levels of waste [11]. From a climate perspective, food loss and waste are estimated to contribute about 8–10% of global anthropogenic greenhouse gas emissions [8].

In 2022, around 59 million tonnes of food were wasted in the European Union, averaging 132 kg per person. Approximately 54% of this originated from households, with a per capita amount of roughly 72 kg. The remaining proportions were processing (19%), food services (11%), retail (8%), and primary production (8%) [12]. These figures underscore that interventions at the consumer level—within households and the food service sector—are crucial for reducing emissions, alleviating resource pressures, and lowering public costs. The extent of the issue and the speed of change differ across countries.

In France, 9.4 million tonnes of food waste were generated in 2022, of which 4 million tonnes were edible fractions, equating to 58 kg per person annually. Between 2016 and 2020, a decrease of approximately 10% was observed [13]. In Germany, food waste in 2022 was estimated at around 10.8 million tonnes, with households accounting for a significant portion [14]. Analyses by the Thünen Institute confirm that households account for about half of the national food waste stream [15]. In the Netherlands, 2.35 million tonnes were reported in 2022, equivalent to 134 kg per person, and saw a decrease of roughly 7% compared to the previous year [16]. In Spain, households generated approximately 1.17 million tonnes of waste in 2022, marking a 6.1% year-on-year decline [17]. In Sweden, total food waste was estimated at approximately 1.42 million tonnes, or 135 kg per person. Recent

https://doi.org/10.3390/su18020880

data revisions have altered sectoral shares and influenced new public policy priorities [18]. In Denmark, total losses amount to approximately 873,000 tonnes annually, with significant contributions from industry and households [19,20]. Ireland generated approximately 750,000 tonnes of waste in 2022 and around 835 tonnes in 2023—roughly 162 kg per person [21,22]. In Portugal, food waste amounted to approximately 1.93 million tonnes in 2022, equivalent to 184 kg per person. National reduction targets for 2030 are being revised based on ongoing monitoring [23]. In the United Kingdom, households waste approximately 6.0 million tonnes of food each year, of which 4.4 million tonnes are edible. This equates to around 88 kg per person and about 16 million tonnes of $CO_2$-equivalent emissions. Interventions targeting consumers in this area provide the most significant benefits for both climate and economic outcomes [24]. WRAP guidelines on date labelling and packaging for meat, bakery products, and fruit and vegetables address the main drivers of household food waste, leading to measurable reductions in the amount of edible food discarded [25].

The most comprehensive estimates for Poland, developed within the PROM project and used in audit reports of the Supreme Audit Office (NIK), indicate that approximately 4.84 million tonnes of food are wasted annually across the entire food supply chain. The distribution across stages is as follows: households, 60% (approximately 2.92 million tonnes); primary production, 15.5%; processing, 15.6%; retail, 6.96%; food services, 1.17%; and transport, 0.65%. These figures clearly show that households constitute the primary area for intervention, whereas waste in retail and food services is smaller in scale but relatively more amenable to regulatory and organisational measures [26–28].

Since 2019, Poland has enacted the Act on Counteracting Food Waste, which introduced, among other measures, an obligation for retailers to establish agreements with non-governmental organisations for the free redistribution of surplus food, a fee payable to the Provincial Funds for Environmental Protection and Water Management (WFOŚiGW) when food is not donated, and compulsory annual reporting [29]. The Federation of Polish Food Banks plays a vital role in this system, bringing together 31 food banks and overseeing the logistics of short-term food redistribution to social aid organisations. It acts as the primary "rescue" channel specified in the Act [27].

The environmental and economic consequences of FLW are well documented. The production of food that ultimately remains unconsumed occupies approximately 1.4 billion hectares of agricultural land each year—around 28% of the global agricultural area. It uses an estimated 250 $km^3$ of freshwater and generates roughly 3.3 billion tonnes of $CO_2$ [30]. Comprehensive full-cost assessments place the combined economic, environmental and social burdens at around USD 2.6 trillion annually, underscoring the apparent cost-effectiveness of preventive measures [31].

At the EU level, work is underway to strengthen binding 2030 reduction targets: approximately 10% in processing and manufacturing, and around 30% per capita in retail and at the consumer level—targets currently analysed by the European Environment Agency [13]. National policies, such as Italy's "Gadda Law," facilitate food donation and reduce waste flows and associated emissions [32].

Taken together, available data and policy instruments indicate that food waste is declining in many countries and across key segments of the food supply chain. The decrease is evident at the household level, although the pace and stability of reductions differ across countries and sectors [12,15,24].

### 2.2. Behavioural Perspective: Habits, Emotions and Norms Across Sociodemographic Groups

Habits, emotions, social norms, and perceived behavioural control influence consumer behaviour within households. Emotional reactions (e.g., guilt) coexist with shopping routines and planning skills, as shown in studies that combine affective and habitual

https://doi.org/10.3390/su18020880

factors in food-related decision-making [33]. In line with this view, Stancu, Haugaard, and Lähteenmäki identify two pathways to food waste: an impulsive pathway and a planning-related pathway [34]. Practically, older adults tend to have more established routines of planning and food management (such as shopping lists, "kitchen FIFO", and leftover utilisation) and higher kitchen self-efficacy, leading to lower self-reported waste levels compared to younger householders [35–38]. An exception is Japan, where per-capita losses increase with age, partly due to a higher proportion of highly perishable items in seniors' baskets [39].

Under limited information and time, consumers rely on heuristics and routines rather than complete optimisation [40], which encourages "stocking up" and large weekend shops—cognitively efficient but not always economically optimal. Prospect theory explains why fear of missing out on a deal (e.g., "buy one, get one free") can outweigh the anticipated cost of discarding surplus food, worsening household finances [41]. Based on the Theory of Planned Behaviour, intentions to reduce waste depend on attitudes, norms and perceived control; when planning, portioning and storage skills are weak, good intentions do not translate into action [34,42,43].

In the Becker–Stigler framework of consumption capital, habits develop through repeated actions that reduce cognitive and transactional costs, stabilising behaviour [44]. In the household context, this results in a "habit pathway": meal planning, shopping lists, proper storage, and systematic use of leftovers. Empirical studies demonstrate that the strength and completeness of this pathway differentiate "low wasters" from "high wasters" [35,45]. Weighing-based and diary studies from Central Europe indicate that the most vulnerable categories are those affected by freshness and portion size, including bread, ready-to-eat meals, fruit, and vegetables [46–49].

From a well-being economics perspective, emotional states during shopping and cooking are highly significant. In stores, positive emotions (such as pleasure, a sense of abundance or security) tend to encourage over-purchasing ("stock gives control"), while in kitchens, negative emotions (like disgust and hygiene-related anxiety) lead to "just-in-case" discarding when uncertainty arises [44,45,47]. Perceived safety risks and strong preferences for freshness strongly influence disposal choices; confusion surrounding date labels ("best before" vs. "use by") also heightens discard rates [48].

Incorporating social norms and suitable information framing (gain/loss) reduces risk aversion and lowers the propensity to waste food. The most effective messages combine practical guidance (such as date checking and storage) with constructive emotions (pride and care) rather than relying solely on guilt [49,50]. At the same time, the "good provider norm"—valuing abundance—can counteract savings-oriented behaviour by promoting oversized purchases and portions [51].

Lifestyle plays a central role. Frugality operates through planning habits, the use of leftovers, storage competencies, and perceived control, reinforced by norms that favour thriftiness [34,35,51]. A value-conscious lifestyle (closely aligned with frugality) is consistently associated with lower waste. In contrast, convenience-oriented and deal-seeking profiles are more likely to result in surpluses and mismatches in package sizes [6]. In practice, "frugality" takes the everyday form of four habits: meal planning, shopping lists, proper storage and consistent use of leftovers—precisely the practices most strongly associated with reduced waste in WRAP data [24,52].

Polish segmentation studies confirm that households with disciplined practices waste less, and that choosing smaller portions or packages, as well as a better understanding of date labels, further reduces risk, especially among younger consumers [48,53]. Findings from Central and Eastern Europe are consistent: in Lithuania, planning and knowledge of date labels have protective effects, reducing the impact of simple demographic patterns [54].

https://doi.org/10.3390/su18020880

In Romania, groups with "careless" routines waste more, suggesting that knowledge alone is not enough without supportive norms and habits [55]. Segmentation studies also show that waste patterns vary between consumer types (e.g., hedonistic vs. pro-environmental), indicating the need for customised communication and interventions [56]. Lastly, exceeding use-by dates and stockpiling are linked to poor planning and household inventory management; therefore, educational tools—such as shopping lists, meal planners, and "eat-first" systems—directly target the key mechanisms behind food waste [57]. The most effective results are achieved when normative components are combined with habit formation and practical support, such as pre-shopping reminders or shopping list apps connected to home inventories [24,58].

To summarise, the evidence suggests that habits, competencies, and norms, interacting with emotions and bounded rationality, provide the most robust explanation for differences in household food waste levels. By contrast, age, education or income tend to exert predominantly indirect effects, operating through these behavioural channels.

### 2.3. Sociodemographic Determinants of Food Waste—A Review of the Evidence

Understanding how sociodemographic characteristics shape household food-waste behaviours is essential for designing targeted and effective interventions. Existing research consistently shows that individual attributes (such as age, education, and income), household composition (notably the presence and number of children), and behavioural patterns related to lifestyle interact in complex ways. These factors influence planning routines, storage practices, perceptions of food safety, and normative beliefs about abundance and thrift. The following section synthesises key empirical findings from Europe, Asia, and North America, highlighting both cross-country patterns and context-specific deviations.

Cross-sectional studies across the EU-27 indicate that age is negatively associated with food waste frequency [36]. Similar trends have been observed in Switzerland and Denmark, where older individuals report wasting less food, a difference attributed to ingrained norms, established habits, and better household management skills [34,35]. In the United Kingdom, respondents aged 55–64 report the highest levels of waste avoidance [37], while WRAP analyses consistently associate lower waste levels with meal planning, shopping lists, proper storage, and the use of leftovers [24]. In the United States, older adults also report less food waste, often driven by economic considerations [47]. Japan stands out as an exception, with per-capita waste increasing with age, partly due to a higher proportion of perishable products in seniors' food baskets [39].

The literature highlights the importance of "habit capital" accumulated with age and reinforced by more stable daily routines, which promotes frugal and predictable household practices [38,59]. Education is not seen as an end in itself but as a means of enhancing skills—such as understanding "use by" versus "best before" labels, planning abilities, and perceived behavioural control (PBC)—which together help to reduce waste [34–36].

In Poland, households that plan meals, store food properly, and utilise leftovers tend to have higher levels of education [48,53]. WRAP findings [24,25] similarly indicate that these practices reduce waste and are more common among individuals with stronger skills. Conversely, segmentation studies in Romania identify, alongside a typically less educated "Careless" segment, an "Ignorant" segment—well-educated individuals who still waste food due to weak habits and poorly internalised norms [55]. Cross-segment comparisons (e.g., hedonic vs. pro-environmental orientations) confirm the need for targeted interventions [56]. Studies of young adults in Spain and Italy also reveal distinct behavioural patterns in this age group [60], aligning with broader conclusions regarding the combination of structural and competence-based factors [43].

https://doi.org/10.3390/su18020880

Household structure plays an equally important role. The presence of children—particularly more than one child—typically increases food waste. A national study in China (N = 1746) found that having "a child or a senior" in the household was positively associated with both waste frequency and the share of wasted food in total expenditure. The effect was strongest for perishable products and snacks, driven by preference volatility and precautionary purchasing [61]. In Japan (N = 5240), having children also increased waste, especially in large cities (faster pace of life, meal-planning constraints). Crucially, reductions were linked to specific practices (action checklists, inventory management), not merely to higher levels of "knowledge" [62].

A Spanish study of 806 families found that planning—particularly shopping lists—was the strongest predictor of lower household waste. Children's age differentiated waste mechanisms: families with younger children reported more plate leftovers, whereas families with older children reported more over-purchasing [60]. Qualitative research supports these findings: parents report children's pressure (requests, pickiness) and norms of abundance ("a good parent provides choice"), which encourage overbuying. Possible responses include offering smaller package sizes and developing practices for reusing leftovers [63]. Evidence from Central Europe corroborates this view: in a diary-based study of over 400 Czech households, the average waste was 2.6 kg per household per week (1.1 kg per capita), with planning and storage habits identified as key factors [64].

In Poland, a representative study (N = 500; 7-day diaries and CAPI) identified the presence and number of children as significant predictors of waste, alongside household size, income, and characteristics of the main cook [65]. In Canada (N = 68; waste weighing plus survey), families with children generated higher total waste, although not always more waste per person (a scale effect), and exhibited greater diversity in food waste types [66]. In rural China, household size is positively correlated with waste (fractional models), primarily due to difficulties in portioning and planning in large families [67]. Further studies indicate that food safety concerns and dietary preference complexity are stronger in larger households (often families with children), which raises the likelihood of discarding food "close to the date" and highlights the importance of label education and storage practices [68].

Income effects are typically indirect and conditional. When mediators—planning, label knowledge, PBC, and norms—are added, the direct effect of socioeconomic status weakens or disappears [54,61,62]. Income operates through habits, competencies, and norms, with its influence shaped by lifestyle orientation (frugality vs. convenience/promotion-seeking), household structure, and food category [24,43,58].

Broader national and segmentation evidence in Poland suggests that younger decision-makers more frequently prefer convenience, freshness, and large packages, thereby increasing the risk of waste [48]. Segmentation analyses similarly divide households into "high" and "low wasters" based on planning, storage, and leftover practices [65]. Evidence from Latvia and Romania aligns with these results: impulsive buying and avoidance of leftovers correlate with higher waste, especially among younger consumers [55,69–71]. Consumer orientations further differentiate outcomes: value-conscious styles (akin to frugality) are associated with lower waste, whereas price- and convenience-oriented styles (including promotion-driven buying) increase waste due to excessive package sizes and over-purchasing [24,52,72] findings reinforce this framework.

Finally, methodological considerations are vital. Differentiating between "low" and "high wasters" is most effective when diary techniques are combined with waste weighing. Diaries record behavioural processes (e.g., planning, transitions from purchase to storage), but are susceptible to selective reporting and observer effects; weighing offers accurate quantity measurements but lacks contextual details. Their combination enables the precise

https://doi.org/10.3390/su18020880

identification of key points along the waste pathway and facilitates targeted, practical interventions where errors most frequently occur [46,53,54,64,73].

From an implementation standpoint, the evidence points to the effectiveness of weekly menu templates, pre-shopping reminders, inventory-linked shopping-list application product-specific storage practices (e.g., freezing bread, correct refrigerator zoning), and normalising leftover reuse—supported by clear label communication ("best before" does not mean "use by") and norm-anchored messaging [24,37,38,49,50,57,64].

Taken together, the reviewed literature demonstrates that household food waste is shaped by a combination of sociodemographic characteristics, everyday routines, and contextual factors, yet the empirical evidence remains fragmented and often country-specific. In particular, studies conducted in Central and Eastern Europe are still relatively limited and frequently rely on small samples or descriptive approaches, providing mixed results regarding the role of variables such as age, education, and household composition. Moreover, existing research rarely integrates sociodemographic analyses with an explicit consideration of national institutional contexts, despite their relevance for shaping household practices. Against this background, the present study addresses this gap by systematically examining the associations between key sociodemographic variables and household food waste in Poland using a large nationwide survey. The hypotheses formulated in the Section 3 are directly derived from these unresolved findings in the literature and aim to test whether commonly assumed relationships—particularly those related to age, household structure, and education—hold in the Polish context.

## 3. Materials and Methods

### 3.1. Data

Understanding household food security and food-waste behaviours requires a measurement strategy that is both standardised and sensitive to the demographic and economic heterogeneity of households. To this end, the study employed a structured survey based on the HFSSM, augmented with modules that captured socio-demographic characteristics, health self-assessment, and self-reported food waste intensity. The survey instrument was translated into Polish and slightly refined following a pilot study conducted prior to the main survey. The pilot phase was used to assess the internal consistency and basic psychometric properties of the instrument, including reliability (Cronbach's $\alpha$).

The data were collected through an online questionnaire administered to a large, nationwide sample of 1000 adults aligned with the age and gender structure of the Polish population.

The study employed a non-probability, purposive quota sampling design implemented by an external research and development centre (Biostat, Rybnik–Poland) specialising in nationwide socio-economic surveys. Quota variables were defined by age and gender to ensure consistency with the demographic structure of the adult population in Poland. Respondents were recruited from the contractor's established respondent panels, with participation being voluntary and anonymous. No probabilistic random sampling was applied, and checks of sample alignment were limited to age and gender distributions.

Eligibility was determined through a screening question asking whether the respondent was personally responsible for food shopping in their household. Only individuals who answered affirmatively proceeded to the full questionnaire. The final sample comprised 1000 individuals responsible for food purchasing (65% women, 35% men). The study employed a cross-sectional design, utilising a Computer-Assisted Web Interviewing (CAWI) survey conducted in December 2022. Its core component was the HFSSM, expanded to include a socio-demographic module and items measuring self-rated health and perceived levels of household food waste.

https://doi.org/10.3390/su18020880

The questionnaire comprised 27 items numbered 0–26, of which 16 referred to experiences within the past 12 months. It included seven conditional questions administered exclusively in households with children. The introductory screening item verified whether the respondent was responsible for purchasing household food. The demographic section consisted of three categorical variables: gender, age group and educational level. The economic status block captured subjective evaluations of lifestyle/financial situation, the main source of income, and net household income. Household composition was assessed with a single numeric question that asked about the number of household members, including adult children (18 years and older) and seniors aged 65 and above. The section on health and dietary practices included two measures: self-rated health and perceived intensity of food waste.

Household food-waste levels were measured using the item: "How much food is wasted in your household?" with response options: 1—very little; 2—little; 3—moderate amount; 4—a lot; 5—very much.

Self-reported measures of household food waste are commonly used in large-scale survey research, as they allow for the efficient collection of comparable data across diverse population groups. Although such measures do not provide precise quantitative estimates, previous studies show that they are suitable for identifying relative differences and patterns in food-waste behaviour, particularly when diary- or weighing-based methods are not feasible. Accordingly, the self-reported indicator applied in this study is intended to capture comparative variation between households rather than exact levels of food waste.

Explanatory variables included age, gender, education, household size (including the number of children and adults aged 65 and above), net income and its source, and lifestyle indicators (Table 1).

**Table 1.** Variables—codes and types.

| Variable | Code | Type |
|---|---|---|
| Level of food waste | FW | Ordinal (1–5) |
| Age | M2 | Ordinal |
| Educational attainment | M3 | Ordinal |
| Lifestyle/perceived financial situation | M4 | Ordinal |
| Main source of income | M5 | Nominal/ordinal |
| Net household income | M6 | Ordinal |
| Household size | M7 | Count/ordinal |
| Presence of children in household (yes/no) | MA8 | Binary |
| Number of children in household | M8 | Count/ordinal |

Source: own study.

### 3.2. Statistical Procedure

Based on literature review we formulated the following hypotheses:

**H1.** *Age and Food Waste—Older adults are expected to report lower levels of food waste than younger individuals (a negative association between age and food waste);*

**H2.** *Education and Food Waste—higher educational attainment is expected to be associated with lower levels of household food waste (a negative association between education and food waste);*

**H3.** *Economic Status and Food Waste—individuals whose primary income derives from pensions, disability benefits or social-assistance transfers are expected to waste less food than those whose income comes from employment or self-employment (a negative association);*

https://doi.org/10.3390/su18020880

**H4.** *Household Structure and Food Waste—larger households are expected to generate more food waste, and the presence of children—particularly a higher number of children—is expected to increase the likelihood and volume of food discarded (a positive association).*

To examine associations between household food-waste levels and sociodemographic characteristics, the analysis employed a set of non-parametric statistical techniques suitable for ordinal data and variables measured on categorical or non-continuous scales. The methodological approach is centred on three complementary procedures:

Chi-square test of independence, supplemented with Yates' continuity correction for 2 × 2 contingency tables, was used to assess whether the distribution of responses on the food-waste scale differed across categories of key sociodemographic variables. Effect sizes for these associations were expressed using the φ-Yule coefficient, which provides a standardised measure of the strength of the relationship for binary or near-binary structures.

Spearman's rank-order correlation was applied to evaluate monotonic relationships between ordinal variables (e.g., age group, household size, educational attainment) and the ordinal food-waste measure. This method is appropriate when assumptions of normality or equal variance are not met, and it is widely recommended in research involving self-reported behavioural scales.

Mann–Whitney U test was used to compare distributions of the food-waste variable across selected categorical groups (e.g., employment types, educational categories, or age brackets). As a rank-based alternative to the independent-samples *t*-test, the Mann–Whitney procedure is suitable for non-normally distributed dependent variables and for Likert-type scales.

## 4. Results

At the beginning, chi-square independence tests with Yates correction and corresponding φ coefficients were carried out, assessing the relationship between selected socio-demographic characteristics and self-assessment of the level of food waste in households. The results are shown for the full sample as well as separately for women and men.

For the overall sample (Table 2), only age (M2) shows a statistically significant relationship with the reported level of food waste ($\varphi = 0.108$; $\chi^2 = 11.686$; $p = 0.001$). Although the effect size is small ($\varphi \approx 0.108$), the result suggests that food waste levels differ significantly across age groups, with younger respondents tending to report higher waste levels. For all other variables—education (M3), lifestyle/financial situation (M4), main income source (M5), net household income (M6), and number of children (M8)—the *p*-values exceed 0.05, suggesting no statistically significant associations in the total sample. When the analysis is stratified by gender, the pattern for women reflects the overall findings. Age again emerges as the only significant correlate of food-waste declarations ($\varphi = 0.117$; $\chi^2 = 9.245$; $p = 0.002$), with a similarly small effect size. A weak, non-significant trend is observed for net income (M6) among women ($p = 0.090$), but it does not meet the conventional threshold for significance. No other variable shows a meaningful association. Among men, none of the examined characteristics display a statistically significant relationship with food-waste declarations. Even age ($\varphi = 0.087$; $\chi^2 = 2.672$; $p = 0.102$), which was significant in the full sample and among women, does not reach significance here.

Across all respondents, and specifically among women, age is the only factor significantly associated with self-reported food waste, although the strength of this relationship is relatively weak. For men, no sociodemographic variable included in the table shows a statistically significant link with food waste levels. This pattern suggests that age-related differences in food-management routines may be more pronounced or more consistently

https://doi.org/10.3390/su18020880

reported among women than among men, while other structural factors (education, income, household composition) do not differentiate food-waste behaviour in this dataset.

**Table 2.** Gender differences in declared household food-waste levels.

| Statistic | M2 | M3 | M4 | M5 | M6 | M8 |
|---|---|---|---|---|---|---|
| | | | Overall | | | |
| φ | 0.108 | −0.001 | 0.048 | −0.054 | 0.036 | −0.036 |
| $\chi^2$ | 11.686 | 0.001 | 2.317 | 2.483 | 1.080 | 1.091 |
| *p*-value | 0.001 | 0.973 | 0.128 | 0.115 | 0.299 | 0.296 |
| | | | Women | | | |
| φ | 0.117 | 0.036 | 0.045 | −0.057 | 0.071 | −0.035 |
| $\chi^2$ | 9.245 | 0.719 | 1.138 | 1.838 | 2.876 | 0.690 |
| *p*-value | 0.002 | 0.397 | 0.286 | 0.175 | 0.090 | 0.406 |
| | | | Men | | | |
| φ | 0.087 | −0.075 | 0.060 | −0.046 | −0.030 | −0.038 |
| $\chi^2$ | 2.672 | 1.718 | 1.283 | 0.603 | 0.282 | 0.427 |
| *p*-value | 0.102 | 0.190 | 0.258 | 0.438 | 0.596 | 0.513 |

Source: own study.

The results show that younger respondents report higher food waste, while those with higher education or more children tend to waste more. In contrast, pensioners and households relying on social benefits declare lower waste levels. Net income itself was not related to food waste, but household composition—especially the number of children—was the strongest predictor of increased waste.

Table 3 presents Spearman's rank-order correlations examining the relationships between selected socio-economic characteristics and the self-reported level of household food waste in Poland.

**Table 3.** Spearman's Rank Correlations: Associations Between Socio-Economic Characteristics and Declared Food-Waste Levels in Poland.

| | R | $t_{(n-2)}$ | *p* |
|---|---|---|---|
| M2 | −0.223818 | −7.25469 | 0.0000 |
| M3 | 0.081304 | 2.57703 | 0.0100 |
| M4 | −0.070563 | −2.23473 | 0.0254 |
| M5 | −0.121532 | −3.86800 | 0.0001 |
| M6 | 0.046767 | 1.47903 | 0.1391 |
| M7 | 0.173854 | 5.57717 | 0.0000 |
| MA8 | −0.203976 | −6.58223 | 0.0000 |
| M8 | 0.205276 | 6.62601 | 0.0000 |

Source: own study.

Age showed a moderate negative correlation (R = −0.224, $p < 0.001$), indicating that younger respondents reported higher levels of food waste. Educational attainment was positively associated with food waste (R = 0.081, $p < 0.010$), while lifestyle/financial self-evaluation showed a weak negative association (R = −0.071, $p < 0.026$).

A significant negative correlation was also observed for the main source of household income (R = −0.121, $p < 0.001$), indicating lower waste among respondents who relied on pensions or social benefits. Net household income did not correlate significantly with food waste.

Household composition demonstrated clear patterns: the presence of children (M8A) correlated negatively with the food-waste scale (R = −0.204, $p < 0.001$), whereas the

https://doi.org/10.3390/su18020880

number of children showed the strongest positive correlation in the dataset (R = 0.205, $p < 0.001$), indicating that food-waste levels increased along with household size, driven by childcare needs. This pattern is not uncommon: the mere presence of children differentiates households in one direction. But among households with children, the number of children shifts the relationship in the opposite direction. This reflects a combination of a threshold effect (the presence of children alters the baseline level of food waste) and an intensity effect (food waste increases in proportion to the number of children). Together. these mechanisms suggest that initial household constraints introduced by childcare may operate differently from the cumulative pressures associated with larger family size.

To examine whether the declared level of household food waste differed across key socio-economic groups, a series of Mann–Whitney U tests was conducted. This non-parametric procedure is appropriate for comparing ordinal self-assessment scales across two independent groups.

The Mann–Whitney U test results demonstrate consistent and statistically significant age-related differences in self-reported food-waste levels (Table 4). In all comparisons involving the youngest group (25–34), this group reported significantly higher waste than respondents aged 35–44, 45–54, 55–64 and 65+ (Z ranging from 2.268 to 7.221; all $p < 0.050$).

**Table 4.** Mann–Whitney U test values for the self-reported level of household food waste with respect to respondents' age.

| Age | Z | *p* |
|---|---|---|
| 25–34/35–44 | 2.268 | 0.023 |
| 25–34/45–54 | 5.139 | 0.000 |
| 25–34/55–64 | 7.221 | 0.000 |
| 25–34/65+ | 5.528 | 0.000 |
| 35–44/45–54 | 3.312 | 0.001 |
| 35–44/55–64 | 5.804 | 0.000 |
| 35–44/65+ | 3.672 | 0.000 |
| 45–54/55–64 | 2.657 | 0.008 |
| 45–54/65+ | 0.189 | 0.850 |
| 55–64/65+ | −2.583 | 0.010 |

Source: own study.

Similarly, respondents aged 35–44 reported higher waste generation than those aged 45–54, 55–64, and 65+ (Z = 3.312–5.804; $p < 0.010$). The same pattern—higher waste among the younger cohort—was observed when comparing the 45–54 group with those aged 55–64 (Z = 2.657; $p = 0.008$). No significant difference emerged between the 45–54 and 65+ groups ($p = 0.850$), suggesting convergence in behaviour in these later life stages.

A significant contrast was observed between respondents aged 55–64 and those aged 65+ (Z = −2.583; $p = 0.010$), indicating that individuals aged 65+ reported even lower levels of waste.

Overall, the results reveal a clear gradient: food waste declines with age, with the largest behavioural gap observed between the youngest adults and all older groups. This is fully consistent with earlier findings on the role of habit stability, frugality, and more predictable daily routines among older respondents.

The Mann–Whitney U test results also indicate statistically significant differences in the self-reported level of household food waste between pensioners/disability pension recipients and respondents from selected income-source groups (Table 5).

A highly significant result was obtained for individuals employed under an employment contract (Z = 4.801; $p < 0.001$). This suggests that, compared with pensioners, respondents in stable wage employment report noticeably higher levels of food waste. The magnitude and significance of this difference point to systematic behavioural or lifestyle

https://doi.org/10.3390/su18020880

distinctions—potentially reflecting time constraints, higher purchasing power, or less stringent planning routines among working adults.

**Table 5.** Mann–Whitney U test values for the self-reported level of household food waste among pensioners/disability pension recipients and respondents from other occupational groups.

| Source of Income | Z | $p$ |
|---|---|---|
| Employment contract | 4.801 | 0.000 |
| Contract of mandate/contract for specific work | 0.918 | 0.359 |
| Self-employment | 2.482 | 0.013 |
| Social welfare benefits | 2.630 | 0.009 |

Source: own study.

A significant difference was also observed for the self-employed (Z = 2.482; $p$ = 0.013). Here as well, food waste levels reported by this group exceed those of pensioners, although the effect is weaker than in the case of standard employment contracts. A plausible explanation may involve irregular working hours, variable income patterns or more fragmented meal-planning practices.

For respondents receiving social welfare benefits, the difference is likewise statistically significant (Z = 2.630; $p$ = 0.009). This finding suggests that, despite constrained budgets, this group reports higher levels of food waste than pensioners. Possible contributors include instability of income, less predictable household needs or limited competencies related to food storage and planned consumption.

In contrast, no statistically significant difference was found for those working under a contract of mandate/contract for specific work (Z = 0.918; $p$ = 0.359). The self-reported waste levels in this group appear comparable to those of pensioners, suggesting that their food-management routines do not differ substantially in this regard.

Overall, the analysis reveals that pensioners and disability pension recipients consistently report lower levels of food waste than most other income-source groups. This pattern aligns with earlier research highlighting the role of established routines, frugality, and more predictable daily rhythms among older adults. The results underscore the heterogeneity of waste-generating behaviours across labour-market groups and point to the importance of tailoring food-waste interventions to the socio-economic context of specific populations.

As expected, households with children reported significantly higher levels of food waste compared with childless households (Z = −6.447; $p$ < 0.001). The results of research on differences in the level of food waste in households depending on the education of respondents are presented in Table 6.

**Table 6.** Mann–Whitney U test values for the self-reported level of household food waste with respect to respondents' education.

| Education | Z | $p$ |
|---|---|---|
| Vocational/Secondary | −1.662 | 0.097 |
| Vocational/Higher | −2.809 | 0.005 |
| Secondary/Higher | −1.954 | 0.051 |

Source: own study.

Finally, the Mann–Whitney U test results reveal that educational attainment differentiates self-reported levels of household food waste only in one comparison. Respondents with higher education declared significantly higher food-waste levels than those with vocational education (Z = −2.809; $p$ = 0.005).

No statistically significant differences were observed between vocational and secondary education ($p$ = 0.097), nor between secondary and higher education ($p$ = 0.051).

https://doi.org/10.3390/su18020880

These borderline values suggest that, although education shows certain tendencies in the correlation analysis, clear group-level differences emerge only between the lowest and highest educational categories.

Overall, the findings suggest that higher-educated respondents tend to report more food waste; however, this effect is not uniformly present across all educational levels.

Table 7 provides a consolidated assessment of hypotheses H1–H4. Three hypotheses (H1, H3, H4) are supported by the empirical results: younger respondents, working adults, and households with children consistently report higher levels of food waste. Only H2 is not supported—higher educational attainment does not reduce food waste and, in some analyses, is associated with slightly higher waste levels. Overall, age, household structure, and economic status emerge as meaningful predictors of food-waste behaviour, while education plays only a marginal role.

**Table 7.** Hypotheses summary.

| Hypothesis | Expected Direction | Empirical Findings | Conclusion |
|---|---|---|---|
| **H1.** *Age and Food Waste—older adults waste less food* | Negative association | Age significantly correlated with lower waste (Spearman R = −0.224, $p < 0.001$); Mann–Whitney tests show younger groups report consistently higher waste; $\chi^2$ significant for total sample | Supported |
| **H2.** *Education and Food Waste—higher education reduces waste* | Negative association | Positive correlation (R = 0.081, $p < 0.010$); significant difference only between vocational vs. higher education, with higher education reporting more waste | Not supported |
| **H3.** *Economic Status and Food Waste—pensioners/benefit recipients waste less* | Negative association | Lower waste among pensioners relative to employed and self-employed (Z = 4.801 and 2.482; $p < 0.050$); negative correlation for income source (R = −0.122, $p < 0.001$) | Supported |
| **H4.** *Household Structure and Food Waste—households with children/larger size waste more* | Positive association | Strongest positive correlation for number of children (R = 0.205, $p < 0.001$); presence of children significantly increases waste (Z = −6.447, $p < 0.001$) | Supported |

Source: own study.

## 5. Discussion

The results of this study confirm the significant influence of demographic and lifestyle factors on household food waste levels. This clearly indicates that educational and preventive measures should be strategically targeted at larger households, particularly those with children. Such initiatives should focus on developing skills related to meal planning, creating rational and needs-based shopping lists, adopting proper food storage practices, and maximising the use of leftovers. Strengthening these skills is likely to help reduce consumer-level food waste over the long term.

The analysis reveals that age is the strongest factor linked to food-waste behaviour. Reported waste levels tend to fall consistently with age, indicating that older adults generally waste less food, while the youngest respondents report the highest waste levels. Educational attainment exhibits a small but positive correlation with food waste; individuals with higher education often report slightly more waste, suggesting that education alone may not always lead to more efficient food management routines.

Lifestyle and perceived financial situation also play a significant role. A more restricted financial situation appears to have a protective effect: respondents who report greater fru-

https://doi.org/10.3390/su18020880

gality and tighter household budgets tend to waste less food. The main source of household income also influences outcomes: lower waste levels are observed among respondents relying on pensions or social assistance benefits compared with those whose primary income comes from employment, contract work, or self-employment. This suggests that both economic stability and purchasing power influence household food management behaviours.

The lack of a significant association between educational attainment and household food waste merits closer consideration. Although higher education is often assumed to enhance environmental awareness and pro-environmental behaviour, the present findings suggest that education alone does not translate into lower levels of food waste. This result is consistent with research emphasising the persistence of everyday routines and habitual practices, which tend to be weakly responsive to information or abstract knowledge. Food-related behaviours are largely embedded in daily time constraints, household organisation, and meal-planning routines, which may limit the behavioural relevance of formal education.

The lack of a significant association between educational attainment and household food waste warrants closer consideration. Although higher education is often assumed to enhance environmental awareness and pro-environmental behaviour, the present findings indicate that education alone does not translate into lower food waste. This result aligns with research emphasising the persistence of everyday routines and habitual practices, which tend to be weakly responsive to information or abstract knowledge. Food-related behaviours are largely embedded in daily time constraints, household organisation, and meal-planning routines, which may limit the behavioural relevance of formal education [33–35].

Moreover, the absence of an educational effect may reflect a well-documented knowledge–action gap in pro-environmental behaviour. While more educated individuals may have greater awareness of sustainability issues, this does not necessarily translate into reduced waste if structural and situational factors—such as household composition, caregiving responsibilities, or time pressure—dominate decision-making. In this sense, education appears to be a less decisive predictor than life stage and household structure, which showed stronger and more consistent associations in the present study.

Importantly, net household income does not show a statistically significant relationship with declared food waste. By contrast, household composition—specifically, the number of children—demonstrates a clear and robust pattern. The number of children is positively and significantly associated with food-waste levels: the more children in the household, the greater the amount of food discarded. This aligns with the broader literature, which links childcare demands, food-preference variability, and planning difficulties to elevated food-waste risks in families with multiple children.

Overall, the results highlight the combined impact of demographic structure, life-stage characteristics, and everyday consumption practices on household food waste, pointing to the importance of tailored, behaviour-focused interventions.

These results are consistent with existing literature. According to CBOS data [38] individuals aged 55 and above report discarding food less frequently in the preceding week. In contrast, a study by Przezbórska-Skobiej et al. [74] found that students declare higher levels of food waste than employed adults—a pattern the authors attribute to shopping routines and leftover management behaviours. Segmentation analyses by Bilska et al. [53] show that Polish households are primarily differentiated by practices (such as meal planning and storage) rather than by demographic traits, whereas Szymkowiak et al. [48] demonstrate that product attributes (including pack size, labelling, and portioning) interact with lifestyle patterns and increase the risk of food waste, particularly among younger consumers. International evidence supports this perspective: in Lithuania, individuals aged 66 and above are more likely to belong to the "low-waster" group, while impulsive

https://doi.org/10.3390/su18020880

purchases and avoidance of leftovers increase waste, and meal planning, combined with label knowledge, exerts a protective effect [54].

In the Czech Republic, survey data indicate that the lowest waste levels are observed among respondents aged 60 and above, as well as among lower-income households [59]. Diary studies conducted in more than 400 households revealed an average of 2.6 kg per household per week (1.1 kg per capita), documenting substantially higher waste levels than self-reports [64]. More recent surveys identify dates, portion sizes, and storage practices as the main drivers of waste, with younger respondents reporting higher levels [46].

In Romania, consumer segmentation identifies distinct clusters that differ in their habits and emotions [55]. Newer findings show that age and behavioural competencies in food management strongly shape waste patterns, with younger individuals more frequently committing planning errors [70].

In Latvia, Beitāne et al.'s [69] large-scale survey (N = 1336) provides raw behavioural data enabling age-differentiated modelling with demographic controls, and the representative RSU dataset (N = 1007) captures awareness, intentions, and actions. Earlier studies similarly indicated higher waste in households with children and lower waste among adult-only households [69,71].

In Hungary, waste-weighing measurements estimate that approximately 68 kg of waste is generated per person annually, of which around 33 kg is avoidable; ready-to-eat meals and bread are the most frequently discarded items [73,75]. Importantly, NÉBIH/Wasteless monitoring indicates ~65.5 kg/person/year and a 24% reduction in "real wastage" compared to 2016. Qualitative interviews highlight the role of personal and social norms, responsibility, and awareness of consequences of the Norm Activation Model/Theory of Planned Behaviour, with clear demographic differentiation [76].

In Slovakia, triangulation using surveys and product weighing reveals discrepancies between self-reports and measured waste, as well as the importance of habits and demographic factors [77]. Additional analyses indicate that age and income differentiate waste levels, while the most frequent causes include spoilage/mould, over-purchasing, and poor storage [78]. Fieldwork further suggests that older rural residents tend to "respect bread," reducing bread waste [79].

The relationship between income and food waste is not universal; it depends on context and measurement strategy. In Lithuanian data, higher income increased the likelihood of being a "high waster" (relative to low-income: OR = 1.969; $p = 0.023$; relative to medium: OR = 1.839; $p = 0.022$). However, once planning and label knowledge were taken into account, the explanatory power of simple socioeconomic traits diminished markedly [54].

In a national Chinese dataset, income did not significantly differentiate whether a household wasted food at all; logistic models showed no significant effect. Tobit models for waste shares revealed only weak positive associations for selected categories—for example, meat ($\beta = 0.054$, $p < 0.10$) and starchy products ($\beta = 0.047$, $p < 0.10$)—indicating borderline signals. Daily practices and household structure were far more influential, suggesting that interventions should prioritise planning, storage, and meal organisation rather than income alone [61]. In rural China, households with village-cadre members (stable, higher incomes) wasted significantly more food—confirmed by fractional logit and PSM ($\beta = 0.33$; $p < 0.05$; ATT = 0.44–0.76 p.p.; for rice 2.45–3.60 p.p.)—and income was positively associated with waste ($\beta = 0.04$; $p < 0.01$) [67].

In Japan, national ordered probit models show no significant link between income and overall waste levels. Instead, practices proved decisive: the "actions" index (sum of waste-reducing behaviours) substantially lowered waste ($\beta = -0.380$, $p < 0.01$), and food-management skills, including storage, had a similar effect ($\beta = -0.056$, $p < 0.01$). Conversely, "checking labels" was positively associated with waste ($\beta = 0.076$, $p < 0.01$), which

https://doi.org/10.3390/su18020880

the authors interpret as precautionary discarding driven by a misinterpretation of "best before" vs. "use by" dates [62]. Systemically, demographic effects—especially age—are strong in Japan, but operate primarily through basket composition (fruits/vegetables) and preparation practices rather than income [39].

The practical implications are clear. Interventions should prioritise larger households and families with children, consistent with model M8 and with findings by Li, Jiang & Qing [61], Nakamura et al. [62], and Tonini et al. [60]. The scope of interventions should include meal planning, shopping-list preparation, inventory management (including "first to eat" principles), product-appropriate storage, systematic leftover reuse, and clear date labelling. Implementation can be supported through simple tools—such as meal planners, list-management, and inventory apps—supplemented with pre-shopping reminders and norm-based messaging reinforced by constructive emotional cues. Priority should be given to categories with short freshness windows, such as bread, ready-to-eat meals, and fruits and vegetables, as highlighted in Kasza et al. [73], Szabó Bódi et al. [75], and WRAP recommendations [24].

At the same time, discrepancies between self-reported data and results from diaries or weighing studies call for cautious interpretation. Combined measurement techniques enable a more comprehensive identification of weak points in everyday food management routines [64,77]. When interpreting correlations, it is essential to avoid causal inference; demographic effects often operate primarily indirectly—through habits, competencies, and norms [43,54,65].

Overall, both the present findings and the international evidence indicate that age (negatively) and the presence of children (positively) remain the most stable correlates of food waste. However, everyday practices—such as planning, storage, and managing leftovers—ultimately determine the levels of waste. Therefore, effective interventions should integrate demographic profiling with the strengthening of habitual food-management behaviours as the primary pathway for behavioural change.

## 6. Conclusions

The results confirm clear demographic differences in household food waste. Age is the most consistent predictor—young adults report significantly higher levels of food waste, consistent with weaker planning routines and shorter "habit capital". Household structure is equally influential: the presence and number of children systematically increase waste levels, reflecting varying preferences, planning difficulties, and greater uncertainty in family food management. Economic status, as measured by income source, distinguishes groups with systematically lower waste (pensioners and disability-benefit recipients) from employed and self-employed individuals. Net household income itself shows no significant impact, indicating that financial resources mainly matter indirectly. Education—contrary to common assumptions—does not reduce food waste, underscoring the superiority of practical skills over formal educational attainment.

These findings should be interpreted against the broader institutional and social context of food-waste reduction in Poland, which shapes individual behaviours but remains outside the scope of direct measurement in this study. Studies conducted among Polish households indicate that food is most frequently discarded due to expired dates, spoilage, and the preparation of excessive quantities, which supports the rationale for information- and awareness-based interventions [80–82]. In practice, however, the effectiveness of such measures may be limited if they are not accompanied by local social norms, trust, and social acceptance of sharing food surpluses rather than disposing of them [34,83].

In Poland, food-waste reduction therefore increasingly takes the form of a collective endeavour. The 2019 Act on Counteracting Food Waste shifts part of the responsibility to the

https://doi.org/10.3390/su18020880

retail sector and institutionalises cooperation with social organisations through mandatory agreements on free food donation, educational campaigns, and a fee imposed on wasted food, thereby creating a systemic incentive for redistribution instead of disposal [84]. This mechanism appears to function most effectively where cooperation between key actors (retailers–NGOs–end beneficiaries) is well coordinated and where bottom-up initiatives, such as community fridges ("jadłodzielnie"), reinforce the "pass it on" norm through local engagement and shared responsibility [85–87].

From an interpretative perspective, this implies that explaining policy effectiveness solely through formal regulatory instruments is incomplete. Public policies aimed at reducing food waste require not only clearly defined targets and monitoring frameworks, but also complementary actions that support cooperation, trust, and social participation. Neglecting this collective-action dimension may weaken the explanatory power of empirical findings based exclusively on individual- or household-level characteristics [83,88].

Overall, the findings highlight that effective food-waste reduction strategies require precise targeting of high-risk household type—particularly younger adults and families with children—and a focus on strengthening daily food management habits, including meal planning, stock control, appropriate storage, and the systematic use of leftovers. These insights provide a solid empirical basis for SCP policies aligned with SDG 12.3 and emphasise the need to combine educational efforts with practical tools to support everyday household organisation.

Future research could build on the present findings by adopting mixed-method approaches that combine large-scale survey data with behavioural or observational measures, such as food diaries or weighing studies, to better capture actual levels of household food waste and to complement self-reported assessments. While the results of this study are embedded in the Polish institutional and socio-demographic context, the observed associations—particularly those related to life stage and household structure—may be transferable to other national settings, especially within Central and Eastern Europe, provided that differences in institutional frameworks and consumption practices are taken into account.

## 7. Limitations

This study has several methodological limitations that should be acknowledged when interpreting the results. First, the data were collected using a non-probability, purposive quota sampling design based on age and gender. Although the sample structure was aligned with the demographic distribution of the adult population in Poland for these two characteristics, no probabilistic random sampling was applied, and no post-stratification, calibration, or weighting procedures were performed for other socio-demographic variables.

Second, the survey instrument was adapted for use in the Polish context and refined following a pilot study conducted prior to the main data collection. While the pilot phase allowed for the assessment of internal consistency and basic psychometric properties, the scope of validation remains limited to the procedures implemented by the survey contractor. As a result, the measurement properties of the adapted instrument should be interpreted with caution, particularly in comparison with tools that have undergone extensive national validation.

Third, household food waste was measured using a single self-reported ordinal item. Although such measures are commonly applied in large-scale survey research and are suitable for identifying relative differences between households, they may be affected by recall bias and social desirability bias. Previous research indicates that self-reported assessments tend to underestimate actual food waste levels compared with diary- or

https://doi.org/10.3390/su18020880

weighing-based methods. Accordingly, the results should be understood as reflecting relative patterns of food waste rather than precise quantitative estimates.

Finally, the cross-sectional design of the study limits the ability to draw causal inferences about the relationships observed. The reported associations capture correlations at a single point in time and should not be interpreted as evidence of causal mechanisms underlying household food-waste behaviour.

**Author Contributions:** Conceptualization, P.U.-J. and A.B.; methodology, P.U.-J. and M.S.; software, M.S.; validation, P.U.-J., M.S. and A.B.; formal analysis, P.U.-J. and M.S.; investigation, A.B.; resources, A.B. and P.P.; data curation, P.U.-J. and M.S.; writing—original draft preparation, P.U.-J. and A.B.; writing—review and editing, P.U.-J., P.P. and M.S.; visualization, P.U.-J. and A.B.; supervision, P.U.-J.; project administration, P.U.-J.; funding acquisition, M.S. All authors have read and agreed to the published version of the manuscript.

**Funding:** This research was financed under the grant awarded by the Faculty of Economics and Finance at Wroclaw University of Economics and Business, grant no. B701111.

**Institutional Review Board Statement:** Ethical review and approval were waived for this study by Institution Committee due to Legal Regulations (Article 47 of the Constitution of the Republic of Poland (Konstytucja Rzeczypospolitej Polskiej Konstytucja Rzeczypospolitej Polskiej z dnia 2 kwietnia 1997 r.)).

**Informed Consent Statement:** Informed consent was obtained from all subjects involved in the study.

**Data Availability Statement:** The data presented in this study are available on request from the corresponding author due to confidentiality agreements and the need to protect sensitive information.

**Conflicts of Interest:** The authors declare no conflicts of interest.

## References

1. *Think Eat Save Tracking Progress to Halve Global Food Waste*; UNEP: Nairobi, Kenya, 2024; ISBN 9789280741391.
2. Gunawan, J.; Permatasari, P.; Tilt, C. Sustainable Development Goal Disclosures: Do They Support Responsible Consumption and Production? *J. Clean. Prod.* **2020**, *246*, 118989. [CrossRef]
3. van Herpen, E.; van Geffen, L.; Nijenhuis-de Vries, M.; Holthuysen, N.; van der Lans, I.; Quested, T. A Validated Survey to Measure Household Food Waste. *MethodsX* **2019**, *6*, 2767–2775. [CrossRef]
4. Quested, T.E.; Palmer, G.; Moreno, L.C.; McDermott, C.; Schumacher, K. Comparing Diaries and Waste Compositional Analysis for Measuring Food Waste in the Home. *J. Clean. Prod.* **2020**, *262*, 121263. [CrossRef]
5. Aitken, J.A.; Sprenger, A.; Alaybek, B.; Mika, G.; Hartman, H.; Leets, L.; Maese, E.; Davoodi, T. Surveys and Diaries and Scales, Oh My! A Critical Analysis of Household Food Waste Measurement. *Sustainability* **2024**, *16*, 968. [CrossRef]
6. Aschemann-Witzel, J.; de Hooge, I.; Amani, P.; Bech-Larsen, T.; Oostindjer, M. Consumer-Related Food Waste: Causes and Potential for Action. *Sustainability* **2015**, *7*, 6457–6477. [CrossRef]
7. Dou, Z.; Toth, J.D. Global Primary Data on Consumer Food Waste: Rate and Characteristics—A Review. *Resour. Conserv. Recycl.* **2021**, *168*, 105332. [CrossRef]
8. Shukla, P.R.; Skea, J.; Slade, R.; van Diemen, R.; Haughey, E.; Malley, J.; Pathak, M.; Pereira, J.P. IPCC-Technical Summary. In *Climate Change and Land: An IPCC Special Report on Climate Change, Desertification, Land Degradation, Sustainable Land Management, Food Security, and Greenhouse Gas Fluxes in Terrestrial Ecosystems*; Shukla, P.R., Skea, J., Calvo Buendia, E., Masson-Delmotte, V., Pörtner, H.-O., Roberts, D.C., Zhai, P., Slade, R., Connors, S., van Diemen, M., Eds.; IPCC: Geneva, Switzerland, 2019.
9. Łazorko, K. Promotion of Sustainable Consumption of Food by Vitrual Communities. *Acta Sci. Polonorum Oecon.* **2015**, *14*, 95–105.
10. United Nations. Transforming Our World: The 2030 Agenda for Sustainable Development United Nations. 2015. Available online: https://sdgs.un.org/2030agenda (accessed on 19 September 2025).
11. The State of Food Security and Nutrition in the World 2024. 2024. Available online: https://doi.org/10.4060/cd1254en (accessed on 19 September 2025).
12. Eurostat Key Figures on the European Food Chain 2024 Edition. 2024. Available online: https://ec.europa.eu/eurostat/documents/15216629/20555393/KS-01-24-000-EN-N.pdf (accessed on 19 September 2025).
13. European Environmental Agency. *Waste Prevention Country Profile France*; European Environmental Agency: Copenhagen, Danmark, 2023.

https://doi.org/10.3390/su18020880

14. Hafner, G.; Slsshsdl, J. *Ermittlung Der Weggeworfenen Lebensmittelmengen Und Vorschläge Zur Verminderung Der Wegwerfrate Bei Lebensmitteln in Deutschland-Kurzfassung*; Universität Stuttgart: Stuttgart, Germany, 2012.
15. Kuntscher, M.; Schmidt, T.G. Pakt Gegen Lebensmittelverschwendung Im Groß-Und Einzelhandel: Ergebnisbericht Zum Monitoring 2023 Standard-Nutzungsbedingungen. 2023. Available online: https://www.openagrar.de/receive/openagrar_mods_00104710 (accessed on 19 September 2025).
16. Vollebregt, M. Monitor Voedselverspilling: Wat Is Aan Voedsel Verloren En Verspild in de Nederlandse Retail in 2023. 2024. Available online: https://research.wur.nl/en/publications/monitor-voedselverspilling-wat-is-aan-voedsel-verloren-en-verspil/ (accessed on 19 September 2025).
17. Informe Sobre El Desperdicio Alimentario En Los Hogares 2023. 2023. Available online: https://www.mapa.gob.es/dam/mapa/contenido/alimentacion/temas/estrategia-desperdicios/nueva-web-2022/2.-desperdicio-hogares/otros-documentos/informesobreeldesperdicioalimentarioenloshogares2023.pdf (accessed on 19 September 2025).
18. Naturvårdsverket. Livsmedelsavfall i Sverige 2022. 2024. Available online: https://www.naturvardsverket.se/49501f/globalassets/media/publikationer-pdf/8900/978-91-620-8908-5.pdf (accessed on 19 September 2025).
19. Miljøstyrelsen Affaldskortlaegning Af Husstandsindsamlet Affald. 2023. Available online: https://www2.mst.dk/Udgiv/publikationer/2023/02/978-87-7038-490-2.pdf (accessed on 19 September 2025).
20. Miljøstyrelsen Madaffald i Detailhandel Og Anden Fødevaredistribution 2023. 2023. Available online: https://www2.mst.dk/Udgiv/publikationer/2025/04/978-87-7564-001-0.pdf (accessed on 19 September 2025).
21. EPA. Circular Economy and Waste Statistics Highlight Report 2022. 2024. Available online: https://indd.adobe.com/view/a9b03f72-c6ff-4919-83b3-3a08db60eedf (accessed on 19 September 2025).
22. Department of Climate, Energy and the Environment; Sustainable Energy Authority of Ireland; Environmental Protection Agency. *Public Sector Climate Action Roadmap Guidance*; Final V1.2; Government of Ireland: Dublin, Ireland, 2025.
23. CNCDA. Plataforma Food Waste e Food Loss Reuniu Em Bruxelas o Seu Subgrupo Doação; Notícias Da CNCDA 18a Edição-Newsletter Novembro 2023. 2024. Available online: https://www.cncda.gov.pt/images/DocumentosLegislacao/Newsletters/NewsletterCNCDA_18aEd.pdf (accessed on 19 September 2025).
24. WRAP. WRAP UK Food Waste & Food Surplus-Key Facts. 2025. Available online: https://www.wrap.ngo/resources/report/uk-food-waste-food-surplus-key-facts#download-file (accessed on 19 September 2025).
25. WRAP. WRAP The Food Waste Reduction Roadmap. Progress Update 2023. 2023. Available online: https://www.wrap.ngo/resources/tool/food-waste-reduction-roadmap-toolkit (accessed on 19 September 2025).
26. NIK. Przeciwdziałanie Marnowaniu Żywności. 2021. Available online: https://www.nik.gov.pl/kontrole/P/21/085/ (accessed on 19 September 2025).
27. Waste Not Food Waste in Poland and Perspectives of Reducing the Phenomenon by 2030. 2021. Available online: https://4cf.pl/wp-content/uploads/pdf/banki-zywnosci-eng.pdf (accessed on 19 September 2025).
28. KOWR. Strategia Racjonalizacji Strat i Ograniczania Marnotrawstwa Żywności. 2021. Available online: https://bankizywnosci.pl/wp-content/uploads/2022/04/STRATEGIA-PROM.pdf (accessed on 19 September 2025).
29. Obwieszczenie Marszałka Sejmu Rzeczypospolitej Polskiej z Dnia 16 Września 2020 r.w Sprawie Ogłoszenia Jednolitego Tekstu Ustawy o Przeciwdziałaniu Marnowaniu Żywności. 2020. Available online: https://isap.sejm.gov.pl/isap.nsf/download.xsp/WDU20200001696/U/D20201696Lj.pdf (accessed on 19 September 2025).
30. FAO. *Food Wastage Footprint: Impacts on Natural Resources: Summary Report*; FAO: Rome, Italy, 2013; ISBN 9789251077528.
31. FAO. *Food Wastage Footprint Full-Cost Accounting: Final Report*; Food and Agriculture Organization of the United Nations: Rome, Italy, 2014; ISBN 9789251085134.
32. Disposizioni Concernenti La Donazione e La Distribuzione Di Prodotti Alimentari e Farmaceutici a Fini Di Solidarieta' Sociale e per La Limitazione Degli Sprechi. 2016. Available online: https://www.normattiva.it/uri-res/N2Ls?urn:nir:stato:legge:2016;166 (accessed on 19 September 2025).
33. Russell, S.V.; Young, C.W.; Unsworth, K.L.; Robinson, C. Bringing Habits and Emotions into Food Waste Behaviour. *Resour. Conserv. Recycl.* **2017**, *125*, 107–114. [CrossRef]
34. Stancu, V.; Haugaard, P.; Lähteenmäki, L. Determinants of Consumer Food Waste Behaviour: Two Routes to Food Waste. *Appetite* **2016**, *96*, 7–17. [CrossRef] [PubMed]
35. Visschers, V.H.M.; Wickli, N.; Siegrist, M. Sorting out Food Waste Behaviour: A Survey on the Motivators and Barriers of Self-Reported Amounts of Food Waste in Households. *J. Environ. Psychol.* **2016**, *45*, 66–78. [CrossRef]
36. Secondi, L.; Principato, L.; Laureti, T. Household Food Waste Behaviour in EU-27 Countries: A Multilevel Analysis. *Food Policy* **2015**, *56*, 25–40. [CrossRef]
37. Office for National Statistics. *A Review of Household Behaviour in Relation to Food Waste, Recycling, Energy Use and Air Travel*; Office for National Statistics: London, UK, 2021.
38. CBOS. Deklaracje Polaków Dotyczące Marnowania Żywności; Komunikat z badań Nr126/2023, 2023. Available online: https://www.cbos.pl/SPISKOM.POL/2023/K_126_23.PDF (accessed on 19 September 2025).

https://doi.org/10.3390/su18020880

39. Shigetomi, Y.; Ishigami, A.; Long, Y.; Chapman, A. Curbing Household Food Waste and Associated Climate Change Impacts in an Ageing Society. *Nat. Commun.* **2024**, *15*, 8806. [CrossRef] [PubMed]
40. Simon, H.A. A Behavioral Model of Rational Choice. *Q. J. Econ.* **1955**, *69*, 99–118. [CrossRef]
41. Kahneman, D.; Tversky, A. Prospect Theory: An Analysis of Decision under Risk. In *Choices, Values, and Frames*; Cambridge University Press: Cambridge, UK, 2019.
42. Ajzen, I. The Theory of Planned Behavior. *Organ. Behav. Hum. Decis. Process.* **1991**, *50*, 179–211. [CrossRef]
43. Graham-Rowe, E.; Jessop, D.C.; Sparks, P. Predicting Household Food Waste Reduction Using an Extended Theory of Planned Behaviour. *Resour. Conserv. Recycl.* **2015**, *101*, 194–202. [CrossRef]
44. Kahneman, D.; Deaton, A. High Income Improves Evaluation of Life but Not Emotional Well-Being. *Proc. Natl. Acad. Sci. USA* **2010**, *107*, 16489–16493. [CrossRef]
45. Motoki, K.; Sugiura, M. Disgust, Sadness, and Appraisal: Disgusted Consumers Dislike Food More than Sad Ones. *Front. Psychol.* **2018**, *9*, 76. [CrossRef]
46. Hanzlová, R. Amount and Causes of Food Waste in Households from Perspective of Consumers—The Case Study of the Czech Republic. *Waste Forum* **2023**, 195–205.
47. Neff, R.A.; Spiker, M.L.; Truant, P.L. Wasted Food: U.S. Consumers' Reported Awareness, Attitudes, and Behaviors. *PLoS ONE* **2015**, *10*, e0127881. [CrossRef]
48. Szymkowiak, A.; Borusiak, B.; Pierański, B.; Kotyza, P.; Smutka, L. Household Food Waste: The Meaning of Product's Attributes and Food-Related Lifestyle. *Front. Environ. Sci.* **2022**, *10*, 918485. [CrossRef]
49. Zheng, H.; Chen, K.; Ma, Z. Interactive Effects of Social Norms and Information Framing on Consumers' Willingness of Food Waste Reduction Behavior. *J. Retail. Consum. Serv.* **2023**, *75*, 103525. [CrossRef]
50. Liu, R.W.; Lapinski, M.K. Cultural Influences on the Effects of Social Norm Appeals. *Philos. Trans. R. Soc. B Biol. Sci.* **2024**, *379*, 20230036. [CrossRef]
51. La Barbera, F.; Amato, M.; Riverso, R.; Verneau, F. Social Emotions and Good Provider Norms in Tackling Household Food Waste: An Extension of the Theory of Planned Behavior. *Sustainability* **2022**, *14*, 9681. [CrossRef]
52. Quested, T.; Luzecka, P. Household Food and Drink Waste: A People Focus. 2014. Available online: https://open.barnet.gov.uk/download/e67p2/fdr/D2%20WRAP%20Household%20Food%20and%20Drink%20Watse%20People-focused%20report.pdf (accessed on 20 July 2025).
53. Bilska, B.; Tomaszewska, M.; Kołozyn-Krajewska, D.; Piecek, M. Segmentation of Polish Households Taking into Account Food Waste. *Foods* **2020**, *9*, 379. [CrossRef]
54. Eičaitė, O.; Alenčikienė, G.; Pauliukaitytė, I.; Šalaševičienė, A. Eat or Throw Away? Factors Differentiating High Food Wasters from Low Food Wasters. *Sustainability* **2021**, *13*, 10741. [CrossRef]
55. Pocol, C.B.; Pinoteau, M.; Amuza, A.; Burlea-Schiopoiu, A.; Glogovețan, A.I. Food Waste Behavior among Romanian Consumers: A Cluster Analysis. *Sustainability* **2020**, *12*, 9708. [CrossRef]
56. Aschemann-Witzel, J.; de Hooge, I.E.; Almli, V.L. Suboptimal Food? Food Waste at the Consumer-Retailer Interface. In *Saving Food: Production, Supply Chain, Food Waste and Food Consumption*; Academic Press: Cambridge, MA, USA, 2019.
57. Farr-Wharton, G.; Foth, M.; Choi, J.H.J. Identifying Factors That Promote Consumer Behaviours Causing Expired Domestic Food Waste. *J. Consum. Behav.* **2014**, *13*, 393–402. [CrossRef]
58. van Geffen, L.; van Herpen, E.; Sijtsema, S.; van Trijp, H. Food Waste as the Consequence of Competing Motivations, Lack of Opportunities, and Insufficient Abilities. *Resour. Conserv. Recycl. X* **2020**, *5*, 100026. [CrossRef]
59. Macková, M.; Hazuchová, N.; Stávková, J. Czech Consumers' Attitudes to Food Waste. *Agric. Econ.* **2019**, *65*, 314–321. [CrossRef]
60. Tonini, P.; Odina, P.M.; Durany, X.G. Predicting Food Waste in Households with Children: Socio-Economic and Food-Related Behavior Factors. *Front. Nutr.* **2023**, *10*, 1249310. [CrossRef]
61. Li, X.; Jiang, Y.; Qing, P. Estimates of Household Food Waste by Categories and Their Determinants: Evidence from China. *Foods* **2023**, *12*, 776. [CrossRef] [PubMed]
62. Nakamura, K.; Kojima, D.; Ando, M. What Reduces Household Food Waste in Japan? Nation-Wide and Region-Specific Contributing Factors in Urban and Rural Areas. *Sustainability* **2022**, *14*, 3174. [CrossRef]
63. Kansal, M.; Ananda, J.; Mitsis, A.; Karunasena, G.G.; Pearson, D. Food Waste in Households: Children as Quiet Powerhouses. *Food Qual. Prefer.* **2022**, *98*, 104524. [CrossRef]
64. Nováková, P.; Hák, T.; Janoušková, S. An Analysis of Food Waste in Czech Households—A Contribution to the International Reporting Effort. *Foods* **2021**, *10*, 875. [CrossRef]
65. Bilska, B.; Tomaszewska, M.; Kołożyn-Krajewska, D. Food Waste in Polish Households—Characteristics and Sociodemographic Determinants on the Phenomenon. Nationwide Research. *Waste Manag.* **2024**, *176*, 30–40. [CrossRef]
66. Parizeau, K.; von Massow, M.; Martin, R. Household-Level Dynamics of Food Waste Production and Related Beliefs, Attitudes, and Behaviours in Guelph, Ontario. *Waste Manag.* **2015**, *35*, 207–217. [CrossRef]

https://doi.org/10.3390/su18020880

67. Luo, Y.; Huang, D.; Cao, F. The Impact of Family Members Serving as Village Cadres on Rural Household Food Waste: Evidence from China. *Sustainability* **2022**, *14*, 2678. [CrossRef]
68. Zhang, L.; Ye, L.; Qian, L.; Zheng, M. The Association Between Safety Preference and Household Food Waste: Evidence from Chinese Households. *Sustainability* **2024**, *16*, 10929. [CrossRef]
69. Beitane, I.; Iriste, S.; Sabovics, M.; Krumina-Zemture, G.; Jenzis, J. Dataset on Food Waste in Households: The Case of Latvia. *Data* **2025**, *10*, 65. [CrossRef]
70. Moroșan, E.; Dărăban, A.; Popovici, V.; Rusu, A.; Ilie, E.I.; Licu, M.; Karampelas, O.; Lupuliasa, D.; Ozon, E.A.; Maravela, V.M.; et al. Socio-Demographic Factors, Behaviors, Motivations, and Attitudes in Food Waste Management of Romanian Households. *Nutrients* **2024**, *16*, 2738. [CrossRef] [PubMed]
71. Tokareva, T.; Eglite, A. Food Waste in Latvian Households: Amounts, Economic Aspects. In Proceedings of the Economic Science for Rural Development, Jelgava, Latvia, 27–28 April 2017; Volume 46.
72. Aschemann-Witzel, J.; Giménez, A.; Ares, G. Convenience or Price Orientation? Consumer Characteristics Influencing Food Waste Behaviour in the Context of an Emerging Country and the Impact on Future Sustainability of the Global Food Sector. *Glob. Environ. Change* **2018**, *49*, 85–94. [CrossRef]
73. Kasza, G.; Dorkó, A.; Kunszabó, A.; Szakos, D. Quantification of Household Food Waste in Hungary: A Replication Study Using the FUSIONS Methodology. *Sustainability* **2020**, *12*, 3069. [CrossRef]
74. Przezbórska-Skobiej, L.; Wiza, P.L. Food Waste in Households in Poland—Attitudes of Young and Older Consumers Towards the Phenomenon of Food Waste as Demonstrated by Students and Lecturers of Puls. *Sustainability* **2021**, *13*, 3601. [CrossRef]
75. Szabó-Bódi, B.; Kasza, G.; Szakos, D. Assessment of Household Food Waste in Hungary. *Br. Food J.* **2018**, *120*, 625–638. [CrossRef]
76. Keller, V.; Gombos, S. Household Food Waste Reduction Determinants in Hungary: Towards Understanding Responsibility, Awareness, Norms, and Barriers. *Foods* **2025**, *14*, 728. [CrossRef]
77. Hazuchová, N.; Tuzová, M.; Macková, M.; Stávková, J. Household Food Waste Behaviour: Subjective and Objective Evidence. *Potravin. Slovak J. Food Sci.* **2019**, *13*, 784–792. [CrossRef] [PubMed]
78. Medved'ová, M.; Kapsdorferová, Z.; Švikruhová, P.; Zábojníková, V. Social Responsibility in Reducing Food Losses and Waste in the Slovak Republic: The Role of Policies—The Responsibility of All. *Potravin. Slovak J. Food Sci.* **2022**, *16*, 219–232. [CrossRef] [PubMed]
79. Kapsdorferová, Z.; Čereš, M.; Švikruhová, P.; Zábojníková, V.; Kataniková, R. Challenges and Innovative Approaches in the Agricultural and Food Industry and Changing Consumer Behaviour in the Milk and Milk Products Market: Case of Slovakia. *Agric. Econ.* **2023**, *69*, 246–254. [CrossRef]
80. Bilska, B.; Tomaszewska, M.; Kołozyn-Krajewska, D. Analysis of the Behaviors of Polish Consumers in Relation to Food Waste. *Sustainability* **2020**, *12*, 304. [CrossRef]
81. Zielińska, D.; Bilska, B.; Marciniak-łukasiak, K.; Łepecka, A.; Trząskowska, M.; Neffe-skocińska, K.; Tomaszewska, M.; Szydłowska, A.; Kołożyn-krajewska, D. Consumer Understanding of the Date of Minimum Durability of Food in Association with Quality Evaluation of Food Products After Expiration. *Int. J. Environ. Res. Public Health* **2020**, *17*, 1632. [CrossRef]
82. Siedlecka, A.; Krzyżanowska, K.; Kuflewska, W. Food waste in polish households as an economic problem. *Econ. Environ.* **2024**, *90*, 704. [CrossRef]
83. Schanes, K.; Dobernig, K.; Gözet, B. Food Waste Matters—A Systematic Review of Household Food Waste Practices and Their Policy Implications. *J. Clean. Prod.* **2018**, *182*, 978–991. [CrossRef]
84. Ustawa z Dnia 19 Lipca 2019 r. O Przeciwdziałaniu Marnowaniu Żywności. 2019. Available online: https://isap.sejm.gov.pl/isap.nsf/download.xsp/WDU20190001680/U/D20191680Lj.pdf (accessed on 19 September 2025).
85. Drozłowska, E.; Sobieraj, M. Food Sharing as Part of the Sharing Economy in the Context of Food Waste. *Zesz. Nauk. Uniw. Szczecińskiego Probl. Transp. I Logistyki* **2019**, *48*, 17–24. [CrossRef]
86. Jastrzębska, E. Konsument w Gospodarce o Obiegu Zamkniętym. Studia i Prace Kolegium Zarządzania i Finansów 2019. Available online: https://econjournals.sgh.waw.pl/SiP/article/view/658 (accessed on 19 September 2025).
87. Morrow, O. Sharing Food and Risk in Berlin's Urban Food Commons. *Geoforum* **2019**, *99*, 202–212. [CrossRef]
88. Sokołowski, Ł.M. Przeciwdziałanie Marnowaniu Żywności Jako Wyzwanie Współczesności—Studium Porównawcze Prawa Polskiego i Niemieckiego. *Przegląd Prawa Rolnego* **2022**, *2*, 99–117. [CrossRef]



https://doi.org/10.3390/su18020880

MDPI AG
Grosspeteranlage 5
4052 Basel
Switzerland
Tel.: +41 61 683 77 34

*Sustainability* Editorial Office
E-mail: sustainability@mdpi.com
www.mdpi.com/journal/sustainability

www.ingramcontent.com/pod-product-compliance
Lightning Source LLC
Chambersburg PA
CBHW040837250926
51850CB00007BA/1335
* 9 7 8 3 7 2 5 8 7 7 8 9 8 *